Topics in
ALGEBRA

AF652892

By

DR. KULBHUSHAN PARKASH
Head, Deptt. of Mathematics
D.A.V. College (Lahore) Ambala City
Haryana, India

OM P. CHUG
Formerly, Head, Deptt. of Mathematics
Hindu College, Sonepat
Haryana, India

PARMANAND GUPTA
M.Sc. M. Phil (Maths)
M.A. (Economics)
Head, Deptt. of Mathematics
Indira Gandhi National College, Ladwa
KURUKSHETRA UNIVERSITY
Haryana, India

FIREWALL MEDIA

(An Imprint of Laxmi Publications Pvt. Ltd.)

BANGALORE • CHENNAI • COCHIN • GUWAHATI • HYDERABAD
JALANDHAR • KOLKATA • LUCKNOW • MUMBAI • RANCHI
NEW DELHI • BOSTON, USA

Published by :

FIREWALL MEDIA

(*An Imprint of Laxmi Publications Pvt. Ltd.*)

113, Golden House, Daryaganj,

New Delhi-110002

Phone : 011-43 53 25 00

Fax : 011-43 53 25 28

www.laxmipublications.com

info@laxmipublications.com

Copyright © 2008 by Laxmi Publications Pvt. Ltd. All rights reserved. No part of this publication may be reproduced, stored in a retrieval system, or transmitted in any form or by any means, electronic, mechanical, photocopying, recording or otherwise without the prior written permission of the publisher.

Price : **Rs. 195.00** *Only.* *First Edition : 2008*

OFFICES

India

✆ **Bangalore**	080-26 61 15 61
✆ **Chennai**	044-24 34 47 26
✆ **Cochin**	0484-239 70 04
✆ **Guwahati**	0361-254 36 69, 251 38 81
✆ **Hyderabad**	040-24 65 23 33
✆ **Jalandhar**	0181-222 12 72
✆ **Kolkata**	033-22 27 37 73, 22 27 52 47
✆ **Lucknow**	0522-220 95 78
✆ **Mumbai**	022-24 91 54 15, 24 92 78 69
✆ **Ranchi**	0651-230 77 64

USA

Boston

11, Leavitt Street, Hingham,

MA 02043, USA

FAL-2972-195-TOPICS IN ALGEBRA **C—15231/08/01**

Typeset at : Goswami Associates, Delhi. *Printed at* : Pack Printers, Delhi.

PREFACE

The present book on 'Algebra' has been written as a textbook according to the latest guidelines and syllabus in Mathematics issued by the U.G.C. for various universities. The text of the book has been prepared with the following salient features :

(*i*) The language of the book is simple and easy to understand.

(*ii*) Each topic has been presented in a **systematic, simple, lucid** and **exhaustive manner**.

(*iii*) A large number of important solved examples properly selected from the previous university question papers have been provided to enable the students to have a clear grasp of the subject and to equip them for attempting problems in the university examination without any difficulty.

(*iv*) Apart from providing a large number of examples, different type of questions in ample quantity have been provided for a thorough practice to the students.

(*v*) A large number of 'notes' and 'remarks' have been added for better understanding of the subject.

A serious effort has been made to keep the book free from mistakes and errors.

In fact no pains have been spared to make the book interesting and useful.

Suggestions and comments for further improvement of the book will be welcomed.

—AUTHORS

CONTENTS

Chapters			Pages
	SECTION I		
1.	Mapping, Equivalence Relations and Partitions	...	3
2.	Congruence Modulo n	...	38
3.	Matrices	...	46
4.	Rank, Row Rank and Column Rank of a Matrix and Linear Equations	...	72
5.	Characteristic Roots and Characteristic Vectors	...	133
	SECTION II		
6.	Relations between the Roots and Co-efficients of an Equation	...	160
7.	Transformations of Equations	...	192
8.	Descarte's Rule of Signs	...	216
9.	Solutions of General Cubic and Bi-quadratic Equations	...	219
	SECTION III		
10.	Groups	...	238
11.	Subgroups	...	264
12.	Normal Subgroups	...	282
13.	Cyclic Groups	...	290
14.	Permutation Groups	...	296
15.	Rings	...	306

SECTION I

1

Mappings, Equivalence Relations and Partitions

1. ORDERED PAIR

We know that the sets $\{a, b\}$ and $\{b, a\}$ are the same because they have exactly the same elements and the order in which the elements are written down is a set is immaterial.

Sometimes we wish to take the order of occurrence of the elements into account. An **ordered pair** is a pair of elements written according to a specified order. Thus, if a and b are two elements, then the ordered pair in which a occupies the first place and b occupies the second place is written as (a, b), whereas the ordered pair in which b occupies the first place and a occupies the second place is written as (b, a).

The order in which the elements occur in an ordered pair is important.

$$(a, b) = (b, a) \text{ if an only if } a = b$$

$$(a, b) = (c, d) \text{ if and only if } a = c \text{ and } b = d$$

While writing the co-ordinates (x, y) of a point in a plane, the first element x is always the abscissa and the second element is always the ordinate. Thus, $(2, 5) \neq (5, 2)$. The two ordered pairs represent distinct points in the plane.

2. CARTESIAN PRODUCT OF TWO SETS

Let A and B be two sets. The set of all ordered pairs (a, b) where $a \in$ A and $b \in$ B is called the **Cartesian product** of the sets A and B and is denoted by A × B (read as 'A cross B'). In symbols, $A \times B = \{(a, b) \mid a \in A, b \in B\}$ and $B \times A = \{(b, a) : b \in B \text{ and } a \in A\}$

For example, if A = {1, 3} and B = {2, 4, 5}, then

$$A \times B = \{(1, 2), (1, 4), (1, 5), (3, 2), (3, 4), (3, 5)\}$$

Note 1. Take one element of A. Keep it in the first place and one by one write all the elements of B in the second place. Take another element of A. Keep it in the first place and, one by one, write all the elements of B in the second place. Continue like this till all the elements of A are exhausted. In this way, we construct the set A × B.

Similarly, B × A = {(2, 1), (2, 3), (4, 1), (4, 3), (5, 1), (5, 3)}

Note 2. $(a, b) \in A \times B \Rightarrow a \in A$ and $b \in$ B.

Note 3. A × B ≠ B × A unless A = B.

Note 4. If one or both of A and B are empty then $A \times B = \phi$.

Note 5. If n(A) = p and n(B) = q, then

$$\mathbf{n(A \times B) = n(A) \times n(B) = pq = qp = n(B) \times n(A) = n(B \times A)}$$

Thus A × B and B × A are equivalent sets.

(Two *finite* sets are said to be **equivalent** if they have the same number of elements).

If R is the set of all real numbers, then $R \times R = \{(x, y) \mid x \in R, y \in R\}$ is the set of all points in the Cartesian plane.

3. RELATION

A relation R from a set A to a set B is a subset of A × B. Thus, $R \subset A \times B$.

In fact, every subset of A × B is a relation from A to B.

If $(a, b) \in R$, then we write it as a R b and read it as 'a is in relation R to b' or 'a is R-related to b'.

If $(a, b) \notin R$, then 'a is not in relation R to b' or 'a is not R-related to b'.

If B = A, then the relation is called a **relation in A. Thus a relation R in A is a subset of A × A.**

The **domain of relation** R from a set A to a set B is the set of all first elements of the ordered pairs which belong to R. Thus

$$\text{Domain of } R = \{a \mid a \in A \text{ and } (a, b) \in R\}.$$

The **range of a relation** R is the set of all second elements of the ordered pairs which belong to R. Thus

Range of $R = \{b \mid b \in B \text{ and } (a, b) \in R\}$

Obviously, domain of $R \subset A$ and range of $R \subset B$.

For example, (*i*) If A = {2, 3, 4, 5, 6}, B = {2, 4, 6, 8} and a R $b \Rightarrow a$ divides b, *i.e.*, $a \mid b$, then the relation R from A to B is given by

$$R = \{(2, 2), (2, 4), (2, 6), (2, 8), (3, 6), (4, 4), (4, 8), (6, 6)\}$$

Here domain of R = Set of all first elements of ordered pairs which $\in$ R = {2, 3, 4, 6} (Dropping the repetitions) and Range of R is the set of all second elements of the ordered pairs which belong to R.

= {2, 4, 6, 8} (Dropping the repetitions)

(*ii*) If A = {1, 2, 3, 4} and R is the relation 'is less than' in A (*i.e.*, a R $b \Rightarrow a < b$), then

$$R = \{(1, 2), (1, 3), (1, 4), (2, 3), (2, 4), (3, 4)\}$$

Domain of $R = \{a : a \in A \text{ and } (a, b) \in R\} = \{1, 2, 3\}$

and Range of $R = \{b : b \in B \text{ and } (a, b) \in R\} = \{2, 3, 4\}$

4. IDENTITY RELATION IN A SET

The identity relation in a set A is the set of all order pairs $(a, b) \in A \times A$ such that $a = b$ ***i.e.*, (a, a) ∈ A × A for all a ∈ A**. The identity relation in A is denoted by I_A. Thus,

$$I_A = \{(a, b) \mid a \in A, b \in A \text{ and } a = b\}.$$

i.e., $\mathbf{I_A = \{(a, a) \mid \text{for all } a \in A\}}$.

For example, if A = {1, 2, 3, 4}, then

$$I_A = \{(1, 1), (2, 2), (3, 3), (4, 4)\}$$

Clearly domain of I_A = range of I_A. (= A)

5. INVERSE RELATION

Let R be a relation from A to B. The inverse relation of R, denoted by R^{-1}, is a relation from B to A defined by

$$R^{-1} = \{(b, a) \mid (a, b) \in R\}$$

Thus R^{-1} consists of those ordered pairs which, when reversed, belong to R.

Obviously, Domain of R^{-1} = Range of R and Range of R^{-1} = Domain of R.

For example, let A = {1, 2, 4}, B = {2, 3, 5} and R be the relation 'is less than' from A to B, then

$$R = \{(x, y) : x \in A, y \in B \text{ and } x < y\}$$

$\therefore$ $R = \{(1, 2), (1, 3), (1, 5), (2, 3), (2, 5), (4, 5)\}$

and hence $R^{-1} = \{(y, x) : (x, y) \in R\}$

$\therefore$ $R^{-1} = \{(2, 1), (3, 1), (5, 1), (3, 2), (5, 2), (5, 4)\}$

Clearly, R^{-1} is the relation 'is greater than' from B to A.

6. REFLEXIVE RELATION

A relation R in a set A is said to be **reflexive** if **a R a for all a $\in$ A,** *i.e.*, if **(a, a) $\in$ R for all a $\in$ A.**

A relation R in a set A is not reflexive if there exists at least one element $a \in$ A such that $(a, a) \notin$ R.

For example, (*i*) Let A be the set of all triangles in a plane. The relation R in A defined by 'is similar to' is reflexive since every triangle is similar to itself. Similarly, the relations in A defined by 'is congruent to' and 'is equal in area to' are also reflexive.

(*ii*) Let N be the set of all natural numbers. The relation R in N defined by 'is a divisor of' is reflexive since every natural number is a divisor of itself, *i.e.*, $a \text{ R } a \ \forall\ a \in$ N.

The relation R in N defined by 'is less than' is not reflexive, since a is not less than a for any natural number a.

But the relation R in N defined by 'is less than or equal to' *i.e.*, $a \text{ R } b \Rightarrow a \le b$ is reflexive since $a \le a$ for every $a \in$ N.

(*iii*) Let A = {1, 2, 3, 4}. Then the relation

$R_1 = \{(1, 1), (2, 4), (3, 3), (4, 1), (4, 4)\}$ in A is not reflexive since 2 $\in$ A and $(2, 2) \notin R_1$.

The relation $R_2 = \{(1, 1), (1, 3), (2, 2), (3, 3), (3, 4), (4, 4)\}$ in A is reflexive since $(a, a) \in R_2$ for every $a \in$ A.

7. SYMMETRIC RELATION

A relation R in a set A is said to be symmetric if $a, b \in$ A and $a \text{ R } b \Rightarrow b \text{ R } a$, *i.e.*, **if (a, b) $\in$ R $\Rightarrow$ (b, a) $\in$ R.**

Thus a relation R in A is symmetric if and only if R = R^{-1}. A relation R in A is not symmetric if there exist two distinct elements $a, b \in$ A such that

$$(a, b) \in R \text{ but } (b, a) \notin R.$$

For example, (*i*) Let L be the set of all straight lines in a plane. The relation R in L defined by 'is perpendicular to' is symmetric, since if $a, b \in$ L than $a \perp b \Rightarrow b \perp a$. Thus, $a \text{ R } b \Rightarrow b \text{ R } a$. The relation in L defined by 'is parallel to' is also symmetric.

(*ii*) Let A be the set of all triangles in a plane. The relation R in A defined by 'is similar to' is symmetric, since if, $a, b \in A$ then a is similar to b implies b is similar to a. Thus $a \text{ R } b \Rightarrow b \text{ R } a$.

(*iii*) Let N be the set of all natural numbers. The relation R in N defined by 'is a divisor of' is not symmetric since 2 divides 4, but 4 does not divide 2.

$\Rightarrow$ 2 R 4 is right but 4 R 2 is wrong.

(*iv*) Let $A = \{1, 2, 3, 4\}$ and $R = \{(1, 2), (3, 4), (2, 1), (3, 3)\}$.

Then $R^{-1} = \{(2, 1), (4, 3), (1, 2), (3, 3)\}$

Since $(3, 4) \in R$ but $(3, 4) \notin R^{-1}$, therefore, $R \neq R^{-1}$. Hence, R is not symmetric. In other words, we can say that $(3, 4) \in R$ but $(4, 3) \notin R$ and hence R is not symmetric.

The relations in A defined by

$$R_1 = \{(1, 1), (2, 4), (4, 2), (1, 3), (3, 1)\}$$
$$R_2 = \{(2, 2), (3, 3), (4, 4)\}$$
$$R_3 = \{(3, 3), (2, 4), (4, 2)\}$$

are all symmetric.

8. TRANSITIVE RELATION

A relation R in a set A is said to be **transitive** if $a \text{ R } b$ and $b \text{ R } c \Rightarrow a \text{ R } c$, *i.e.*, if

$$\mathbf{(a, b) \in R \text{ and } (b, c) \in R} \Rightarrow (a, c) \in R.$$

For example, (*i*) Let L be the set of all lines in a plane and R be the relation in L defined by 'is parallel to'. If $a, b, c \in L$ such that $a \parallel b$ and $b \parallel c$ then $a \parallel c$. Therefore, R is transitive.

(*ii*) Let N be the set of all natural numbers and let R be the relation in N defined by 'is less than'.

Since $a < b$ and $b < c \Rightarrow a < c$, therefore R is transitive.

Let R be the relation in N defined by 'is a divisor of'. Since a divides b and b divides $c \Rightarrow a$ divides c, therefore, R is transitive.

(*iii*) Let L be the set of all lines in a plane and R be the relation in L defined by 'is perpendicular to'.

If $a \perp b$ and $b \perp c$ then $a \parallel c$, *i.e.*, a is not perpendicular to c. Hence, R is not transitive.

(*iv*) Let $A = \{1, 2, 3, 4\}$ and $R = \{(1, 2), (2, 2), (2, 1), (4, 3)\}$, then R is not transitive, since $(1, 2) \in R$, $(2, 1) \in R$ but $(1, 1) \notin R$.

9. EQUIVALENCE RELATION

A relation R in a set A is said to be an **equivalence relation** if **R is reflexive, symmetric and transitive.** Thus R is an equivalence relation if

(*i*) $a \text{ R } a \ \forall \ a \in A$ [R is reflexive]

(*ii*) $a \text{ R } b \Rightarrow b \text{ R } a \ \forall \ a, b \in A$ [R is symmetric]

(*iii*) $a \text{ R } b$ and $b \text{ R } c \Rightarrow a \text{ R } c \ \forall \ a, b, c \in A$ [R is transitive]

Whenever any one of the three properties (*viz.*, reflexive, symmetric, transitive) is absent, the relation is not an equivalence relation.

For example, (*i*) Let A be the set of all triangles in a plane. Let R be the relation in A defined by 'is congruent to', then

a R a, for every $a \in$ A since every triangle is congruent to itself. Thus, R is reflexive.

a R $b \Rightarrow b$ R a, since if triangle a is congruent to triangle b, then b is congruent to a. Thus, R is symmetric.

a R b and b R $c \Rightarrow a$ R c since if triangle a is congruent to triangle b and triangle b is congruent to triangle c, then a is congruent to c. Thus, R is transitive.

Hence, R is an equivalence relation.

Similarly, if R be the relation in A defined by 'is similar to', then also R is an equivalence relation.

(*ii*) The relation R of equality between real numbers is an equivalence relation

For, (*i*) $a = a$

(*ii*) $a = b \;\Rightarrow\; b = a$

(*iii*) $a = b,\; b = c \;\Rightarrow\; a = c.$

(*iii*) **The Identity Relation I_A on a set A is an equivalence relation.**

10. PARTITION OF A SET

Def. *Let S be a non-empty set. A set* $P = \{A_1, A_2, ..., A_n\}$, *(where* A_i*'s are non-empty subsets of S) is called a partition of set S if*

(*i*) $S = A_1 \cup A_2 \cup ... \cup A_n$

(*ii*) $A_i \cap A_j = \phi$ for $i \neq j$ *i.e., the sets* A_i $(i = 1, 2, ... n)$ *are pairwise disjoint sets.*

For example, (*i*) Let S = {1, 2, 3, 4, ..., 8}.

Let $A_1 = \{1, 3\}, A_2 = \{2, 4, 5\}, A_3 = \{6, 7\}, A_4 = \{8\}$

Then $P = \{A_1, A_2, A_3, A_4\}$ is a partition of S.

[$\because$ These sets A_1, A_2, A_3, A_4 satisfy both above conditions (*i*) and (*ii*) of Article 10)]

(*ii*) If $P = \{A_1, A_2, A_3\}$ where $A_1 = \{1, 3, 5\}, A_2 = \{2, 7\}, A_3 = \{4, 8\}$, then P is not a partition of S because $6 \notin A_1 \cup A_2 \cup A_3$ and therefore, $A_1 \cup A_2 \cup A_3 \neq S$

(*iii*) If $A_1 = \{1,2, 5\}, A_2 = \{2, 3, 4\}, A_3 = \{5, 6, 7, 8\}$, then A_1, A_2, A_3 do not form a partition of S because $A_1 \cap A_3 = \{5\} \neq \phi$.

11. EQUIVALENCE CLASSES

An important property of an equivalence relation, on any set S is that it gives rise to a **partition** of the set S (as defined in Art. 10) *i.e.,* it (equivalence relation) divides the given set S into pairwise disjoint sets A_i $(i = 1, 2, 3, ..., n$ (say) (or $i = 0, 1, 2, ...)$ whose union is S.

Each of these sets A_i is called an **equivalence class.**

Note. (*i*) Union of all equivalence classes is S.

(*ii*) Equivalence classes are pairwise disjoint.

We illustrate it by an example.

In the set N of all natural numbers, let us define a relation R as follows :

For $a, b \in$ N, a R $b \Rightarrow a$ and b leave the same remainder when divided by 5 (so that the remainder is one of the numbers 0, 1, 2, 3 and 4).

Since a and a leave the same remainder when divided by 5

$\therefore$ $a \, R \, a \; \forall \; a \in N$ $\therefore$ **R is reflexive.**

If a and b leave the same remainder when divided by 5, then b and a also leave the same remainder when divided by 5.

$\therefore$ $a \, R \, b \Rightarrow b \, R \, a$ $\therefore$ **R is symmetric.**

If a and b leave the same remainder when divided by 5, and b and c also leave the same remainder when divided by 5, then a and c also leave the same remainder when divided by 5.

$\therefore$ $a \, R \, b$ and $b \, R \, c \Rightarrow a \, R \, c$ $\therefore$ **R is transitive.**

Hence R is an equivalence relation.

Now, let
$A_0 = \{n \mid n \in N$ and on division by 5, n leaves the remainder 0$\}$
$A_1 = \{n \mid n \in N$ and on division by 5, n leaves the remainder 1$\}$
$A_2 = \{n \mid n \in N$ and on division by 5, n leaves the remainder 2$\}$
$A_3 = \{n \mid n \in N$ and on division by 5, n leaves the remainder 3$\}$
$A_4 = \{n \mid n \in N$ and on division by 5, n leaves the remainder 4$\}$

Since there can be only five remainders, *viz.*, 0, 1, 2, 3, 4 on division by 5, there are no other sets.

Now
$A_0 = \{5, 10, 15, 20 \ldots\}$
$A_1 = \{1, 6, 11, 16, \ldots\}$, $A_2 = \{2, 7, 12, 17, \ldots\}$
$A_3 = \{3, 8, 13, 18, \ldots\}$, $A_4 = \{4, 9, 14, 19, \ldots\}$

Clearly, $A_0 \cup A_1 \cup A_2 \cup A_3 \cup A_4 = \bigcup_{i=0}^{4} A_i = N$ and the five sets are pairwise disjoint.

Thus the equivalence relation R divides the set N into five pairwise disjoint subsets.

These five sets A_0, A_1, A_2, A_3, A_4, A_5 are five *equivalence classes*.

Remark. *One can define an equivalence relation R in a set S corresponding to a partition P by defining the relation R in S such that a R b iff a and b are in same subset A_i belonging to P.*

12. UNIVERSAL RELATION

Every subset of A × A is a relation in A. Since every set is a subset of itself, A × A is a subset of A × A. **The relation R = A × A is called the universal relation in A.**

The universal relation is the largest equivalence relation which can be defined in a given set.

13. VOID RELATION

Every subset of A × A is a relation in A. Since the empty set ϕ is also a subset of A × A, therefore, ϕ is also a relation in A. The relation R = ϕ is called the void relation or empty relation in A.

For example, if A = {3, 5, 7} and R is a relation in A defined by $R = \{(a, b) \mid a \in A, b \in A, a \neq b$ and $a = b\}$, then $R = \phi$.

$\therefore$ R is a void relation in A.

Remark. Void Relation on a set A is the **smallest** relation on the set A.

14. ANTISYMMETRIC RELATION

A relation R in a set A is said to be an anti-symmetric relation iff $(a, b) \in R$ and $(b, a) \in R \Rightarrow a = b$ for all $a, b \in A$.

For example, (*i*) Let R be a relation in the set N of natural numbers defined by a R $b \Leftrightarrow$ "a divides b" $\forall\, a, b \in N$.

This relation is an antisymmetric relation in N because a divides b and b divides $a \Rightarrow a = b$.

In the above example if we replace N by Z, the set of integers ; then the relation "a divides b" is not antisymmetric because 4 divides – 4 and – 4 divides 4 but $4 \neq -4$.

(*ii*) The relation $\leq$ (*i.e.* less than or equal to) on the set R of real numbers is antisymmetric because $a \leq b$ and $b \leq a$

$$\Rightarrow \quad a = b \text{ for all } a, b \in A$$

(*iii*) Let S be a non-empty set and R be a relation in the power set P(S), the set of all subsets of the set S defined by $(A, B) \in R \Rightarrow A \subset B$ for all sets A and $B \in P(S)$.

This relation is reflexive because $A \subset A$ *i.e.*, every set A is a subset of itself for all $A \in P(S)$.

This relation is Transitive also because $A \subset B$ and $B \subset C \Rightarrow A \subset C$ for all A, B, $C \in P(S)$.

This relation of "subset of" is antisymmetric also because A R B and B R A *i.e.*, $A \subset B$ and $B \subset A \Rightarrow A = B$.

But this relation is not symmetric.

An identity relation in the set A is an anti-symmetric relation.

Example 1. *Let A = {1, 2, 3} and B = {2, 3, 4}.*

(*a*) *Which of the following are relations from A to B ?*

(*i*) *{(1, 3), (2, 2), (3, 4), (1, 4)}* (*ii*) *{(2, 2), (2, 1), (2, 3), (2, 4)}*

(*b*) *Which of the following are relations from B to A ?*

(*i*) *{(3, 1), (3, 3), (4, 2), (2, 3)}* (*ii*) *{(2, 1), (2, 2), (2, 3), (3, 3), (1, 4)}*

Sol. (*a*) (*i*) Every ordered pair in the set {(1, 3), (2, 2), (3, 4), (1, 4)} is an element of $A \times B$.

$\therefore \quad \{(1, 3), (2, 2), (3, 4), (1, 4)\} \subset A \times B$

Hence {(1, 3), (2, 2), (3, 4), (1, 4)} is a relation from A to B.

(*ii*) $(2, 1) \in \{(2, 2), (2, 1), (2, 3), (2, 4)\}$ but $(2, 1) \notin A \times B$

$\therefore \quad \{(2, 2), (2, 1), (2, 3), (2, 4)\} \not\subset A \times B$

Hence {(2, 2), (2, 1), (2, 3), (2, 4)} is not a relation from A to B.

(*b*) (*i*) Every ordered pair in the set {(3, 1), (3, 3), (4, 2), (2, 3)} is an element of $B \times A$.

$\therefore \quad \{(3, 1), (3, 3), (4, 2), (2, 3)\} \subset B \times A$

Hence {(3, 1), (3, 3), (4, 2), (2, 3)} is a relation from B to A.

(*ii*) $(1, 4) \in \{(2, 1), (2, 2), (2, 3), (3, 3), (1, 4)\}$ but $(1, 4) \notin B \times A$

$\therefore \quad \{(2, 1), (2, 2), (2, 3), (3, 3), (1, 4)\} \not\subset B \times A$

Hence {(2, 1), (2, 2), (2, 3), (3, 3), (1, 4)} is not a relation from B to A.

Example 2. *If R is the relation 'is less than' from A = {1, 2, 3, 4, 5} to B = {1, 4, 5}, write down the Cartesian product corresponding to R. Find the inverse relation to R.*

Sol. $R = \{(a, b) \mid a \in A, b \in B \text{ and } a < b\}$

$= \{(1, 4), (1, 5), (2, 4), (2, 5), (3, 4), (3, 5), (4, 5)\}$

Inverse relation to R is

$$R^{-1} = \{(y, x) : (x, y) \in R\} = \{(4, 1), (5, 1), (4, 2), (5, 2), (4, 3), (5, 3), (5, 4)\}$$

which corresponds to 'is greater than' from B to A.

Example 3. *Write all the possible relations from A = {1, 2} to B = {0}.*

Sol. $A \times B = \{(1, 0), (2, 0)\}$

A relation from A to B is a subset of A × B.

∴ The possible relations from A to B are

ϕ (void relation), {(1, 0)}, {(2, 0)}, {(1, 0), (2, 0)}, (universal relation).

Since $n(A \times B) = 2$, the number of possible relations from A to B is equal to number of subsets of set $A \times B = 2^n = 2^2 = 4$.

Example 4. *Describe all the possible relations in the set A = {0, 2}.*

Sol. $A \times A = \{(0, 0), (0, 2), (2, 0), (2, 2)\}$

A relation in the set A means a relation from A to A which is a subset of A × A.

∴ The possible relations in the set A are possible subsets of A × A.

ϕ, {(0, 0)}, {(0, 2)}, {(2, 0)}, {(2, 2)}, {(0, 0), (0, 2)}, {(0, 0), (2, 0)},
{(0, 0), (2, 2)}, {(0, 2), (2, 0)}, {(0, 2), (2, 2)}, {(2, 0), (2, 2)},
{(0, 0), (0, 2), (2, 0)}, {(0, 0), (0, 2), (2, 2)}, {(0, 0), (2, 0), (2, 2)},
{(0, 2), (2, 0), (2, 2)}, {(0, 0), (0, 2), (2, 0), (2, 2)}.

Since $n(A \times A) = 4$, the number of possible relations in the set A is $2^4 = 16$.

Example 5. *How many relations are possible from a set A of m elements to another set B of n elements ? Why ?*

Sol. Number of elements in A = m, and Number of elements in B = n(given)

∴ Number of elements in $A \times B = mn$. ∴ Number of subsets of $A \times B = 2^{mn}$. Since every subset of A × B is a relation from A to B, therefore 2^{mn} relations are possible from A to B.

Example 6. *Let R be a relation on the set of the all lines in a plane defined by $(l_1, l_2) \in R \Leftrightarrow$ line l_1 is parallel to the line l_2. Show that R is an equivalence relation.*

Sol. Let L be the set of all straight lines in a plane.

It is given that R is a relation in this set L defined by $(l_1, l_2) \in R \Leftrightarrow$ Line l_1 parallel to line l_2.

We examine below the following properties on R.

Reflexive. Each $l \in L$ is parallel to itself *i.e.*, $l \parallel l$ ∴ $(l, l) \in R$ for all $l \in L$

∴ R is reflexive. (Here we assume that each line is ∥ to itself.)

Symmetric. Let $l_1, l_2, \in L$ *s.t.* $(l_1, l_2) \in R$

∴ By Def. of R, $l_1 \parallel l_2$.

Now we know that therefore $l_2 \parallel l_1$ ∴ $(l_2, l_1) \in R$

∴ $(l_1, l_2) \in R \Rightarrow (l_2, l_1) \in R$

∴ R is symmetric.

Transitive. Let $l_1, l_2, l_3 \in L$ such that $(l_1, l_2) \in R$ and $(l_2, l_3) \in R$.

$(l_1, l_2) \in R \Rightarrow l_1$ is parallel to l_2.

$(l_2, l_3) \in R \Rightarrow l_2$ is parallel to l_3.

$\therefore$ l_1 must be parallel to l_3 *i.e.*, $l_1 \parallel l_3$ $\therefore (l_1, l_3) \in R$

$\therefore$ $(l_1, l_2) \in R$ and $(l_2, l_3) \in R \Rightarrow (l_1, l_3) \in R$

$\therefore$ R is transitive.

Hence R being Reflexive, Symmetric and Transitive is an equivalence relation in L.

Example 7. *Prove that the relation R on the set Z of all integers defined by $(x, y) \in R \Leftrightarrow (x - y)$ is divisible by n for a fixed $n\ (\neq 0) \in Z$ and for all $x, y \in Z$ is an equivalence relation.*

Sol. It is given that R is a relation in the set Z defined by "$(x - y)$ is divisible by n". ...(1)

We examine below the following properties on R.

Reflexive. Let $x \in Z$

We know that $x - x = 0 = n \times 0$ is divisible by n.

$\therefore$ By (1) $(x, x) \in R$

$\therefore$ R is Reflexive.

Symmetric. Let $(x, y) \in R$.

$\therefore$ By (1) $(x - y)$ is divisible by n.

$\therefore$ There exists an integer m such that $(x - y) = n.m$

Multiplying both sides by -1, $y - x = n(-m)$

$\therefore$ $(y - x)$ is divisible by n [$\because$ m is an integer $\Rightarrow -m$ is also an integer.]

$\therefore$ By (1) $(y, x) \in R$

$\therefore$ R is symmetric.

Transitive. Let $(x, y) \in R$ and $(y, z) \in R$

$\therefore$ By (1) $(x - y)$ is divisible by n and $(y - z)$ is also divisible by n.

$\therefore$ There exist integers p and q such that

$$(x - y) = np \quad ...(2)$$

and

$$(y - z) = nq \quad ...(3)$$

Adding Eqns. (2) and (3) (to eliminate y)

$$x - z = n(p + q)$$

$\therefore$ $x - z$ is divisible by n.

($\because$ p, q are integers, $\therefore p + q$ is also an integer.)

$\therefore$ By (1) $(x, z) \in R$

$\therefore$ $(x, y) \in R$ and $(y, z) \in R \Rightarrow (x, z) \in R$

$\therefore$ R is Transitive.

Hence R being reflexive, symmetric and transitive is an equivalence relation in Z.

Example 8. *If R is the relation in $N \times N$ defind by $(a, b)\ R\ (c, d)$ if and only if $a + d = b + c$. Show that R is an equivalence relation.*

Sol. We know that N denotes the set of Natural numbers.

We know that a relation in N × N means a relation from N × N to N × N.

R is the relation in N × N defined by (a, b) R (c, d) iff $a + d = b + c$ ...(1)

Reflexive. Let $(a, b) \in$ N × N.

R is Reflexive if (a, b) R (a, b) (Putting $c = a$ and $d = b$ in (1))

i.e., if $a + b = b + a$, which is true.

∴ (a, b) R (a, b)

∴ R is reflexive.

Symmetric. Let $(a, b), (c, d) \in$ N × N such that (a, b) R (c, d).

∴ By (1) $a + d = b + c$

or $b + c = a + d$

or $c + b = d + a$

∴ (c, d) R (a, b) [By (1)]

∴ R is symmetric.

Transitive. Let $(a, b), (c, d), (e, f) \in$ N × N such that (a, b) R (c, d) and (c, d) R (e, f).

∴ By (1) $a + d = b + c$ ...(2)

and $c + f = d + e$ ...(3)

Adding Eqns. (2) and (3), we get

$$a + d + c + f = b + c + d + e$$

Cancelling $c + d$ from both sides

$$a + f = b + e$$

∴ (a, b) R (e, f) [By (1)]

∴ R is transitive.

Hence R being reflexive, symmetric and transitive is an equivalence relation in N × N.

Example 9. *R is the relation in N × N defined by (a, b) R (c, d) iff ad = bc. Prove that R is an equivalence relation.*

Sol. R is the relation in N × N defined by (a, b) R (c, d) iff $ad = bc$ ...(1)

Reflexive. Let $(a, b) \in$ N × N

R is **reflexive** if (a, b) R(a, b) (Putting $c = a$, $d = b$ in (1))

i.e., if $ab = ba$, which is true.

∴ (a, b) R (a, b)

∴ R is reflexive.

Symmetric. Let (a, b) and $(c, d) \in N \times N$ such that $(a, b)\ R\ (c, d)$.

$\therefore$ By (1) $\quad ad = bc$

$\therefore$ $\quad bc = ad$

or $\quad cb = da$

$\therefore$ By (1), $(c, d)\ R\ (a, b)$

$\therefore$ R is symmetric.

Transitive. Let (a, b), (c, d) and $(e, f) \in N \times N$ and $(a, b)\ R\ (c, d)$ and $(c, d)\ R\ (e, f)$

$\therefore$ By (1) $\quad ad = bc \quad$ ...(2)

and By (1) $\quad cf = de \quad$ ...(3)

Multiplying Eqns. (2) and (3) $adcf = bcde$

Dividing by cd, $af = be$

$\therefore$ $(a, b)\ R\ (e, f)$ [By (1)]

$\therefore$ R is transitive.

$\therefore$ R is an equivalence relation.

Example 10. *For any relation R in a set A, prove that R is symmetric if and only if* $R = R^{-1}$.

Sol. Let $R = \{(a, b) : a, b \in A\}$.

$\therefore$ By Definition $R^{-1} = \{(b, a) : (a, b) \in R\}$

Let R be symmetric.

To prove $\quad R = R^{-1}$

Let $\quad (a, b) \in R$

$\therefore$ $\quad (b, a) \in R \quad$ [$\because$ R is symmetric]

$\therefore$ $\quad R \subseteq R^{-1} \quad$ [$\therefore$ $(a, b) \in R^{-1}$] ...(1)

Now let $\quad (a, b) \in R^{-1}$

$\therefore$ By def. $\quad (b, a) \in R$

$\therefore$ $\quad (a, b) \in R \quad$ [$\therefore$ R is symmetric]

$\therefore$ $\quad R^{-1} \subseteq R \quad$ ...(2)

From (1) and (2), we have $R = R^{-1}$

Converse. Let $\quad R = R^{-1}$. (given)

We shall prove that R is symmetric.

Let $\quad (a, b) \in R\ (= R^{-1})$

$\therefore$ $\quad (a, b) \in R^{-1}$

$\therefore$ $\quad (b, a) \in R \quad$ [By def. of inverse relation]

$\therefore$ $\quad (a, b) \in R \Rightarrow (b, a) \in R$

$\therefore$ R is a symmetric relation.

Example 11. *In the set of integers, let a relation R be defined as : a R b if and only if a – b is even. Prove that R is an equivalence relation.*

Sol. Reflexive. $\forall\ a \in Z$, $a - a = 0$ = an even integer

Thus $a\ R\ a\ \forall\ a \in Z \quad \therefore$ R is reflexive.

Symmetric. Let $a \text{ R } b \Rightarrow a - b$ is an even integer

$\Rightarrow b - a$ is an even integer $\Rightarrow b \text{ R } a$

$\therefore$ R is symmetric.

Transitive. Let $a \text{ R } b$ and $b \text{ R } c \Rightarrow a - b$ and $b - c$ are both even

$\Rightarrow (a - b) + (b - c)$ is even

$\Rightarrow a - c$ is even $\Rightarrow a \text{ R } c$

$\therefore$ R is transitive.

Since R is reflexive, symmetric and transitive, therefore, R is an equivalence relation.

(Second Solution)

Proceed as in Example 7 replacing n by 2.

Example 12. *Let the relation R in the set of real numbers be defined as a R b if and only if 1 + ab > 0. Show that this relation is reflexive and symmetric but not transitive.*

Sol. It is given that R is a relation in the set of real numbers defined by $a \text{ R } b$ if and only if $1 + ab > 0$ ...(1)

We examine below the following properties on R.

Reflexive. Let a be a real number

$a \text{ R } a$ (Putting $b = a$ in (1))

if $1 + a.a > 0$ *i.e.*, $1 + a^2 > 0$ which is true.

$\therefore$ R is reflexive.

Symmetric. Let $(a, b) \in \text{R}$

$\therefore$ By (1) $1 + ab > 0$

or $1 + b.a > 0$

$\therefore$ $(b, a) \in \text{R}$ (By (1))

$\therefore$ $(a, b) \in \text{R} \Rightarrow (b, a) \in \text{R}$

$\therefore$ R is symmetric.

Transitive. Relation R defind in (1) is not transitive.

Consider three real numbers $a = 1, b = -\frac{1}{2}, c = -4$

$$1 + ab = 1 + 1\left(-\frac{1}{2}\right) = 1 - \frac{1}{2} = \frac{1}{2} > 0 \quad \therefore \quad a \text{ R } b$$

$$1 + bc = 1 + \left(-\frac{1}{2}\right)(-4) = 1 + 2 = 3 > 0 \quad \therefore \quad b \text{ R } c$$

But $1 + ac = 1 + 1(-4) = 1 - 4 = -3$ which is not greater than 0.

$\therefore$ a is not R–related to c.

$\therefore$ R is not transitive.

Example 13. *Prove that the intersection of two equivalence relations on a set is an equivalence relation on the set.*

Sol. Let R_1 and R_2 be two equivalence relations on a set A, then each of R_1 and R_2 is reflexive, symmetric and transitive.

Since $R_1 \subset A \times A, R_2 \subset A \times A \Rightarrow R_1 \cap R_2 \subset A \times A$

$\therefore$ **$R_1 \cap R_2$ is a relation on A.**

(*i*) **Reflexive.** R_1 and R_2 are reflexive relations on A.

$\Rightarrow (a, a) \in R_1$ and $(a, a) \in R_2$ for all $a \in A$

$\Rightarrow (a, a) \in R_1 \cap R_2$ for all $a \in A$

$\Rightarrow$ **$R_1 \cap R_2$ is a reflexive relation on A.**

(*ii*) **Symmetric.** Let $(a, b) \in R_1 \cap R_2$

$\Rightarrow (a, b) \in R_1$ and $(a, b) \in R_2$

$\Rightarrow (b, a) \in R_1$ and $(b, a) \in R_2$ [$\because$ R_1 and R_2 are symmetric]

$\Rightarrow (b, a) \in R_1 \cap R_2$

Since $(a, b) \in R_1 \cap R_2 \Rightarrow (b, a) \in R_1 \cap R_2$

$\therefore$ **The relation $R_1 \cap R_2$ is symmetric on A.** ...(2)

(*iii*) **Transitive.** Let $(a, b) \in R_1 \cap R_2$ and $(b, c) \in R_1 \cap R_2$

$\Rightarrow \{(a, b) \in R_1$ and $(a, b) \in R_2\}$ and $\{(b, c) \in R_1$ and $(b, c) \in R_2\}$

$\Rightarrow \{(a, b) \in R_1$ and $(b, c) \in R_1\}$ and $\{(a, b) \in R_2$ and $(b, c) \in R_2\}$

$\Rightarrow (a, c) \in R_1$ and $(a, c) \in R_2$ [$\because$ R_1 and R_2 are transitive]

$\Rightarrow (a, c) \in R_1 \cap R_2$

Since $(a, b) \in R_1 \cap R_2$ and $(b, c) \in R_1 \cap R_2 \Rightarrow (a, c) \in R_1 \cap R_2$ ...(3)

$\therefore$ **The relation $R_1 \cap R_2$ is transitive on A.**

From (1), (2) and (3), we conclude that $R_1 \cap R_2$ is an equivalence relation on A.

Example 14. *Prove that the union of two symmetric relations on a set A is symmetric.*

Sol. Let R and S be two symmetric relations on a set A.

$\therefore R \subset A \times A$ and $S \subset A \times A$

$\therefore R \cup S \subset A \times A$

Let $(a, b) \in R \cup S$

$\therefore (a, b) \in R$ or $(a, b) \in S$

$\therefore (b, a) \in R$ or $(b, a) \in S$ [$\because$ R and S are Symmetric Relations]

$\therefore (b, a) \in R \cup S$

$\therefore R \cup S$ is a symmetric relation.

EXERCISE 1

1. Let A = {– 1, 0, 1} and B = {0, 1, 2, 3}.

   (*a*) Which of the following are relations from A to B ?

   (*i*) {(0, 0), (0, 2), (– 1, 3), (1, 0), (1, 3)}

   (*ii*) {(1, 3), (3, 0), (0, 1), (0, 2), (– 1, 1)}

   (*iii*) {(– 1, 0), (– 1, 1), (1, 0), (1, 1), (0, 2)}.

(*b*) Which of the following are relations from B to A ?

(*i*) $\{(0, 1), (0, -1), (2, 0), (-1, 2)\}$ (*ii*) $\{(2, 1), (2, 0), (3, -1), (0, -1), (0, 0)\}$.

2. In each of the following, state which of the ordered pairs belongs to the given relations :

(*i*) $\{(x, y) \mid y = x + 1\}$; (0, 1), (1, 2), (2, 3), (2, 4), (3, 5).

(*ii*) $\{(x, y) \mid x > y + 5\}$; (1, 0), (8, 2), (0, 1), (2, 8), (9, 3).

3. Find the domain and range of the following relations :

(*i*) $\{(x, y) \mid x, y$ are natural numbers and $x + y = 9\}$.

(*ii*) $\{(x, x^2) \mid x$ is a prime number all $10 < x < 20\}$.

(*iii*) $\{(x, |x|) \mid x$ is an integer and $-2 \le x \le 2\}$.

4. Write down the relation 'is the square of' in the set $\{-2, -1, 0, 1, 2, 3, 4\}$.

5. Let A = {2, 3, 4, 5, 6} and R = $\{(x, y) \mid x, y$ are relatively prime$\}$ be a relation in R. Write R as the set of ordered pairs in roster form.

[**Hint. Def.** Two numbers x and y are said to be relatively prime (or Co-prime) if 1 is the only common factor of x and y].

6. Let R be the relation in the set {2, 3, 5, 6, 8} defined by R = $\{(x, y) \mid 3$ divides $|x - y|\}$. Write R as the set of ordered pairs in roster form.

7. If R is the relation 'is greater than' from A = {4, 5, 6, 7} to B = {1, 3, 5, 7}, write R as a set of ordered pairs. Also find the inverse relation to R.

8. Describe all the possible relations from A = {2, 3} to B = {3, 4}.

9. Describe all the possible relations in the set {1, 4}.

10. If a set A has n elements, how many relations are possible in the set A ?

11. Let S = {1, 2, 3, 4, 5} and R = {(1, 3), (2, 4), (3, 5), (1, 1), (2, 2), (4, 2), (3, 1)}. Is R an equivalence relation in S ?

12. In the set of human beings, is the relation 'is brother of' an equivalence relation ?

13. Prove that the relation of similarity in the set of all triangles in a plane is an equivalence relation.

14. Prove that the relation of congruence in the set of all triangles in a plane is an equivalence relation.

15. Let O be a fixed point in a plane. Two points P and Q in the plane are related if OP = OQ. Show that the relation is an equivalence relation.

16. In the set of integers, a relation R is defined by a R b if and only if $a = \pm b$. Prove that R is an equivalence relation.

17. In the set N of natural numbers, let a relation R be defined by R = $\{(x, y) \mid x \in N, y \in N, x - y$ is divisible by 7$\}$. Prove that R is an equivalence relation.

18. (*a*) If R is an equivalence relation on a set A, then R^{-1} is also an equivalence relation on A.

(*b*) The Inverse of an equivalence relation is an equivalence relation.

19. If R and S are relations on a set A and R is reflexive, prove that $R \cup S$ is reflexive.

20. Let N denote the set of all natural numbers and R be the relation on N × N defined by (a, b) R$(c, d) \Leftrightarrow ad(b + c) = bc(a + d)$. Check whether R is an equivalence relation on N × N.

$$\left(\textbf{Hint. } ad\,(b + c) = bc(a + d) \Leftrightarrow \frac{a + d}{ad} = \frac{b + c}{bc} \text{ or } \frac{1}{a} + \frac{1}{d} = \frac{1}{b} + \frac{1}{c}\right).$$

21. Show that the relation R in the set N of natural numbers defined by a R b if and only if a divides b, is reflexive and transitive but not symmetric.

22. Show that the relation 'is greater than' in the set N of natural numbers is transitive but neither reflexive nor symmetric.

23. Prove that the relation R in the set Z of integers defined by a R b if and only if a and b, both odd, is symmetric and transitive but not reflexive.

[**Hint**. If a is even, then $(a, a) \notin R$].

24. Which of the following relations in the set of real numbers are equivalence relations :

(*i*) a R b if $|a| = |b|$ (*ii*) a R b if $a - b \geq 0$?

ANSWERS

1. (*a*) (*i*), (*iii*) (*b*) (*ii*)

2. (*i*) (0, 1), (1, 2), (2, 3) (*ii*) (8, 2), (9, 3)

3. (*i*) Domain = {1, 2, 3, 4, 5, 6, 7, 8} ; Range = {8, 7, 6, 5, 4, 3, 2, 1}

(*ii*) Domain = {11, 13, 17, 19,} ; Range {121, 169, 289, 361}

(*iii*) Domain = {– 2, – 1, 0, 1, 2} ; Range {0, 1, 2}

4. {(0, 0), (1, – 1), (1, 1), (4, – 2), (4, 2)}

5. {(2, 3), (3, 2), (2, 5), (5, 2), (3, 4), (4, 3), (3, 5), (5, 3), (4, 5), (5, 4), (5, 6), (6, 5)}.

6. {(2, 5), (5, 2), (2, 8), (8, 2), (3, 6), (6, 3), (5, 8), (8, 5)}

7. R = {(4, 1), (4, 3), (5, 1), (5, 3), (6, 1), (6, 3), (6, 5), (7, 1), (7, 3), (7, 5)}.

R^{-1} = {(1, 4), (3, 4), (1, 5), (3, 5), (1, 6), (3, 6), (5, 6), (1, 7), (3, 7) (5, 7)} which corresponds to 'is less than' from B to A.

8. ϕ {(2, 3)}, {(2, 4)}, {(3, 3)}, {(3, 4)}, {(2, 3), (2, 4)}, {(2, 3)}, (3, 3)}, {(2, 3)}, (3, 4)}, {(2, 4), (3, 3)}, {(2, 4), (3, 4)}, {(3, 3), (3, 4)}, {(2, 3), (2, 4), (3, 3)}, {(2, 3), (2, 4), (3, 4)}, {(2, 3), (3, 3), (3, 4)}, {(2, 4), (3, 3), (3, 4)}, {(2, 3), (2, 4), (3, 3), (3, 4)}.

9. ϕ {(1, 1)}, {(1, 4)}, {(4, 1)}, {(4, 4)}, {(1, 1), (1, 4)}, {(1, 1), (4, 1)}, {(1, 1), (4, 4)}, {(1, 4), (4, 1)}, {(1, 4), (4, 4)}, {(4, 1), (4, 4)}, {(1, 1), (1, 4), (4, 1)},

{(1, 1), (1, 4), (4, 4)}, {(1, 1), (4, 1), (4, 4)}, {(1, 4), (4, 1), (4, 4)}, {(1, 1), (1, 4), (4, 1), (4, 4)}.

10. 2^{n^2} **11.** No

12. No. Suppose x is a female human being. Then x is not a brother of x and therefore $x \not R x$. Thus, the relation is not reflexive. Note that this relation is also not symmetric.

20. Yes **24.** (*i*).

15. FUNCTIONS (OR MAPPINGS)

A function is a special case of a relation.

A function *f* from a set A to a set B is a relation (*i.e.*, a subset of A × B) which associates, to **each** element of A, **one and only one** element of B.

Thus, a function is a single-valued association of elements of the set A with the elements of the set B.

A function f from A to B is a subset of A × B(*i.e.*, a relation from A to B) such that

(*i*) every element of A is necessarily the first entry of some ordered pair.

(*ii*) no two ordered pairs have the same first entry.

Note 1. If one or more elements of A are not the first entry of any ordered pair, then *f* is not a function. If two ordered pairs have the same first entry, then also *f* is not a function.

∴ **Every relation need not be a function.**

Note 2. It must be noted that there may be some elements of the set B which are not associated to any element of the set A. Each element of the set A must be associated to one and only one element of the set B. Two or more elements of the set A may be associated to the same element of the set B.

'f is a function (or map) from A to B' is denoted by $f : A \to B$ or $A \xrightarrow{f} B$.

The set A is called the **domain** of the function f and is denoted by **dom f or D_f**. The set B is called the **co-domain** of f.

The unique element $b \in B$ associated by f to an element $a \in A$ is denoted by $f(a)$ and is called the **f-image** of a or value of f and we write $b = f(a)$. Here a is called the **pre-image** of b under f. The set of all those elements in B which are the f-image of at least one element in A, is called the range of f. Hence, we often speak of the range of a function as the **image of its domain.** The range of f is denoted by **ran f** or $\mathbf{R_f}$ or $f(\mathbf{A})$.

Thus, ran $f = f(A) = \{(f(a) \mid a \in A\}$. Obviously, $f(\mathbf{A}) \subset \mathbf{B}$.

If $f : A \to B$ is a function, then by the **graph of f** is meant the subset $\{(a, f(a)) \mid a \in A\}$ of $A \times B$.

A function $f : A \to B$ is called a **constant function** if every element of A has the same image in B. *i.e.*, if $f(a) = b$ for every $a \in A$ and for a fixed $b \in B$. The range of **a constant function is a singleton.**

Remark. When a function is defined by a formula, its domain is the set of all those real numbers for which formula is meaningful (*i.e.*, $f(x)$ is Real, Finite).

Example 1. *Let* $X = \{1, 2, 3, 4\}$ *and* $Y = \{2, 4, 6, 8, 10\}$.

(*i*) *Consider* $R_1 = \{(1, 2), (2, 4), (3, 6), (4, 8)\}$.

Here every element of X is the first entry of some ordered pair and no two ordered pairs have the same first entry.

$\therefore$ R_1 is a function from X to Y.

Domain of $R_1 = \{1, 2, 3, 4\} = X$; Range of $R_1 = \{2, 4, 6, 8\}$; Co-domain = Y.

Clearly, range $\subset$ co-domain.

(*ii*) *Consider* $R_2 = \{(1, 4), (2, 8), (4, 10)\}$.

Here $3 \in X$ but 3 is not the first entry of any ordered pair. $\therefore$ R_2 is not a function from X to Y.

(*iii*) *Cosider* $R_3 = \{(1, 2), (2, 4), (2, 10), (3, 6)\}$.

Here the first entry of the two ordered pairs (2, 4) and (2, 10) is the same.

$\therefore$ R_3 is not a function from X to Y.

(*iv*) *Consider* $R_4 = \{(1, 6), (2, 6), (3, 6), (4, 6)\}$.

Here every element of X is the first entry of some ordered pair and no two ordered pairs have the same first entry.

$\therefore$ R_4 is a function from X to Y. Range of $R_4 = \{6\}$.

Also all the elements of X are associated with the same element of Y, namely 6. Such functions are called *constant functions. The range of a constant function is a singleton.*

Example 2. *Consider the following arrow diagrams :*

(*i*)

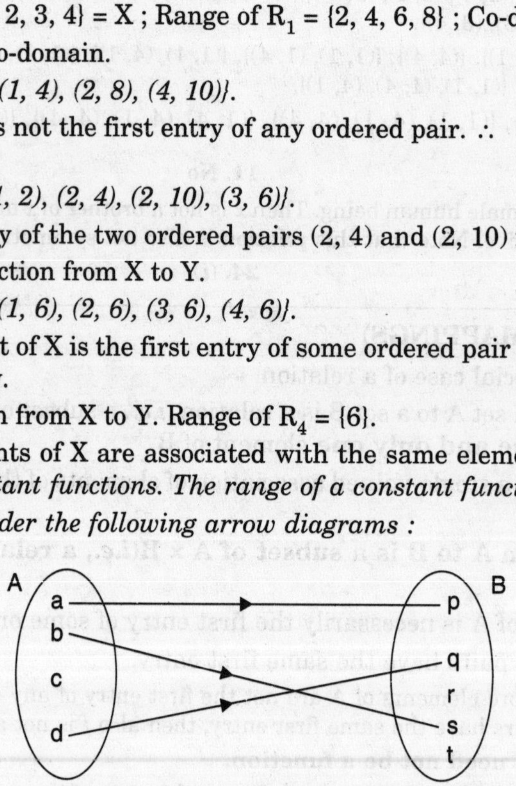

Sol. Here $c \in A$ but c is not associated with any element of B.

$\therefore$ **The arrow diagram does not represent a function from A to B.**

(*ii*)

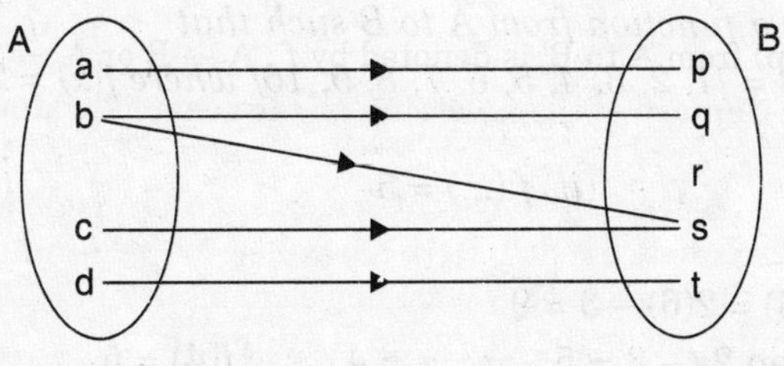

Sol. Here $b \in$ A and b is associated with two elements of B, namely q and s.
Thus b is not associated with a unique element of B.

$\therefore$ The arrow diagram does not represent a function from A to B.

(*iii*)

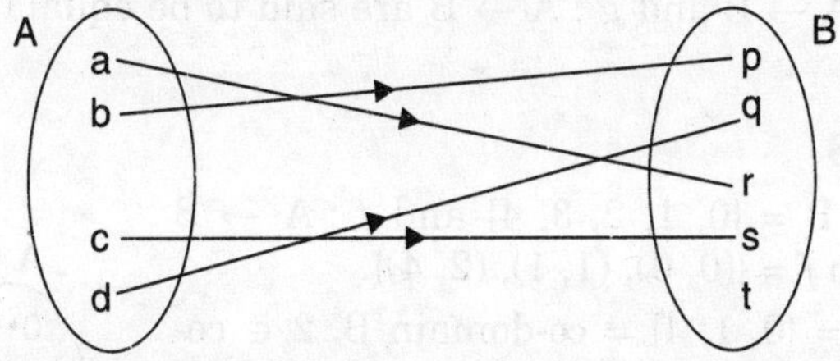

Sol. Here every element of A is associated with a unique element of B.
Thus the arrow diagram represents a function from A to B.
Domain of the function = $\{a, b, c, d\}$ = A ; Co-domain of the function = $\{p, q, r, s, t\}$ = B ;
Range of the function = $\{p, q, r, s\}$.

(*iv*)

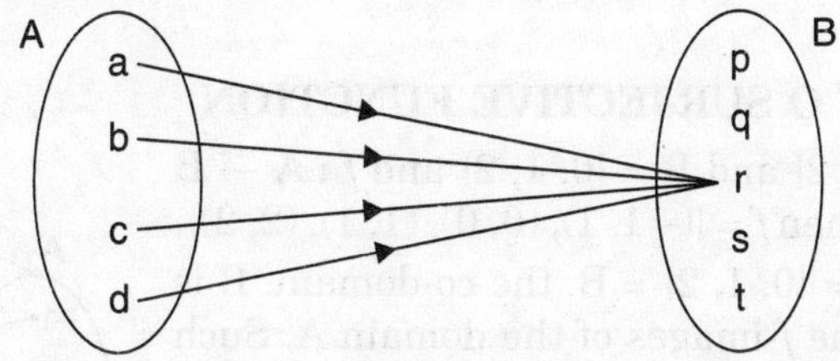

Sol. Here every element of A is associated with a unique element of B.

$\therefore$ The arrow diagram represents a function from A to B.

Since every element of A is associated with the *same* element of B, namely r, the function is a constant function whose range = $\{r\}$.

Example 3. *Consider X = {1, 2, 3, 4} and Y = {1, 2, 3, 4, 5, 6, 7, 8, 9, 10, 11, 12, 13, 14, 15, 16, 17}. Let $f : X \to Y$ be defined as (i) $f(x) = 2x + 5$ (ii) $f(x) = x^2 + 1$. Find domain and range of f.*

Sol. (*i*) Here, $f(x) = 2x + 5$, and $f(1) = 2(1) + 5 = 7$. $\therefore$ $1 \to 7$ $\Rightarrow$ $(1, 7) \in f$.

Similarly, $f(2) = 9, f(3) = 11, f(4) = 13$

$\therefore$ $f = \{(1, 7), (2, 9), (3, 11), (4, 13)\}$

Domain of f = {1, 2, 3, 4} = X, Range of f = {7, 9, 11, 13}.

(*ii*) Here, $f(x) = x^2 + 1$, and

$f(1) = 1^2 + 1 = 2, f(2) = 5, f(3) = 10, f(4) = 17.$

$\therefore$ $f = \{(1, 2), (2, 5), (3, 10), (4, 17)\}$

Domain of f = {1, 2, 3, 4} = X, Range of f = {2, 5, 10, 17}.

Example 4. *If f is a function from A to B such that A = {2, 4, 6} and B = {1, 2, 3, 4, 5, 6, 7, 8, 9, 10} where f(x) = 2x – 3, fill up the following blanks.*

(*i*) $f(6) = ...$ (*ii*) $f(...) = 5$ (*iii*) $f(...) = 9$.

Sol. $f(x) = 2x - 3$

(*i*) When $x = 6$, $f(6) = 2(6) - 3 = 9$

(*ii*) Let $f(x) = 5$, then $2x - 3 = 5 \Rightarrow x = 4 \quad \therefore \quad f(4) = 5$.

(*iii*) Let $f(x) = 9$, then $2x - 3 = 9 \Rightarrow x = 6 \quad \therefore \quad f(6) = 9$.

16. EQUAL FUNCTIONS

Two functions $f : A \to B$ and $g : A \to B$ are said to be equal (or the same) if and only if $f(a) = g(a) \ \forall \ a \in A$.

17. INTO FUNCTIONS

Let A = {0, 1, 2}, B = {0, 1, 2, 3, 4} and $f : A \to B$ defined by $f(x) = x^2$. Then f = {(0, 0), (1, 1), (2, 4)}.

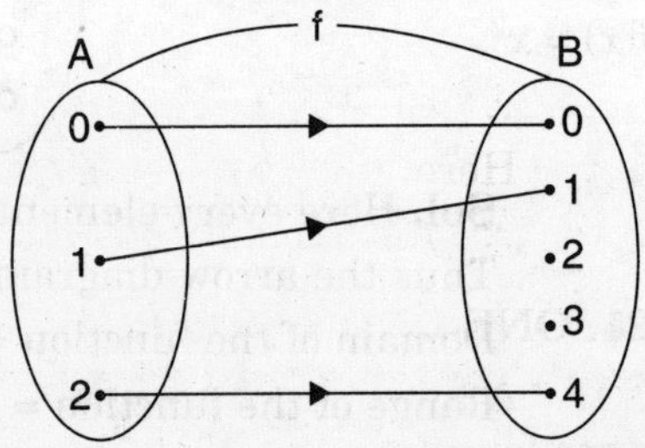

Range of $f = f(A) = \{0, 1, 4\} \neq$ co-domain B. $2 \in$ co-domain B, $3 \in$ B but 2 and 3 are not the image of any element of the domain A. Such functions are called **into functions.**

If the function $f : A \to B$ is such that there is at least one element in B which is not the f-image of any element in A *i.e.*, $f(A) \neq B$, then we say 'f is a function from A into B' or 'f maps A into B'.

18. ONTO FUNCTION O SURJECTIVE FUNCTION

Let A = {– 1, 0, 1, 2} and B = {0, 1, 2} and $f : A \to B$ defined by $f(x) = |x|$. Then f = {(– 1, 1), (0, 0), (1, 1), (2, 2)}.

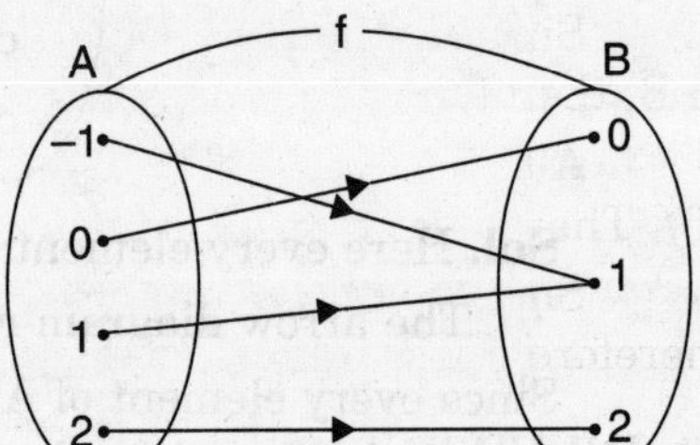

Range of $f = f(A) = \{0, 1, 2\} = B$, the co-domain. B is completely covered by the f-images of the domain A. Such functions are called **onto functions** or **surjective functions** or **surjections.**

If the function $f : A \to B$ is such that each element in B is the f-image of at least one element in A *i.e.*, **f(A) = B** *i.e.*, **Range of f = co-domain of f,** then we say 'f' is a function from A **onto** B' or 'f maps A onto B'.

19. ONE-ONE FUNCTION OR INJECTIVE FUNCTION

If the function $f : A \to B$ is such that distinct elements of A have distinct f-images in B *i.e.* $\mathbf{x_1, x_2 \in A, x_1 \neq x_2 \Rightarrow f(x_1) \neq f(x_2)}$, **then f is called a one-one function** or 1 – 1 function or an injective function or an injection.

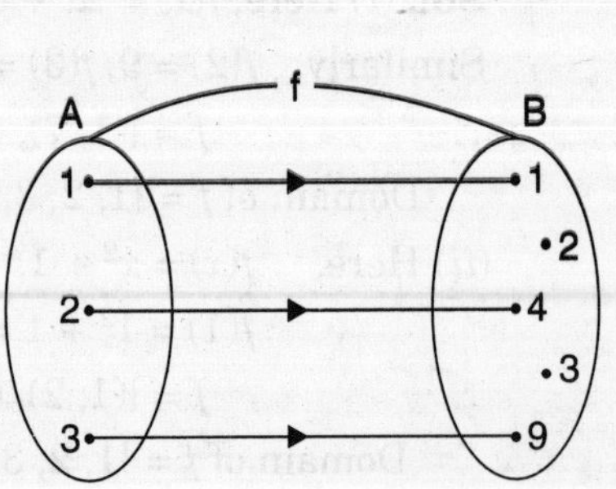

For example, if A = {1, 2, 3}, B = {1, 2, 3, 4, 9} and $f : A \to B$ defined by $f(x) = x^2$; then $f(1) = 1, f(2) = 4, f(3) = 9$. Distinct elements of A are taken to distinct elements of B by f. $\therefore$ f is an injection.

Second Definition of one-one function

A function $f : A \to B$ is said to be **one-one** if $x_1, x_2 \in A$ and $f(x_1) = f(x_2) \Rightarrow x_1 = x_2$

Remark 1. This second definition is more useful for proving a function to be one-one for which the rule is given.

Remark 2. If $x_1, x_2 \in A$ and $x_1 = x_2$, then $f(x_1) = f(x_2)$ is always true wherever f is a function.

Caution. Therefore we should never start with $x_1 = x_2$ and prove $f(x_1) = f(x_2)$ for proving a function to be one-one.

20. MANY-ONE FUNCTION

A function $f : A \to B$ is said to be a many-one function if two or more elements of set A have the same image in B. Thus $f : A \to B$ is a many-one function if there exist $x_1, x_2 \in A$ *s.t.* $x_1 \neq x_2$ but $f(x_1) = f(x_2)$.

Remark. A function which is not a one-one function is a many-one function.

For example, Let A = {0, – 1, 1} and B = {0, 1}. Let us define a function $f : A \to B$ as $f(x) = x^2$.

$\therefore \quad f(0) = 0^2 = 0, f(-1) = (-1)^2 = 1, f(1) = 1^2 = 1$

Here $\quad -1 \neq 1$ but $f(-1) = f(1)\ (= 1)$.

$\therefore$ f is a many-one function.

21. ONE-ONE ONTO FUNCTION OR BIJECTIVE FUNCTION

A function which is **both one-one** (injective) and **onto** (surjective) is said to be a one-one onto function or a bijective function or simply a bijection.

For example, if A = {1, 2, 3}, B = {2, 4, 6} and $f : A \to B$ defined by $f(x) = 2x$, then $f(1) = 2, f(2) = 4, f(3) = 6$.

Since distinct elements of A have distinct f-images in B, therefore, f is an injection.

Also $f(A) = \{2, 4, 6\} = B$ *i.e.*, Range of f = Co-domain of f. Therefore f is a surjection.

Since f is both **injection** as well as **surjection,** therefore, f is a **bijection.**

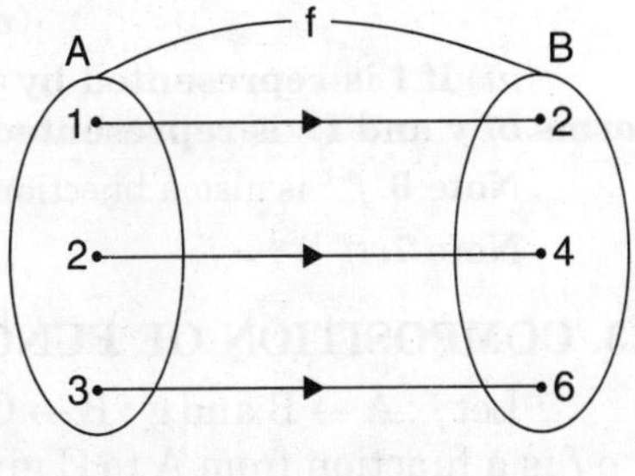

One-One onto function (Bijection)

22. INVERSE FUNCTION

Consider a **bijection** $f : A \to B$. Since f is one-one, distinct elements of A are taken to distinct elements of B by f. Also f is onto, therefore, every element of B has pre-image in A. Combining, for any $x \in B$, there exists one and only one $a \in A$ such that $f(a) = x$.

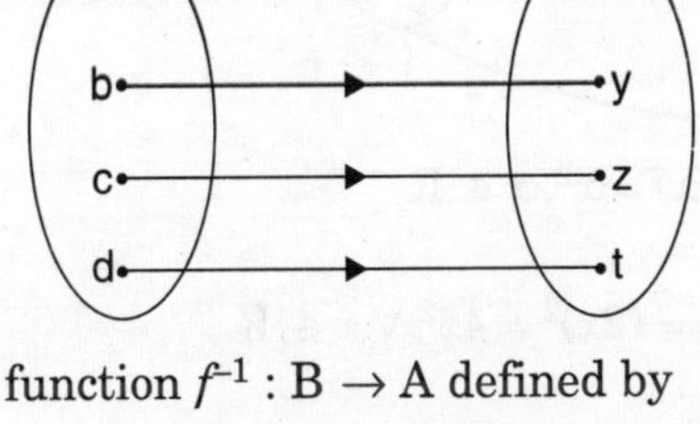

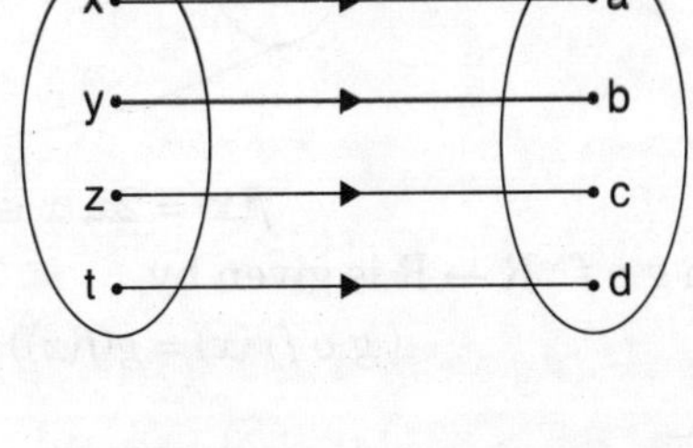

The function $f^{-1} : B \to A$ defined by

$f^{-1}(x) = a$ if and only if $f(a) = x$. f^{-1} is called the *inverse function* of f.

Thus, if $f : A \to B$ where $f = \{(a, x), (b, y), (c, z), (d, t)\}$, then $f^{-1} = \{(x, a), (y, b), (z, c), (t, d)\}$.

Note 1. If f : A → B, then f^{-1} : B → A exists if and only if f is one-one and onto i.e., iff f is a bijection.

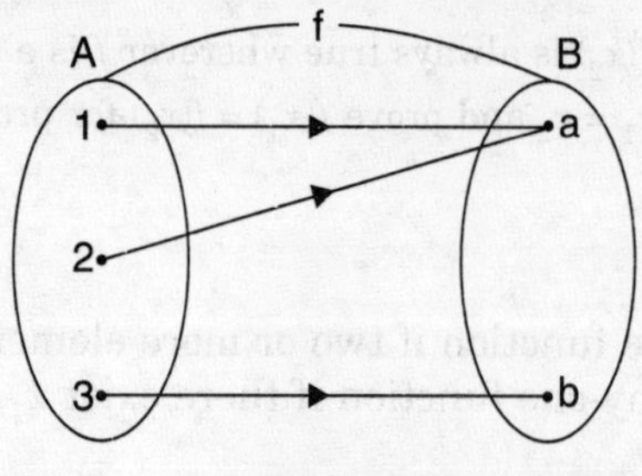

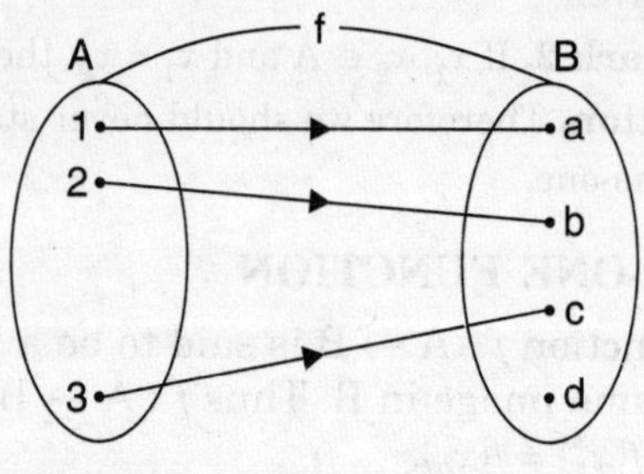

[f is onto but not one-one. f^{-1} does not exit.] [f is one-one but not onto. f^{-1} does not exit]

Remark. If R is a relation from A to B ; then R^{-1} always exist.

Note 2. Domain of f^{-1} = Range of f : Range of f^{-1} = Domain of f.

Note 3. $f(a) = b \Leftrightarrow a = f^{-1}(b)$.

Note 4. f^{-1} exists $\Rightarrow$ f is invertible.

Note 5. Rule to find f^{-1} when it exists :

(*i*) If f is represented by an arrow diagram, then f^{-1} is obtained by reversing the arrows in the diagram.

(*ii*) If f is represented by ordered pairs (x, y), then f^{-1} is obtained by interchanging x and y, *i.e.*,

$$(x, y) \in f \Rightarrow (y, x) \in f^{-1} \text{ provided } f \text{ is a bijection.}$$

(*iii*) **If f is represented by y = f(x), then we solve the equation y = f(x) so as to get x in terms of y and f^{-1} is represented by x = f^{-1} (y).**

Note 6. f^{-1} is also a bijection and f^{-1} is unique.

Note 7. $(f^{-1})^{-1} = f$.

23. COMPOSITION OF FUNCTIONS

Let $f : A \to B$ and $g : B \to C$ be two functions. Then the composition of f and g, denoted by $g \circ f$ is a function from A to C given by $g \circ f : A \to C$ such that **(g o f) (x) = g(f(x)) ∀ x ∈ A.**

Note. Domain of $g \circ f$ is the domain of f and co-domain of $g \circ f$ is the co-domain of g.

For example, (*i*) Let $f : R \to R$ and $g : R \to R$ be two functions defined respectively by

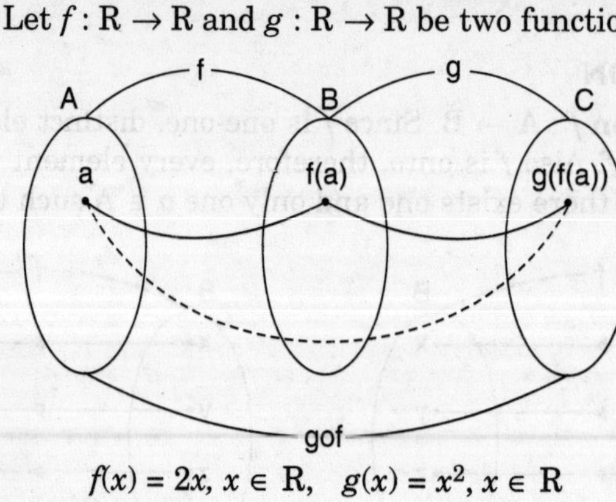

$$f(x) = 2x, x \in R, \quad g(x) = x^2, x \in R$$

Then $g \circ f : R \to R$ is given by

$$(g \circ f)(x) = g(f(x)) = g(2x) = (2x)^2 = 4x^2 \ \forall\ x \in R.$$

(*ii*) Let $f : A \to B$ and $g : B \to C$ be two functions defined by

$$f(x) = |x|, x \in A \text{ and } g(x) = 2x + 1, x \in B$$

where $\quad A = \{-1, 0, 1\}, B = \{0, 1\}, C = \{1, 2, 3, 4\}$

Then $\quad f(-1) = |-1| = 1, f(0) = 0, f(1) = 1$

$$(g \circ f)(-1) = g(f(-1)) = g(1) = 2(1) + 1 = 3$$

$$(g \circ f)(0) = g(f(0)) = g(0) = 2(0) + 1 = 1$$

$$(g \circ f)(1) = g(f(1)) = g(1) = 3$$

$$(g \circ f)(x) = g(f(x)) = g(|x|) = 2|x| + 1. \qquad [\because \quad g(x) = 2x + 1]$$

Thus, $g \circ f : A \to C$ defined by

$$(g \circ f)(x) = 2|x| + 1, x \in A$$

Note. g o f is defined only when range of f ⊂ domain of g.

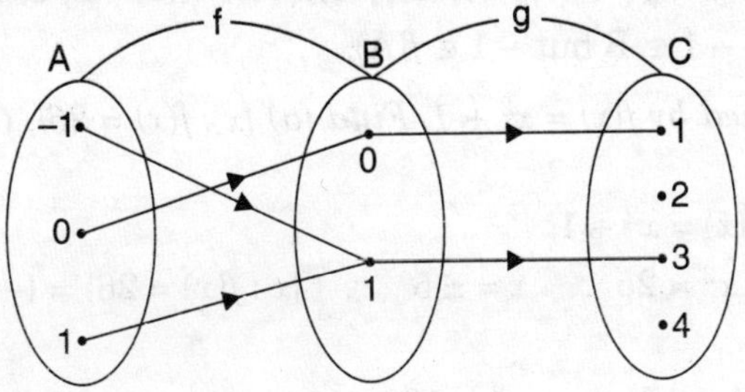

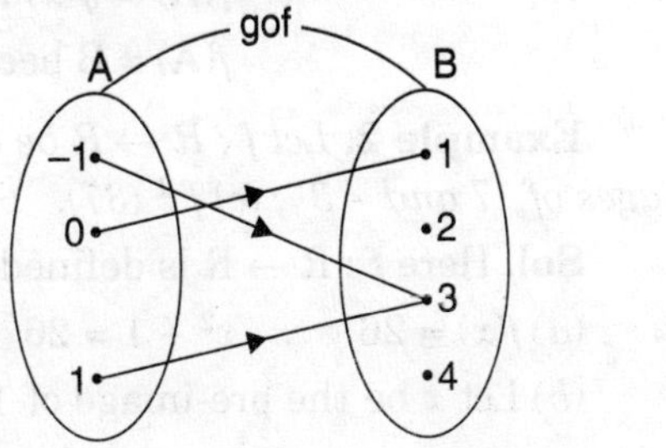

24. IDENTITY FUNCTION

The identity function (or map) of a set A onto itself is denoted by I_A and defined as

$$I_A(x) = x \;\forall\; x \in A$$

i.e., the identity function I_A if denoted by f is given by

$$y = f(x) = x.$$

Clearly, I_A is a bijection (*i.e.*, **one-one and onto**) and **is its own inverse.**

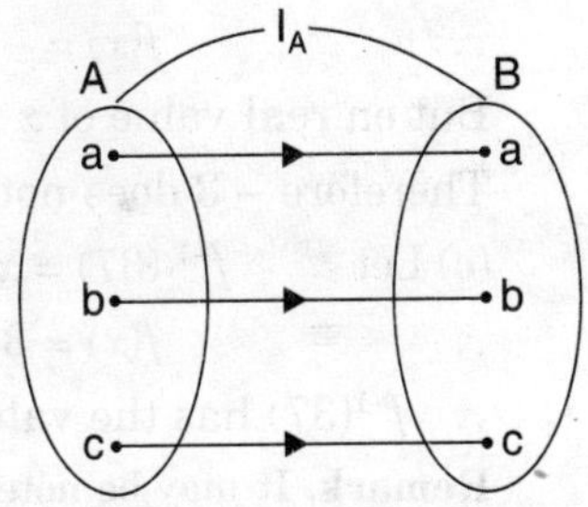

Remark. If $f : A \to A$ is a bijection from A onto itself ;

then $\quad \mathbf{f^{-1} \circ f = f \circ f^{-1} = I_A}$

25. EVEN AND ODD FUNCTIONS

A function $f(x)$ is said to be an **even function** of x if **f(– x) = f(x)** and an odd function if **f(– x) = – f(x).**

For example, (*i*) $f(x) = x^2$ is an even function.

$$(\because \quad f(-x) = (-x)^2 = x^2 = f(x))$$

(*ii*) $g(x) = x + \dfrac{1}{x}$ $(x \neq 0)$ is an odd function.

$$\left(\because \quad g(-x) = -x - \frac{1}{x} = -\left(x + \frac{1}{x}\right) = -g(x)\right)$$

Every function need not be even or odd.

The function $f(x) = x + 1$ is neither even nor odd

$$[\because \quad f(-x) = -x + 1 \neq f(x). \text{ Also } f(-x) \neq f(x)].$$

Example 1. *Given* $A = \{-1, 0, 2, 5, 6\}$ *and* $B = \{-2, -1, 0, 18, 28\}$ *and* $f(x) = x^2 - x - 2$ *; find f(A). Is f(A) = B ?*

Sol. Set $A = \{-1, 0, 2, 5, 6\}$.

$$f(x) = x^2 - x - 2 \qquad ...(1) \text{ (given)}$$

Putting $x = -1, 0, 2, 5, 6$ successively in Eqn. (1),

$$f(-1) = 1 + 1 - 2 = 0$$
$$f(0) = 0^2 - 0 - 2 = -2$$
$$f(2) = 2^2 - 2 - 2 = 0$$
$$f(5) = 5^2 - 5 - 2 = 18$$
$$f(6) = 6^2 - 6 - 2 = 28$$

$\therefore$ $f(A) = \{f(x) : x \in A\} = \{f(-1), f(0), f(2), f(5), f(6)\} = \{0, -2, 0, 18, 28\}$

$f(A) \neq B$ because $-1 \in B$ but $-1 \notin f(A)$.

Example 2. *Let* $f : R \to R$ *be defined by* $f(x) = x^2 + 1$. *Find (a)* $\{x : f(x) = 26\}$, *(b) the pre-images of 17 and – 3 ; (c)* $f^{-1}(37)$.

Sol. Here $f : R \to R$ is defined by $f(x) = x^2 + 1$.

(*a*) $f(x) = 26$ $\therefore$ $x^2 + 1 = 26$ or $x^2 = 25$ $\therefore$ $x = \pm 5$ $\therefore$ $\{x : f(x) = 26\} = \{-5, 5\}$

(*b*) Let x be the pre-image of 17.

$\therefore$ $f(x) = 17$ or $x^2 + 1 = 17$ or $x^2 = 16$

$\therefore$ $x = \pm 4$

$\therefore$ Pre-images of 17 are – 4 and 4.

Let x be the pre-image of – 3.

$\therefore$ $f(x) = -3$ or $x^2 + 1 = -3$ or $x^2 = -4$

But on real value of x satisfies $x^2 = -4$.

Therefore – 3 does not have any pre-image under f.

(*c*) Let $f^{-1}(37) = x$

$\therefore$ $f(x) = 37$ or $x^2 + 1 = 37$ or $x^2 = 36$ $\therefore$ $x = \pm 6$.

$\therefore$ $f^{-1}(37)$ has the values – 6 and 6.

Remark. It may be noted here that $f^{-1}(x)$ *i.e.*, f^{-1} (an element of co-domain) may exist without f being one-one and onto. Here f is neither one-one ($\because f(-x) = f(x)$) nor onto ($\because$ In (*b*) part above $-3 \in$ co-domain R does not have any pre-image in R).

But the function f^{-1} exists iff f is one-one and onto.

Example 3. *If* $f(x) = \dfrac{1}{1-x}$, *find* $f\left(f\left(\dfrac{1}{2}\right)\right)$.

Sol.
$$f(x) = \frac{1}{1-x} \qquad ...(1) \text{ (given)}$$

Putting $x = \dfrac{1}{2}$ in (1), $f\left(\dfrac{1}{2}\right) = \dfrac{1}{1-\dfrac{1}{2}} = \dfrac{1}{\left(\dfrac{1}{2}\right)} = 2$

$$\therefore \quad f\left(f\left(\frac{1}{2}\right)\right) = f(2) = \frac{1}{1-2} \qquad \text{(Putting } x = 2 \text{ in (1))}$$

$$= \frac{1}{-1} = -1.$$

Example 4. *If $f : R \to R$ is defined by $f(x) = x^2 - 3x + 2$; find $f(f(x))$.*

Sol. $$f(x) = x^2 - 3x + 2 \qquad \text{...(1) (given)}$$

$$\therefore \quad f(f(x)) = f(x^2 - 3x + 2) \qquad \text{(By (1))}$$

Changing x to $x^2 - 3x + 2$ in (1),

$$= (x^2 - 3x + 2)^2 - 3(x^2 - 3x + 2) + 2$$

$$= x^4 + 9x^2 + 4 - 6x^3 + 4x^2 - 12x - 3x^2 + 9x - 6 + 2$$

$$[\because \quad (a + b + c)^2 = a^2 + b^2 + c^2 + 2ab + 2ac + 2bc]$$

$$= x^4 - 6x^3 + 10x^2 - 3x.$$

Example 5. *If $f(x) = \log \dfrac{1+x}{1-x}$ show that $f\left(\dfrac{2x}{1+x^2}\right) = 2f(x)$.*

Sol. $$f(x) = \log \frac{1+x}{1-x} \qquad \text{...(1)}$$

Putting $x = \dfrac{2x}{1+x^2}$ in Eqn. (1),

We have $$f\left(\frac{2x}{1+x^2}\right) = \log\left(\frac{1+\dfrac{2x}{1+x^2}}{1-\dfrac{2x}{1+x^2}}\right)$$

Multiplying by L.C.M. $= 1 + x^2$

$$= \log\left(\frac{1+x^2+2x}{1+x^2-2x}\right) = \log\frac{(1+x)^2}{(1-x)^2}$$

$$= \log\left(\frac{1+x}{1-x}\right)^2 = 2\log\frac{1+x}{1-x} = 2f(x). \qquad \text{[By (1)]}$$

Example 6. *If A = {a, b, c, d} and f corresponds to the Cartesian product {(a, b), (b, d), (c, a), (d, c)}, show that f is one-to-one from A onto A. Find f^{-1}.*

Sol. Because f corresponds to the Cartesian product {(a, b), (b, d), (c, a), (d, c)}, therefore

$$f(a) = b,\ f(b) = d,\ f(c) = a,\ f(d) = c$$

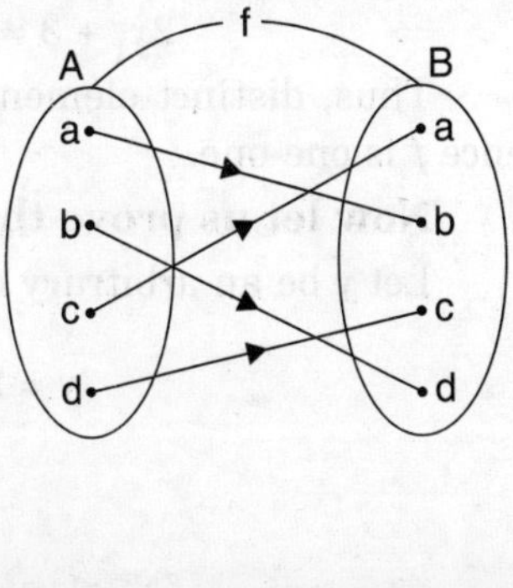

$\therefore$ Distinct elements of A have distinct f-images in A.

$\Rightarrow$ f is a one-to-one function.

Also $f(A) = \{b, d, a, c\} = A$

$\therefore$ f is an onto function.

Hence f is one-to-one from A onto A.

Reversing the entries in the ordered pairs, we have

$$f^{-1} = \{(b, a), (d, b), (a, c), (c, d)\}.$$

Example 7. *Given an example of a map :*

(i) which is one-to-one but not onto *(ii) which is not one-to-one but onto*

(iii) which is neither one-to-one nor onto.

Sol. (*i*) Consider the map $f : N \to N$, defined by $f(x) = 2x$, N being the set of all natural numbers.

Let x_1, x_2 be two distinct elements of the domain N.

$$x_1 \neq x_2 \Rightarrow 2x_1 \neq 2x_2 \Rightarrow f(x_1) \neq f(x_2)$$

$\therefore$ Distinct elements of the domain have distinct f-images. $\therefore$ f is one-to-one.

Range of $f = \{2, 4, 6, ...\}$ = the set of all even natural numbers $\neq$ N, the co-domain.

$\therefore$ f is not onto. Hence f is one-to-one but not onto.

(*ii*) Consider the map $f : A \to B$, defined by $f(x) = x^2$, where

$$A = \{1, -1, 2\} \text{ and } B = \{1, 4\}.$$

The element $1 \in B$ has more than one pre-images, *viz.*, 1 and – 1 belonging to A.

$$[\because \ f(1) = (1)^2 = 1 \text{ and } f(-1) = (-1)^2 = 1]$$

$\therefore$ f is not one-to-one.

OR $1, -1, \in A$ and $1 \neq -1$ But $f(1) = 1 = f(-1)$

$\therefore$ f is not one-to-one.

Range of $f = f(A) = \{1, 4\} = B$, the co-domain of f. $\therefore$ f is onto.

Hence f is not one-to-one but onto.

(*iii*) Consider the map $f : R \to R$, defined by $f(x) = |x|$, where R is the set of all real numbers.

$$2, -2 \in R, \text{ the domain of } f \text{ and } 2 \neq -2$$

But $f(2) = 2 = f(-2)$ $\therefore$ f is not one-to-one.

Range of $f = f(R)$ = the set of non-negative real numbers $(\because \ f(x) = |x| \geq 0) \neq R$

$\therefore$ f is not onto.

Hence f is neither one-to-one nor onto.

Example 8. *Let Q be the set of rational numbers. Let $f : Q \to Q$ be defined by $f(x) = 2x + 3$, $x \in Q$. Show that f is one-to-one and onto. Also find a formula which defines the inverse function f^{-1}.*

Sol. Let $x_1, x_2 \in Q$, the domain of f, and $x_1 \neq x_2$.

Now $x_1 \neq x_2 \Rightarrow 2x_1 \neq 2x_2$

$\Rightarrow$ $2x_1 + 3 \neq 2x_2 + 3 \Rightarrow f(x_1) \neq f(x_2)$.

Thus, distinct elements in the domain Q have distinct f-images in the co-domain Q. Hence f is one-one.

Now let us prove that f is onto.

Let y be an arbitrary element of the co-domain Q. If $y = f(x)$, then

$$y = 2x + 3 \quad \text{or} \quad x = \frac{y-3}{2}$$ which is also a rational number.

Thus, $f\left(\dfrac{y-3}{2}\right) = x$, *i.e.*, every element y of the co-domain Q has pre-image $\dfrac{y-3}{2}$ in the domain Q. Hence f is onto.

Since $f : Q \to Q$ is one-one and onto, therefore, f has the inverse function $f^{-1} : Q \to Q$.

Now let us find the rule for f^{-1}.

Let y be the image of x under the function f. Then $y = f(x) = 2x + 3$.

Consequently, x will be the image of y under the inverse function f^{-1} *i.e.*, $x = f^{-1}(y)$.

Solving for x in terms of y in the equation $y = 2x + 3$, we get $x = \dfrac{y-3}{2}$

Hence $\quad (x =)\ f^{-1}(y) = \dfrac{y-3}{2}$.

Remark. For the above example, here below we give another method to prove that $f : Q \to Q$ defined by

$$f(x) = 2x + 3,\ x \in Q \qquad \text{...(1)}$$

is one-one.

Let $x_1, x_2 \in Q$ such that $\quad f(x_1) = f(x_2)$.

$\Rightarrow \quad 2x_1 + 3 = 2x_2 + 3 \quad$ (Putting $x = x_1$ and $x = x_2$ in (1))

$\Rightarrow \quad 2x_1 = 2x_2$ or $x_1 = x_2$

$\therefore$ f is one-one.

Example 9. *If* $f = \{(5, 2), (6, 3)\}$, $g = \{(2, 5), (3, 6)\}$, *find* $f \circ g$.

Sol. Here Domain of $f = \{5, 6\} = A$ (say), Range of $f = \{2, 3\} = B$(say)

Domain of $g = \{2, 3\} = B$, Range of $g = \{5, 6\} = A$

$\therefore \quad f : A \to B$ and $g : B \to A \Rightarrow f \circ g : B \to B$

Now $\quad f \circ g(2) = f(g(2)) = f(5) = 2$, $f \circ g(3) = f(g(3)) = f(6) = 3$

$\therefore \quad f \circ g = \{(2, 2), (3, 3)\}$.

Example 10. *If* $f, g : R \to R$ *are defined respectively by* $f(x) = 3x + 1$ *and* $g(x) = x^2 + 2$, *find formulae for*

(*i*) $f \circ g$ (*ii*) $g \circ f$ (*iii*) $f \circ f$ (*iv*) $g \circ g$.

Sol. $f, g : R \to R$ are defined by

$$f(x) = 3x + 1 \qquad \text{...(1)}$$

and

$$g(x) = x^2 + 2 \qquad \text{...(2)}$$

(*i*) $\quad (f \circ g)\,x = f(g(x)) = f(x^2 + 2) \quad$ [By (2)]

Putting $\quad x = x^2 + 2$ in (1) $= 3(x^2 + 2) + 1 = 3x^2 + 7$

(*ii*) $\quad (g \circ f)\,x = g(f(x)) = g(3x + 1) \quad$ [By (1)]

Putting $\quad x = 3x + 1$ in (2)

$$= (3x + 1)^2 + 2 = 9x^2 + 1 + 6x + 2 = 9x^2 + 6x + 3$$

(*iii*) $\quad (f \circ f)x = f(f(x)) = f(3x + 1) \quad$ [By (1)]

Putting $\quad x = 3x + 1$ in (1) $= 3(3x + 1) + 1 = 9x + 4$

(iv) $(g \circ g)x = g(g(x)) = g(x^2 + 2)$ [By (2)]

Putting $x = x^2 + 2$ in (2)

$$= (x^2 + 2)^2 + 2 = x^4 + 4 + 4x^2 + 2 = x^4 + 4x^2 + 6$$

Example 11. *If $f, g : R \to R$ are defined respectively by $f(x) = x^2 + 3x + 1$, $g(x) = 2x - 3$, find formulae for*

(i) $f \circ g$ *(ii)* $g \circ f$ *(iii)* $f \circ f$ *(iv)* $g \circ g$.

Sol. *(i)*
$$f \circ g(x) = f(g(x)) = f(2x - 3) = (2x - 3)^2 + 3(2x - 3) + 1$$
$$= (4x^2 - 12x + 9) + 6x - 9 + 1 = 4x^2 - 6x + 1$$

(ii)
$$g \circ f(x) = g(f(x)) = g(x^2 + 3x + 1) = 2(x^2 + 3x + 1) - 3 = 2x^2 + 6x - 1$$

(iii)
$$f \circ f(x) = f(f(x)) = f(x^2 + 3x + 1) = (x^2 + 3x + 1)^2 + 3(x^2 + 3x + 1) + 1$$
$$= (x^4 + 9x^2 + 1 + 6x^3 + 2x^2 + 6x) + 3x^2 + 9x + 3 + 1$$
$$= x^4 + 6x^3 + 14x^2 + 15x + 5$$

(iv)
$$g \circ g(x) = g(g(x)) = g(2x - 3) = 2(2x - 3) - 3 = 4x - 9.$$

Example 12. *If $f : A \to B$ is a function, prove that $f \circ I_A = f = I_B \circ f$.*

Sol. We know that $I_A : A \to A$ and $I_B : B \to B$

$\therefore$ $f \circ I_A : A \to B$ and $I_B \circ f : A \to B$

Let a be any element of A.

$$(f \circ I_A)(a) = f(I_A(a)) = f(a) \ \forall \ a \in A.$$

$\Rightarrow$ $f \circ I_A = f$

$$(I_B \circ f)(a) = I_B(f(a)) = f(a) \ \forall \ a \in A.$$

$\Rightarrow$ $I_B \circ f = f$

Hence $f \circ I_A = f = I_B \circ f$.

Example 13. *Prove that the composite function of two one-one functions is a one-one function.*

Sol. Let $f : A \to B$ and $g : B \to C$ be two one-one functions.

Then $g \circ f : A \to C$

We are to prove that $g \circ f : A \to C$ is one-one

Let $x_1, x_2 \in A$ such that $(g \circ f)(x_1) = (g \circ f)(x_2)$

$\Rightarrow$ $g(f(x_1)) = g(f(x_2))$

$\Rightarrow$ $f(x_1) = f(x_2)$ | $\because$ g is one-one (given)

$\Rightarrow$ $x_1 = x_2$ | $\because$ f is one-one (given)

$\therefore$ $g \circ f$ is a one-one function.

Example 14. *Prove that the composite function of two surjections (i.e., onto functions) is a surjection.*

Sol. Let $f : A \to B$ and $g : B \to C$ be two onto functions.

Then $g \circ f : A \to C$

Let $z \in C$.

Because $g : B \to C$ is onto ; therefore there exists $y \in B$ such that $g(y) = z$

Because $y \in B$ and $f : A \to B$ is onto, therefore there exists $x \in A$ such that $y =$

Putting $y = f(x)$ from (2) in (1), we have

$$z = g(f(x)) \text{ or } z = (g \circ f)\, x$$

$\therefore$ Co-domain of $g \circ f$ namely C is also range of $g \circ f$.

$\therefore$ $g \circ f$ is onto.

Example 15. *Prove that composite of two one-one and onto functions is one-one and onto.*

Sol. Reproduce examples (13) and (14).

Example 16. *If $f : A \to B$ is a bijection, prove that $f^{-1} \circ f = I_A$ and $f \circ f^{-1} = I_B$.*

Sol. $f : A \to B$ is a bijection

$\Rightarrow$ f^{-1} exists and $f^{-1} : B \to A$

$\therefore$ $f^{-1} \circ f : A \to A$ and $f \circ f^{-1} : B \to B$

Let $a \in A$, then there exists a unique element $b \in B$ such that

$$f(a) = b \quad \text{and} \quad f^{-1}(b) = a$$

Now $\quad (f^{-1} \circ f)(a) = f^{-1}(f(a)) = f^{-1}(b) = a = I_A(a) \quad \therefore \quad f^{-1} \circ f = I_A.$

Also $\quad (f \circ f^{-1})(b) = f(f^{-1}(b)) = f(a) = b = I_B(b) \quad \therefore \quad f \circ f^{-1} = I_B.$

Example 17. *Find the domains of the following functions :*

(i) $\dfrac{x}{x^2 - 3x + 2}$ *(ii)* $\sqrt{2x + 1}$ *(iii)* $\dfrac{1}{\sqrt{x + 3}}$ *(iv)* $\dfrac{1}{\sqrt{(x - 1)(2 - x)}}$.

Sol. Domain is the set of those values of x for which $f(x)$ is real and finite.

(i) Let $\quad f(x) = \dfrac{x}{x^2 - 3x + 2}$

$f(x) \to \infty$ (*i.e.*, is not finite) when $x^2 - 3x + 2 = 0$

or $\quad (x - 1)(x - 2) = 0$

$\therefore \quad x = 1, 2$

$\therefore \quad$ Domain $f = R - \{x : x^2 - 3x + 2 = 0\} = R - \{1, 2\}$

(ii) Let $\quad f(x) = \sqrt{2x + 1}$

For $f(x)$ to be real and finite, we must have $2x + 1 \geq 0$

i.e., $\quad 2x \geq -1 \quad \text{or} \quad x \geq -\dfrac{1}{2}$

$\therefore \quad$ Domain of $f = \left\{x : x \geq -\dfrac{1}{2}\right\}$

(iii) Let $\quad f(x) = \dfrac{1}{\sqrt{x + 3}}$

For $f(x)$ to be real and finite, we must have

$$x + 3 > 0 \quad i.e., \quad x > -3.$$

$\therefore \quad$ Domain of $f = \{x : x > -3\}$.

(iv) Let $\quad f(x) = \dfrac{1}{\sqrt{(x - 1)(2 - x)}}$

For $f(x)$ to be real and finite, $(x-1)(2-x) > 0$

or $$-(x-1)(x-2) > 0$$

$\therefore$ $$(x-1)(x-2) < 0$$

Case I. $x - 1 > 0$ and $x - 2 < 0$

$\Rightarrow$ $x > 1$ and $x < 2$ $\Rightarrow$ $1 < x < 2$

Case II. $x - 1 < 0$ and $x - 2 > 0$

$\Rightarrow$ $x < 1$ and $x > 2$

But this is impossible.

$\therefore$ $(x-1)(2-x) > 0$ only when $1 < x < 2$

$\therefore$ Domain of f is $\{x : 1 < x < 2\}$.

Example 18. *Find the range of the following functions :*

(i) $\sin^2 5x + \cos^2 5x$ (ii) $5 - 3 \cos 2x$.

Sol. (i) Let $f(x) = \sin^2 5x + \cos^2 5x = 1 \ \forall \ x \in \mathbf{R}$.

$\therefore$ f is the constant function 1.

$\therefore$ Range of $f = \{1\}$.

(ii) Let $f(x) = 5 - 3 \cos 2x$

We know that for all $x \in \mathbf{R}$, $-1 \le \cos 2x \le 1$

Multiplying by -3, $3 \ge -3 \cos 2x \ge -3$

Adding 5 to all sides $8 \ge 5 - 3 \cos 2x \ge 2$

or $8 \ge f(x) \ge 2$ or $2 \le f(x) \le 8$

$\therefore$ Range of f is $\{f(x) : 2 \le f(x) \le 8\}$.

Example 19. *If* $y = f(x) = \dfrac{2x-3}{5x-2}$, *show that* $f(y) = x$.

Sol. $$y = f(x) = \frac{2x-3}{5x-2} \quad \text{(given) ...(1)}$$

Cross-multiplying

$$y(5x-2) = 2x - 3 \quad \text{or} \quad 5xy - 2y = 2x - 3$$

or $$5xy - 2x = 2y - 3 \quad \text{or} \quad x(5y-2) = 2y - 3$$

$\therefore$ $$x = \frac{2y-3}{5y-2} = f(y)$$

$$\left[\because \text{ Changing } x \text{ to } y \text{ in Eqn. (1), } f(y) = \frac{2y-3}{5y-2}.\right]$$

Example 20. *Find the domain and range of the following functions :*

(i) $\dfrac{x^2-1}{x-1}$ (ii) $\dfrac{3-x}{x-3}$.

Sol. (i) Let $$f(x) = \frac{x^2-1}{x-1}$$

Dom f = the set of all real numbers x for which $f(x)$ is real and finite.

$= \mathbf{R} - \{x : x - 1 = 0\} = \mathbf{R} - \{x : x = 1\} = \mathbf{R} - \{1\}$

Range of *f*.

Since $x \neq 1$, we have $x - 1 \neq 0$

$$f(x) = \frac{(x+1)(x-1)}{x-1} = x + 1$$

Now $\quad x \neq 1 \Rightarrow x + 1 \neq 2 \Rightarrow f(x) \neq 2$

Range f = the set of all $f(x)$ such that $x \in$ dom f = R – {2}.

(*ii*) Let $\quad f(x) = \dfrac{3-x}{x-3}$

Domain of $f(x)$ is the set of all real x for which $f(x)$ is real and finite.

Domain of f = R – {x : x – 3 = 0} = R – {3}

Range of *f*.

Since $x \neq 3$, we have $x - 3 \neq 0$.

$$f(x) = \frac{-(x-3)}{x-3} = -1$$

$\therefore$ Range of f = {– 1}.

EXERCISE 2

1. If A = {1, 2, 3} and f, g are relations corresponding to the subsets of A × A indicated against them, which of f, g is a function ? Why ?

f = {(1, 3), (2, 3), (3, 2)}, g = {(1, 2), (1, 3), (3, 1)}.

2. Given A = {1, 3, 5} and B = {2, 4, 6, 8}, point out which of the following sets of ordered pairs define a function. Find their domain and range.

(*i*) R_1 = {(1, 2), (3, 4), (5, 6)} (*ii*) R_2 = {(1, 4), (3, 6), (5, 8)}

(*iii*) R_3 = {(1, 6), (3, 8), (3, 4), (5, 2)} (*iv*) R_4 = {(1, 8), (3, 8),(5, 8)}.

3. Given A = {1, 2, 3} and B = {1, 2, 3, 4, 5, 6, 7, 8, 9}. Write the following function from A to B as the set of ordered pairs. Also write the domain and range in each case.

(*i*) $f(x) = 3x - 1$ (*ii*) $f(x) = x^2 - x + 2$.

4. If $f(x) = x^2 - 3x + 1$, find x so that $f(x) = f(2x)$.

5. If $f(x) = \dfrac{1}{1-x}$, find $f(f(x))$.

6. What is the fundamental difference between a relation and a function ? Is every relation a function ?

[**Hint.** See Art. 15, Note 1].

7. Let A = {– 2, – 1, 0, 1, 2} and f : A → Z (set of Integers) be given by $f(x) = x^2 - 2x - 3$. Find (*a*) the range ; (*b*) pre-images of 6, – 3, 5.

8. Let f : R → R be given by $f(x) = x^2 + 3$. Find (*a*) {x : $f(x) = 28$} (*b*) the pre-images of 39 and 2 under f.

9. Let A = {12, 13, 14, 15, 16 ; 17} and f : A → Z be a function given by $f(x)$ = Highest prime factor of x. Find range of f.

(**Hint.** $f(12)$ = Highest prime factor of 12 = 3) ($\because$ Factors of 12 are 1, 2, 3, 6, 12).

10. Let A = {1, 2, 3} and B = {2, 4, 6}. A function $f : A \to B$ is defined by

(*i*) {(1, 2), (2, 4), (3, 2)}. Prove that f maps A into B.

(*ii*) {(1, 2), (2, 4), (3, 6)}. Prove that f is a bijection. Find f^{-1}. Is f^{-1} also a bijection ?

11. Show that $f : N \to N$ defined by $f(x) = 3x$, N being the set of all natural numbers, is one-one but not onto.

12. Show that $f : R \to R$ defined by $f(x) = \cos x$, R being the set of all real numbers, is neither one-one nor onto.

(Not one-one as $\cos(-x) = \cos x$. Not onto because range of $\cos x$ is $[-1, 1] \neq R$).

13. Show that $f : R \to Z$ defined by $f(x) = [x]$, where $[x]$ denotes the greatest integer $\leq x$: Z being the set of all integers, is not one-one but onto.

14. Show that $f : R \to R$ defined by $f(x) = x^2$ is neither one-one nor onto.

15. Show that $g : A \to B$ defined by $g(x) = 3x + 1$, where A = {1, 2, 3} and B = {4, 7, 10}, is a bijection.

16. Let R_0 denote the set of all non-zero real numbers. Prove that the map $f : R_0 \to R_0$ Given by $f(x) = \frac{1}{x}$, $x \in R_0$ is both one-to-one and onto.

17. Let N be the set of natural numbers and E be the set of even natural numbers. Let $f : N \to E$ be defined by $f(x) = 2x$, $x \in N$. Show that the map f is a bijection. Find the formula that defines the inverse function f^{-1}.

18. For each of the following functions find whether it is invertible or not. If not, give reason. If invertible, find the inverse function.

(*i*) $f : A \to B$, $f(x) = x^2$, A = {1, 2, 3, 4}, B = {1, 4, 9, 16, 25}.

(*ii*) $g : A \to B$, $g(x) = x^2$, A = {0, 1, 2, 3}, B = {0, 1, 4, 9}.

19. Let A and B be the sets A = {1, 2, 3, 4, 5} and B = {2, 5, 10,17, 26}. Let f be a function from A to B defined by $f(x) = x^2 + 1$. Find f^{-1}, if it exists.

20. Let f = {(1, 3), (2, 5), (3, 3), (4, 1), (5, 2)} and g = {(1, 4), (2, 1), (3, 1), (4, 2), (5, 3)}. Find $f \circ g$ and $g \circ f$.

21. Given $f(x) = x^3$, $g(x) = \sqrt{x} - 1$, find

(*i*) $(g \circ f)(1)$, (*ii*) $(f \circ g)(9)$.

22. If $f(x) = a^x$, show that $f(x) f(y) = f(x + y)$.

23. If $f(x) = \log x$, show that $f\left(\frac{a}{b}\right) = f(a) - f(b)$, where $a > 0$, $b > 0$.

24. If $f(x) = \frac{1}{1-x}$, find $(f \circ f)\left(\frac{1}{2}\right)x$.

25. If $f(x) = \frac{x-1}{x+1}$, show that $(f \circ f)(x) = -\frac{1}{x}$.

26. If $f(x) = \frac{1+x}{1-x}$, show that $(f \circ f)(\tan x) = -\cot x$.

27. If $y = f(x) = \frac{2x-3}{5x-2}$, show that $f(y) = x$.

28. If $f(x) = \frac{1+e^x}{1-e^x}$, prove that $f(-x) = -f(x)$.

29. Find the domain of the functions :

(i) $\dfrac{x^2}{x^2-1}$ (ii) $\dfrac{7}{x^3-3x^2+2x}$

(iii) $\sqrt{(x-2)(4-x)}$ (iv) $\dfrac{1}{\sqrt{(3-x)(x-5)}}$

(v) $\dfrac{1}{\sqrt{16-x^2}}$.

30. Find the range of the following functions :

(i) $11 - 7\sin x$ (ii) $2 + 3\cos x$.

31. If f, g, h are three functions defined from R to R as follows :

(i) $f(x) = x^2$ (ii) $g(x) = \sin x$ (iii) $h(x) = x^2 + 1$.

Find the range of each function.

32. Let $f : R \to R$ and $g : C \to C$ be two functions defined as $f(x) = x^2$ and $g(x) = x^2$ where R and C denote the sets of real numbers and complex numbers respectively. Are they equal functions ?

Classify the following functions as injection, surjection or bijection.

33. $f : \mathrm{R} \to \mathrm{R}$, $f(x) = \sin x$.

34. $f : \mathrm{R} \to \mathrm{R}$, $f(x) = |\, x \,|$.

35. $f : \mathrm{N} \to \mathrm{N}$, $f(x) = x^3$.

36. $f : \mathrm{R} \to \mathrm{R}$, $f(x) = 5x^3 + 4$.

ANSWERS

1. f is a function, g is not a function.

2. (i) Function, domain = A, range = {2, 4, 6}

(ii) Function, domain = A, range = {4, 6, 8}

(iii) Not a function (iv) Function, domain = A, range = {8}

3. (i) {(1, 2), (2, 5), (3, 8)}, domain = A, range = {2, 5, 8}

(ii) {(1, 2), (2, 4), (3, 8)}, domain = A, range = {2, 4, 8}

4. 0, 1 **5.** $\dfrac{x-1}{x}$

6. No

7. (a) Range (f) = {0, 5, – 3, – 4}

(b) No pre-image of 6 and pre-images of – 3 are $x = 0, 2$ and $x = -2$ is the pre-image of 5.

8. (a) {– 5, 5} (b) pre-images of 39 are $x = -6$ and 6 but there is no pre-image of 2.

9. Range of f = {3, 13, 7, 5, 2, 17} **10.** (ii) {(2, 1), (4, 2), (6, 3)} ; Yes.

17. $f^{-1} : \mathrm{E} \to \mathrm{N}$ where $f^{-1}(y) = \dfrac{y}{2}$.

18. (i) Not invertible, f is not onto since 25 has no pre-image.

(ii) Invertible, $f^{-1} : \mathrm{B} \to \mathrm{A}$, $f^{-1}(y) = \sqrt{y}$.

19. f^{-1} = {(2, 1), (5, 2), (10, 3), (17, 4), (26, 5)}, where $f^{-1}(y) = \sqrt{y-1}$.

20. $f \circ g$ = {(1, 1), (2, 3), (3, 3), (4, 5), (5, 3)}, $g \circ f$ = {(1, 1), (2, 3), (3, 1), (4, 4), (5, 1)}.

21. (i) 0 (ii) 8 **24.** – 1

29. (*i*) R – {– 1, 1} (*ii*) R – {0, 1, 2) (*iii*) $\{x : 2 \le x \le 4\}$
(*iv*) $\{x : 3 < x < 5\}$ (*v*) $\{x : -4 < x < 4\}$

30. (*i*) $\{y : 4 \le y \le 18\}$ (*ii*) $\{y : -1 \le y \le 5\}$

31. (*i*) $\{f(x) : f(x) \ge 0\}$ (*ii*) $\{g(x) : -1 \le g(x) \le 1\}$ (*iii*) $\{h(x) : h(x) \ge 1\}$

32. No, because domain R of $f \ne$ domain C of g **33.** neither injective nor surjective

34. neither injective nor surjective

35. injective **36.** bijective.

---

26. THEOREM(COMPOSITION OF FUNCTIONS IS ASSOCIATIVE)

Let $f : X \to Y$, $g : Y \to Z$ and $h : Z \to T$ be three functions. Then $(hog)of = ho(gof)$.

Proof. Clearly, both the functions $(hog)of$ and $ho(gof)$ are functions from $X \to T$.

Now, if $x \in X$, then

$$((hog)of)(x) = (hog)f(x) = h(g(f(x))) = h((gof)(x)) = (ho(gof))(x)).$$

Hence $(hog)of = ho(gof)$.

27. DIRECT IMAGES

Let $f : X \to Y$ be a function. If $A \subseteq X$, then its direct image $f(A)$ is a subset of Y defined by $f(A) = \{f(x) \mid x \in A\}$.

Examples. (*i*) Let $f : R \to R$ be a function defined by

$$f(x) = 2x + 5.$$

Let $A = \{2, 5, 7, 9\}$.

Then $f(A) = \{f(2), f(5), f(7), f(9)\} = \{9, 15, 19, 23\}$.

(*ii*) Let $f : R \to R$ be a function defined by.

$$f(x) = x^2.$$

Let $A = [-1, 0]$.

$f(x)$ dereases from 1 to 0 as x increases from – 1 to 0.

$\therefore$ $f(A) = [0, 1]$.

Note. $x \in A \Rightarrow f(x) \in f(A)$,

but $f(x) \in f(A) \not\Rightarrow x \in A$.

For example, in example (*ii*), $f(\frac{1}{2}) = \frac{1}{4} \in [0, 1]$, but $\frac{1}{2} \notin A$.

28. THEOREM

Let $f : X \to Y$ be a function and $A \subseteq X$, $B \subseteq X$.

Then (i) $A \subseteq B \Rightarrow f(A) \subseteq f(B)$

(ii) $f(A \cup B) = f(A) \cup f(B)$

(iii) $f(A \cap B) \subseteq f(A) \cap f(B)$.

Proof. (*i*) Let $f(x) \in f(A)$, where $x \in A$

Since $A \subseteq B$, we have

$\therefore$ $x \in B$

$\Rightarrow$ $f(x) \in f(B)$.

Thus $f(x) \in f(A) \Rightarrow f(x) \in f(B)$

So $f(A) \subseteq f(B)$.

(*ii*) If $A = \phi$ or $B = \phi$, then the result is obviously true.

So, let $A \neq \phi, B \neq \phi$.

Since $A \subseteq A \cup B$

$\therefore$ by case (*i*),

$$f(A) \subseteq f(A \cup B)$$

Since $B \subseteq A \cup B$

$\therefore$ $f(B) \subseteq f(A \cup B)$

$\therefore$ $f(A) \cup f(B) \subseteq f(A \cup B)$ ...(1)

Let $f(x) \in f(A \cup B)$, where $x \in A \cup B$

$\Rightarrow$ $x \in A$ or $x \in B$

$\Rightarrow$ $f(x) \in f(A)$ or $f(x) \in f(B)$

$\Rightarrow$ $f(x) \in f(A) \cup f(B)$

$\therefore$ $f(A \cup B) \subseteq f(A) \cup f(B)$ ...(2)

$\therefore$ From (1) and (2),

$$f(A \cup B) = f(A) \cup f(B).$$

(*iii*) Since $A \cap B \subseteq A$

$\therefore$ $f(A \cap B) \subseteq f(A)$ (by (*i*)) ...(1)

Since $A \cap B \subseteq B$

$\therefore$ $f(A \cap B) \subseteq f(B)$ (by (*i*)) ...(2)

$\therefore$ From (1) and (2),

$$f(A \cap B) \subseteq f(A) \cap f(B).$$

Remark. Following example shows that inclusion in (*iii*) may be proper.

Let $f : R \to R^+$ be a function defined by

$$f(x) = x^2$$

Let $A = \{-2, 5\}, B = \{2, 5\}$.

Then $A \cap B = \{5\}$

$$f(A) = \{4, 25\}, f(B) = \{4, 25\}$$

$$f(A) \cap f(B) = \{4, 25\}.$$

$$f(A \cap B) = \{25\}$$

Thus, $f(A \cap B) \subset f(A) \cap f(B)$

However, equality exists in (*iii*) iff f is one-one.

Proof. *Let f be one-one.*

Case (*i*) If $A \cap B = \phi$, then $f(A \cap B) = \phi$

Also $f(A) \cap f(B) = \phi$.

$\therefore$ the result is obviously true.

Case (*ii*) Let $A \cap B \neq \phi$

Let $x \in A \cap B$

$\Rightarrow$ $x \in A$ and $x \in B \Rightarrow f(x) \in f(A)$ and $f(x) \in f(B)$

$\Rightarrow$ $f(x) \in f(A) \cap f(B) \therefore f(A) \cap f(B) \neq \phi$

Let $y \in f(A) \cap f(B)$

$\Rightarrow$ $y \in f(A)$ and $y \in f(B)$

$\Rightarrow$ $\exists\, a \in A$ and $b \in B$ such that

$f(a) = y$ and $f(b) = y$

$\Rightarrow$ $f(a) = f(b) \Rightarrow a = b$ [$\because$ f is one-one]

$\Rightarrow$ $a \in A \cap B \Rightarrow f(a) \in f(A \cap B) \Rightarrow y \in f(A \cap B)$

$\therefore$ $f(A) \cap f(B) \subseteq f(A \cap B)$

Also $f(A \cap B) \subseteq f(A) \cap f(B)$

$\therefore$ $f(A) \cap f(B) = f(A \cap B)$.

Generalization. If $f : X \to Y$ is a function and $\{A_i\}$ $1 \leq i \leq n$ is a family of subsets of X, then

(*i*) $f\left(\bigcup_{i=1}^{n} A_i\right) = \bigcup_{i=1}^{n} f(A_i)$ (*ii*) $f\left(\bigcap_{i=1}^{n} A_i\right) \subseteq \bigcap_{i=1}^{n} f(A_i)$

Proof is exactly on same lines.

29. INVERSE IMAGES

Let $f : X \to Y$ be a function and B be a subset of Y. Then inverse image of B is denoted by f^{-1} (B) and is a subset of X defined by

$$f^{-1}(B) = \{x \mid x \in X \text{ and } f(x) \in B\}$$

Example. Let $X = \{-2, -1, 0, 1, 2\}$

and $Y = \{0, 1, 2, 3, 4, 5, 6, 7\}$

Let $f : X \to Y$ such that $f(x) = x^2 \ \forall\, x \in X$

(*i*) If $B = \{0, 1, 2\}$, then $f^{-1}(B) = \{-1, 0, 1\}$

(*ii*) If $B = \{2, 3, 5, 7\}$, then $f^{-1}(B) = \phi$

(*iii*) If $B = \{0, 1, 4\}$, then $f^{-1}(B) = X$.

Note. (*i*) $f^{-1}(\phi) \neq \phi$

(*ii*) $f^{-1}(Y) = X$

(*iii*) $f^{-1}(B) = \phi \Rightarrow B = \phi$ as is clear from example (*ii*) above

(*iv*) $f^{-1}(y) \neq f^{-1}(\{y\})$.

In fact $f^{-1}(y)$ is defined iff f is one-one and onto whereas $f^{-1}\{y\}$ is always defined.

30. THEOREM

Let $f : X \to Y$ be a function

Let $A \subseteq Y$ and $B \subseteq Y$. Then,

(*i*) $A \subseteq B \Rightarrow f^{-1}(A) \subseteq f^{-1}(B)$ (*ii*) $f^{-1}(A \cup B) = f^{-1}(A) \cup f^{-1}(B)$

(*iii*) $f^{-1}(A \cap B) = f^{-1}(A) \cap f^{-1}(B)$.

Proof. (*i*) Let $x \in f^{-1}(A)$

$\Rightarrow \quad f(x) \in A$

$\Rightarrow \quad f(x) \in B \qquad [\because A \subseteq B]$

$\Rightarrow \quad x \in f^{-1}(B)$

$\therefore \quad f^{-1}(A) \subseteq f^{-1}(B)$

(*ii*) Since $\quad A \subseteq A \cup B \quad \text{and} \quad B \subseteq A \cup B$

$\Rightarrow \quad f^{-1}(A) \subseteq f^{-1}(A \cup B) \quad \text{and} \quad f^{-1}(B) \subseteq f^{-1}(A \cup B) \qquad$ [By (*i*)]

$\Rightarrow \quad f^{-1}(A) \cup f^{-1}(B) \subseteq f^{-1}(A \cup B) \qquad ...(1)$

Again, let $x \in f^{-1}(A \cup B)$

$\Rightarrow \quad f(x) \in A \cup B$

$\Rightarrow \quad f(x) \in A \quad \text{or} \quad f(x) \in B$

$\Rightarrow \quad x \in f^{-1}(A) \quad \text{or} \quad x \in f^{-1}(B)$

$\Rightarrow \quad x \in f^{-1}(A) \cup f^{-1}(B)$

$\therefore \quad f^{-1}(A \cup B) \subseteq f^{-1}(A) \cup f^{-1}(B) \qquad ...(2)$

$\therefore$ From (1) and (2),

$$f^{-1}(A \cup B) = f^{-1}(A) \cup f^{-1}(B)$$

(*iii*) Since $\quad A \cap B \subseteq A \quad \text{and} \quad A \cap B \subseteq B$

$\therefore \quad f^{-1}(A \cap B) \subseteq f^{-1}(A) \quad \text{and} \quad f^{-1}(A \cap B) \subseteq f^{-1}(B)$

$\Rightarrow \quad f^{-1}(A \cap B) \subseteq f^{-1}(A) \cap f^{-1}(B) \qquad ...(1)$

Again, let $\quad x \in f^{-1}(A) \cap f^{-1}(B)$

$\Rightarrow \quad x \in f^{-1}(A) \text{ and } x \in f^{-1}(B)$

$\Rightarrow \quad f(x) \in A \text{ and } f(x) \in B$

$\Rightarrow \quad f(x) \in A \cap B$

$\Rightarrow \quad x \in f^{-1}(A \cap B)$

$\therefore \quad f^{-1}(A) \cap f^{-1}(B) \subseteq f^{-1}(A \cap B) \qquad ...(2)$

$\therefore$ From (1) and (2),

$$f^{-1}(A) \cap f^{-1}(B) = f^{-1}(A \cap B)$$

Generalization. If $f : X \to Y$ is a function and $\{A_i\}$ $1 \le i \le n$ is a family of subsets of Y, then

(*i*) $f^{-1}\left(\bigcup_{i=1}^{n} A_i\right) = \bigcup_{i=1}^{n} f^{-1}(A_i)$ $\qquad$ (*ii*) $f^{-1}\left(\bigcap_{i=1}^{n} A_i\right) = \bigcap_{i=1}^{n} f^{-1}(A_i)$

Proof is on the same lines.

2

Congruence Modulo n

1. DEFINITION

Let n be a positive integer. If a and b are integers such that n divides $(a - b)$, which is symbolically written as $n \mid (a - b)$, we say that "a is congruent to b modulo n" and write it as $a \equiv b \pmod{n}$ or $a - b \equiv 0 \pmod{n}$.

This notation for congruence is due to *Gauss*.

If n does not divide $a - b$, we say that a is not congruent to b modulo n and we write $a \not\equiv b \pmod{n}$.

For example, $16 \equiv 11 \pmod{5}$ | $\because$ $5/(16 - 11)$

$3 \equiv -5 \pmod{4}$ | $\because$ $4/(3 - (-5))$ *i.e.,* $4/8$

17 is not congruent to 3 (mod 5). | $\because$ 5 does not divide $(17 - 3)$

2. THEOREM

Theorem 1. *Prove that two integers are congruent modulo n if and only if they leave the same remainder when divided by n.*

Proof. Let $a \equiv b \pmod{n}$.

$\Rightarrow$ n divides $(a - b)$ $\Rightarrow$ $a - b = tn$, where t is an integer. ...(1)

Let r_1 and r_2 be the remainders when a and b are divided by n.

Let $a = na_1 + r_1 \ (0 \le r_1 < n)$ ...(*i*)

$b = nb_1 + r_2 \ (0 \le r_2 < n)$ ...(*ii*)

(*i*)–(*ii*) gives $a - b = n(a_1 - b_1) + r_1 - r_2$

$\Rightarrow$ $tn = n(a_1 - b_1) + r_1 - r_2$ [By (1)]

$\Rightarrow$ $(t - a_1 + b_1)n = r_1 - r_2$

$\Rightarrow$ $n \mid r_1 - r_2$

But $r_1 - r_2 < n$ [$\because$ $r_1 < n, r_2 < n$]

$\therefore$ $r_1 - r_2 = 0$

[$\because$ If a/b and $b < a$, where a ($\neq 0$) and b are non-negative integers, then $b = 0$]

or $r_1 = r_2$.

$\therefore$ a and b leave the same remainder when divided by n.

Converse

Let a and b leave the same remainder r when divided by n.

$\therefore$ $a = na_1 + r$ and $b = nb_1 + r$ (by division alogrithm)

Subtracting, $a - b = n(a_1 - b_1)$

$\therefore$ n divides $(a - b)$.

$\therefore$ $a \equiv b \bmod n$.

Cor. If a and b are incongruent modulo n, then a and b have distinct remainders r_1 and r_2 where $0 \le r_1, r_2 < n$, when divided by n.

Def. When two integers are congruent to each other modulo n, each is said to be a **residue** of the other for that modulo.

Since $12 \equiv 2 \bmod 10$, 2 is a residue of 12 modulo 10 and 12 is a residue of 2 Modulo 10.

The totality of integers congruent to a given integer for the modulus n constitutes a **residue class modulo n** *e.g.* the set of integers {2, 12, 22, 32, 42, ...} is a residue class modulo 10. ($\because$ all the integers in this set are congruent to 2 for the Modulus 10).

Example 1. *Show that every number and its cube when divided by 6 leave the same remainder.*

Sol. Let n be any number.

$$n^3 - n = n(n^2 - 1) = n(n - 1)(n + 1) = (n - 1) \cdot n \cdot (n + 1),$$

which being the product of three consecutive integers is divisible by $3! = 6$.

[$\because$ Product of r consecutive integers is divisible by r !]

$\therefore$ 6 divides $(n^3 - n)$

$\therefore$ $n^3 \equiv n \pmod 6$ | By def. of congruence

$\therefore$ n^3 and n leave the same remainder when divided by 6. (Art. 2)

Example 2. *Prove that* $2.7^n + 3.5^n - 5 \equiv 0 \pmod{24}$.

Sol. Let $f(n) = 2.7^n + 3.5^n - 5$

Put $n = 1$, $\therefore$ $f(1) = 2.7 + 3.5 - 5 = 24$ is divisible by 24

$\therefore$ $f(1) \equiv 0 \pmod{24}$

$\therefore$ The result is true for $n = 1$.

Let $f(n) \equiv 0 \pmod{24}$ *i.e.* $f(n)$ is divisible by 24.

$\therefore$ $f(n) = 2.7^n + 3.5^n - 5 = 24k$

$\therefore$ $2.7^n = 24k - 3.5^n + 5$ ...(1)

Changing n to $n + 1$ in $f(n)$,

$$f(n + 1) = 2.7^{n+1} + 3.5^{n+1} - 5$$

$$= 2.7.7^n + 3.5.5^n - 5 = 7.2.7^n + 15.5^n - 5$$

Putting the value of 2.7^n from (1),

$$= 7(24k - 3.5^n + 5) + 15.5^n - 5$$

$$= 168k - 21.5^n + 35 + 15.5^n - 5$$

$$= 168k - 6.5^n + 30 = 168k - 6(5^n - 5)$$

$$= 168k - 6.5\,(5^{n-1} - 1) = 168k - 30(5^{n-1} - 1^{n-1})$$

$$= 168k - 30(5-1)(5^{n-2} + 5^{n-3} + \ldots + 1^{n-2})$$

$$[\because \quad x^k - y^k = (x-y)(x^{k-1} + x^{k-2}y + x^{k-3}y^2 + \ldots + y^{k-1})]$$

$$= 168k - 120(5^{n-2} + 5^{n-3} + \ldots + 1)$$

or $$f(n+1) = 24[7k - 5(5^{n-2} + 5^{n-3} + \ldots + 1)]$$

$\therefore$ By def. of divisibility, $f(n+1)$ is divisible by 24, whenever $f(n)$ is divisible by 24.

$\therefore$ by P.M.I., $f(n) \equiv 0 \bmod 24$.

Theorem 2. *If r is the remainder on dividing a by n, prove that $a \equiv r \pmod{n}$ i.e., an integer is congruent to its remainder.*

Proof. $\because$ r is the remainder on dividing a by n.

$\therefore$ $a = nq + r$ *i.e.*, $a - r = nq$

$\therefore$ By def. of divisibility, $n \mid (a - r)$.

$\therefore$ By def. of congruence, $a \equiv r \pmod{n}$

Theorem 3. *Show that the relation of congruence modulo n is an equivalence relation.*

Proof. By definition of congruence,

$$a \equiv b \pmod{n} \text{ if } n \mid (a-b)$$

The relation of congruence modulo n is an equivalence relation as it is (*i*) reflexive (*ii*) symmetric and (*iii*) transitive as shown below :

(*i*) **Reflexive.** Since $n \mid a - a \quad \forall a$

$\therefore$ $$a \equiv a \pmod{n}$$

$\therefore$ the relation is reflexive.

(*ii*) **Symmetric.** Let $a \equiv b \pmod{n}$

$\Rightarrow$ $n \mid (a - b)$

$\Rightarrow$ $n \mid (b - a)$

$\Rightarrow$ $b \equiv a \pmod{n}$

$\therefore$ The relation is symmetric.

(*iii*) **Transitive.** Let $a \equiv b \pmod{n}$, $b \equiv c \pmod{n}$

$\Rightarrow$ $n \mid (a - b)$ and $n \mid (b - c)$

$\Rightarrow$ $n \mid (a - b + b - c)$

$\Rightarrow$ $n \mid a - c$

$\Rightarrow$ $a \equiv c \pmod{n}$

$\therefore$ The relation is transitive.

Theorem 4. *If $a \equiv b \pmod{n}$, then*

(*i*) $(a + c) \equiv (b + c) \pmod{n}$ *i.e., any common number can be added to both sides of the congruence.*

(*ii*) $ac \equiv bc \pmod{n}$ *i.e., both numbers of a given congruence can be multiplied by a common number.*

Proof. (*i*) We have $a \equiv b \pmod{n}$.

$\therefore$ n divides $(a - b)$

$\therefore$ n divides $[(a + c) - (b + c)]$ | $\because$ $(a + c) - (b + c) = a - b$

$\therefore$ By def. of congruence, $(a + c) \equiv (b + c) \pmod{n}$.

(*ii*) We have $a \equiv b \pmod{n}$.

$\therefore$ n divides $(a - b)$

$\therefore$ n divides $c(a - b)$ *i.e.*, $(ac - bc)$

$\therefore$ By def. of congruence, $ac \equiv bc \pmod{n}$.

Remark. The converse of Theorem 4 (*ii*) need not be true *i.e.*, it is not always possible to cancel a common number from the two sides of a given congruence.

For example, $36 \equiv 12 \pmod 8$

If we cancel the common factor 12 from 36 and 12, we get $3 \equiv 1 \pmod 8$ which is impossible because 8 does not divide $(3 - 1)$.

The converse of Theorem 4(*ii*) is true only if c and n have no common factor other than 1 *i.e.*, c and n are coprime, which is symbolically written as $(c, n) = 1$ and we are proving this converse in the following theorem.

Theorem 5. *If* $ca \equiv cb \pmod{n}$ *and* $(c, n) = 1$, *then* $a \equiv b \pmod{n}$.

Proof. We have $ca \equiv cb \pmod{n}$.

$\therefore$ n divides $(ca - cb)$ *i.e.*, $n/c(a - b)$

We are given that $(n, c) = 1$.

$\therefore$ n divides $(a - b)$

[$\because$ By Gauss theorem if n/ab *i.e.*, n divides ab and $(n, a) = 1$; then n/b.]

$\therefore$ $a \equiv b \pmod{n}$.

Theorem 6. *If* $a \equiv b \pmod{n}$ *and* $c \equiv d \pmod{n}$ *prove that :*

(i) $(a + c) \equiv (b + d) \pmod{n}$ *i.e., two congruences can be added (with same modulus).*

(ii) $(a - c) \equiv (b - d) \pmod{n}$ *i.e., two congruences can be subtracted (with same modulus).*

Proof. (*i*) We have $a \equiv b \pmod{n}$ and $c \equiv d \pmod{n}$.

$\therefore$ n divides $(a - b)$ and n divides $(c - d)$

$\therefore$ n divides their sum which is $(a - b) + (c - d)$

$\therefore$ n divides $(a + c) - (b + d)$

$\therefore$ $a + c \equiv (b + d) \pmod{n}$.

Generalisation

If $a_i \equiv b_i \pmod{n}$ for $i = 1, 2, \ldots r$

then
$$\sum_{i=1}^{r} a_i \equiv \left(\sum_{i=1}^{r} b_i\right) \bmod n.$$

(*ii*) We have $a \equiv b \pmod{n}$ and $c \equiv d \pmod{n}$.

$\therefore$ n divides $(a - b)$ and n divides $(c - d)$

$\therefore$ n divides their difference which is $[(a - b) - (c - d)]$

$\therefore$ n divides $(a - c) - (b - d)$

$\therefore$ By def. of congruence, $(a - c) \equiv (b - d) \pmod{n}$.

Theorem 7. *If $a \equiv b \pmod{n}$ and $c \equiv d \pmod{n}$, prove that $ac \equiv bd \pmod{n}$ i.e., two congruences can be multiplied.*

Proof. We have $a \equiv b \pmod{n}$ and $c \equiv d \pmod{n}$.

$\therefore$ n divides $(a - b)$ and n divides $(c - d)$

$\therefore$ There exist integers r and s such that

$$a - b = nr \text{ and } c - d = ns.$$

$$\therefore \quad a = nr + b \text{ and } c = ns + d$$

$$\therefore \quad ac = (nr + b)(ns + d) = n^2rs + nrd + nbs + bd$$

$$\therefore \quad ac - bd = n(nrs + rd + bs)$$

$\therefore$ By def. of divisibility, n divides $(ac - bd)$

$$\therefore \quad ac \equiv bd \pmod{n}.$$

Generalisation. *If $b_1, b_2, b_3, \ldots\ldots, b_r$ are respectively congruent to $c_1, c_2, c_3, \ldots\ldots, c_r$ with respect to modulo n, then prove that $b_1.b_2\ldots\ldots b_r$ is congruent to $c_1.c_2\ldots\ldots c_r$ modulo n.*

Proof. We have $b_1 \equiv c_1 \pmod{n}$, $b_2 \equiv c_2 \pmod{n}$, ..., $b_r \equiv c_r \pmod{n}$.

$\therefore$ n divides $(b_1 - c_1), (b_2 - c_2), \ldots\ldots, (b_r - c_r)$.

$\therefore$ There exist integers $d_1, d_2, \ldots\ldots, d_r$ s.t.

$$b_1 - c_1 = nd_1,\ b_2 - c_2 = nd_2,\ \ldots\ldots,\ b_r - c_r = nd_r.$$

$$\therefore \quad b_1 = c_1 + nd_1,\ b_2 = c_2 + nd_2,\ \ldots\ldots,\ b_r = c_r + nd_r$$

$$\therefore \quad b_1.b_2\ldots\ldots b_r = (c_1 + nd_1)(c_2 + nd_2) \ldots\ldots (c_r + nd_r)$$

$$= c_1.c_2 \ldots\ldots c_r + n(\text{an integer})$$

$\therefore$ $b_1.b_2\ldots\ldots b_r - c_1.c_2 \ldots\ldots c_r = n(\text{an integer})$

$\therefore$ By def. of divisibility, n divides $b_1.b_2 \ldots\ldots b_r - c_1.c_2 \ldots\ldots c_r$

$\therefore$ $b_1.b_2 \ldots\ldots b_r \equiv c_1.c_2 \ldots\ldots c_r \pmod{n}$.

Cor. *If $a \equiv b \pmod{n}$, then $a^2 \equiv b^2 \pmod{n}$.*

We have $a \equiv b \pmod{n}$ and $a \equiv b \pmod{n}$

Multiplying the two congruences, we get $a^2 \equiv b^2 \pmod{n}$.

Theorem 8. *If $a \equiv b \pmod{n}$, prove that $a^k \equiv b^k \pmod{n}$ for every positive integer k.*

Proof. We have $a \equiv b \pmod{n}$.

$\therefore$ n divides $(a - b)$

$\therefore$ There exists an integer t such that $a - b = nt$ ...(1)

We know that

$$a^k - b^k = (a - b)[a^{k-1} + a^{k-2} . b + \ldots\ldots + b^{k-1}]$$

or $$a^k - b^k = nt[a^{k-1} + a^{k-2}.b + \ldots\ldots + b^{k-1}]$$ | Replacing $a - b$ by nt from (1)

$\therefore$ By def. of divisibility, n divides $a^k - b^k$.

$$\therefore \quad a^k \equiv b^k \pmod{n}.$$

Example 3. *Find the least positive integer (mod 11) to which 335 is congruent.*

Sol. By Division Algorithm

$$335 = (11)(30) + 5$$

$\therefore$ $335 \equiv 5 \pmod{11}$ | By Theorem 2, Art. 2

$\therefore$ 5 is the least positive integer (mod 11) to which 335 is congruent.

Example 4. *If $a \equiv b \pmod m$ and*

$$f(x) = p_0x^n + p_1x^{n-1} + p_2x^{n-2} + \ldots\ldots + p_{n-1}\,x + p_n$$

is a polynomial with integral coeffs., then prove that

$$f(a) \equiv f(b) \pmod m.$$

Sol. $\because$ $a \equiv b \pmod m$ (given)

$\therefore$ $m \mid (a - b)$

$\therefore$ There exists an integer t such that $a - b = mt$ ...(1)

$$f(x) = p_0x^n + p_1x^{n-1} + p_2x^{n-2} + \ldots.. + p_{n-1}\,x + p_n \quad \text{(given)}$$

Putting $x = a$ and $x = b$ respectively

$$f(a) = p_0a^n + p_1\,a^{n-1} + p_2\,a^{n-2} + \ldots\ldots\, p_{n-1}\,a + p_n$$

$$f(b) = p_0\,b^n + p_1b^{n-1} + p_2\,b^{n-2} + \ldots\ldots + p_{n-1}\,b + p_n$$

Subtracting

$$f(a) - f(b) = p_0\,(a^n - b^n) + p_1\,(a^{n-1} - b^{n-1}) + p_2\,(a^{n-2} - b^{n-2}) + \ldots\ldots + p_{n-1}\,(a - b)$$

$$= p_0(a - b)\,(a^{n-1} + a^{n-2}b + \ldots\ldots + b^{n-1})$$

$$+ p_1\,(a - b)(a^{n-2} + a^{n-3}b + \ldots\ldots + b^{n-2}) \ldots + p_{n-1}\,(a - b)$$

$$[\because\ \mathbf{a^n - b^n = (a - b)(a^{n-1} + a^{n-2}\,b + a^{n-3}\,b^2 + \ldots\ldots + b^{n-1})}]$$

Taking $(a - b)$ common from R.H.S. ;

$$f(a) - f(b) = (a - b)\ [\text{Some Integer say } k]$$

$$= mt\,(k) \quad \text{[By (1)]}$$

$\therefore$ $f(a) - f(b)$ is divisible by m

i.e., $f(a) \equiv f(b) \bmod m.$

Example 5. *Find the remainder if 3^{40} is divided by 23.*

Sol. $3^1 = 3 \equiv 3 \pmod{23}$

$3^2 = 9 \equiv 9 \pmod{23}$

$3^3 = 27 \equiv 4 \pmod{23}$

($\because$ 4 is the remainder obtained on dividing 27 by 23)

or $3^3 \equiv 4 \pmod{23}$ ...(1)

Raising both sides to power 4, | By Theorem 8, Art. 2

$(3^3)^4 \equiv 4^4 \pmod{23}$

or $3^{12} \equiv 256 \pmod{23}$ ...(2)

But $256 \equiv 3 \pmod{23}$ ...(3)

($\because$ 3 is the remainder obtained on dividing 256 by 23)

From (2) and (3),

$3^{12} \equiv 3 \pmod{23}$ | Because of transitive property. Theorem 2 (*iii*)

Cancelling 3, (By Theorem 5, Art. 2) ($\because$ (3, 23) = 1), we have

$$3^{11} \equiv 1 \bmod 23$$

Raising both sides to power 3,

$$3^{33} \equiv 1 \bmod 23 \quad ...(4)$$

Again raising both sides of congruence (1) to power 2

$$(3^3)^2 \equiv 4^2 \bmod 23$$

or $$3^6 \equiv 16 \bmod 23 \quad ...(5)$$

Multiplying congruence (4) and (5)

$$3^{39} \equiv 16 \bmod 23$$

Multiplying both sides by 3

$$3^{40} \equiv 48 \bmod 23 \equiv 2 \bmod 23.$$

[$\because$ 2 is the remainder obtained on dividing 48 by 23]

Example 6. *If x is odd, show that $x^2 \equiv 1$ mod 8.*

Sol. Because x is odd, so let $x = 2k + 1$ where k is an integer.

$\therefore$ $$x^2 - 1 = (2k + 1)^2 - 1 = 4k^2 + 4k + 1 - 1 = 4k^2 + 4k = 4k(k + 1) \quad ...(1)$$

Now $k(k + 1)$ being the product of two consecutive integers is divisible by $2\,! = 2$

[$\because$ Product of r consecutive integers is divisible by r !]

$\therefore$ $\exists$ an integer t s.t. $k(k + 1) = 2t$.

Putting $k(k + 1) = 2t$ in (1),

$$x^2 - 1 = 8t$$

which is divisible by 8.

$\therefore$ $$x^2 \equiv 1 \bmod 8.$$

Example 7. *If $ab \equiv c$ mod m and $b \equiv d$ mod m, prove that $ad \equiv c$ mod m.*

Sol. $\because$ $ab \equiv c \bmod m$ (given), therefore m divides $(ab - c)$

$\therefore$ There exists an integer t such that $ab - c = m.t.$ ...(1)

Because $b \equiv d \bmod m$ (given), therefore m divides $(b - d)$.

$\therefore$ There exists an integer k s.t. $b - d = mk$

$\therefore$ $$b = d + mk.$$

Putting this value of b in Eqn. (1) (To eliminate b),

we have $$a(d + mk) - c = mt$$

or $$ad + amk - c = mt$$

or $$ad - c = mt - amk = m(t - ak)$$

$\therefore$ $ad - c$ is divisible by m *i.e.,* $ad \equiv c \bmod m$.

EXERCISE

1. (*a*) Find the least positive integer (mod 11) to which 282 is congruent. [**Ans.** 7]

(*b*) Find *a* such that $a \equiv 7 \bmod 5$. [**Ans.** 7, 12, 17, 22, ..., 2, – 3, – 8, ...]

2. Show that $3^{2n+2} - 8n \equiv -9 \bmod 64$.

3. Prove that $5^{2n+2} - 24n - 25 \equiv 0 \pmod{576}$.

4. If $a \equiv b \pmod m$ and $c \equiv d \pmod m$, prove that $(ax + cy) \equiv (bx + dy) \pmod m$ for all integers x and y.

5. Show that $a \equiv b \pmod m$ if and only if $ax \equiv bx \pmod{mx}$ for all natural numbers x.

6. Prove that $a^n \equiv a \bmod (a - 1)$, where $a > 1$.

7. **If $a + b \equiv c \bmod m$ and $b \equiv d \bmod m$, show that $(a + d) \equiv c \bmod m$.**

8. **Find the remainder obtained on dividing 2^{20} by 7.** **[Ans.** 4]

[Hint. $2^3 = 8 \equiv 1 \pmod 7$ $\therefore$ $(2^3)^6 \equiv 1^6 \pmod 7$, *i.e.*, $2^{18} \equiv 1 \pmod 7$ $\therefore$ $2^{18}.2^2 \equiv 4 \pmod 7$].

9. **Find the remainder on dividing the sum**

$1! + 2! + 3! + 4! + \ldots\ldots + 99! + 100!$ by 12. **[Ans.** 9 ($= 1! + 2! + 3!$)]

[Hint. **We know that $4! = 24$ is divisible by 12**

$\therefore$ $5! = 5.4!$ is also divisible by 12 etc.

$\therefore$ $4! + 5! + \ldots\ldots + 100! \equiv 0 \bmod 12$

Add $1! + 2! + 3!$ to both sides.]

10. **Using congruence modulo *n*, find the remainder on dividing the sum**

$1! + 2! + 3! + 4! + \ldots\ldots + 99! + 100!$ by 24. **[Ans.** 9]

3

Matrices

1. DEFINITION

An arrangement of mn numbers belonging to a number system F (real or complex) into m rows and n columns is called a *matrix of order* $m \times n$ *over F*.

For example :

(*i*) $\begin{bmatrix} 2 & -3 & i \\ 3 & 8 & 6+2i \end{bmatrix}$ is a matrix of order 2×3, as it has two rows and three columns.

(*ii*) $\begin{bmatrix} 1 & 8 & -7 \\ 2 & 5 & 6 \\ i+2 & 0 & 4 \end{bmatrix}$ is a matrix of order 3×3.

(*iii*) In general a matrix of order $m \times n$ can be written as $\begin{bmatrix} a_{11} & a_{12} \ldots\ldots a_{1n} \\ a_{21} & a_{22} \ldots\ldots a_{2n} \\ \ldots\ldots & \ldots\ldots \\ a_{i1} & a_{i2} \ldots\ldots a_{in} \\ \ldots\ldots & \ldots\ldots \\ \ldots\ldots & \ldots\ldots \\ a_{m1} & a_{m2} \ldots\ldots a_{mn} \end{bmatrix}$

which can be briefly written as $[a_{ij}]_{m \times n}$.

Note 1. We shall denote a matrix by capital letters, A, B, C ...... etc.

2. The element a_{ij} is that which occurs in the ith row and jth col. The first suffix indicates row number, while the second suffix indicates the col. number.

3. Members of the number system F are called scalars relative to the matrix.

4. The elements $a_{11}, a_{22}, a_{33}, \ldots\ldots, a_{nn}$ in which both suffixes are same, are called the **diagonal elements,** all other are called non-diagonal elements.

Thus a_{ij} is a diagonal element if $i = j$ and a_{ij} is non-diagonal elements if $i \neq j$.

5. The line along which the diagonal elements

$a_{11}, a_{22}, \ldots\ldots, a_{nn}$ lie is called the **principal diagonal.**

2. DIFFERENT TYPES OF MATRICES

1. **Zero Matrix or Null Matrix.** A matrix each of whose element is zero is called a zero matrix or null matrix.

e.g., $\begin{bmatrix} 0 & 0 & 0 \\ 0 & 0 & 0 \end{bmatrix}$ or $\begin{bmatrix} 0 & 0 \\ 0 & 0 \\ 0 & 0 \end{bmatrix}$ or $\begin{bmatrix} 0 & 0 & 0 \\ 0 & 0 & 0 \\ 0 & 0 & 0 \end{bmatrix}$

are zero matrices respectively of order 2×3 ; 3×2 and 3×3.

In general, a *zero* matrix of order $m \times n$ is denoted by $O_{m \times n}$.

Note. A matrix which is not a zero-matrix is called a non-zero matrix.

2. **Square matrix.** A matrix in which the number of rows is equal to the number of columns is called a Square matrix.

A square matrix of order $n \times n$ is called square matrix of order n.

A matrix which is not square is called a **rectangular matrix.**

3. **Row-matrix or Row-Vector.** A matrix of type $1 \times n$ *i.e.*, having only one row is called a **row-matrix.** For example, $[1, -3, -7, i, 0]$ is a row- matrix of order 1×5.

4. **Column-matrix or Column-vector.** A matrix of type $m \times 1$ *i.e.*, having only one column is called a *column-matrix*.

For example, $\begin{bmatrix} 1 \\ 7 \\ 8 \end{bmatrix}$ is a column matrix of order 3×1.

5. **Diagonal Matrix.** A square matrix in which all non-diagonal elements are zero is called a *diagonal matrix*.

In symbols. The matrix $A = [a_{ij}]_{n \times n}$ is diagonal matrix if $a_{ij} = 0$ for $i \neq j$. Thus

$$\begin{bmatrix} 2 & 0 \\ 0 & -8 \end{bmatrix}, \begin{bmatrix} 1 & 0 & 0 \\ 0 & -5 & 0 \\ 0 & 0 & 0 \end{bmatrix} \text{ are diagonal matrices.}$$

Note. The diagonal matrix $\begin{bmatrix} x_1 & 0 & 0 \\ 0 & x_2 & 0 \\ 0 & 0 & x_3 \end{bmatrix}$ can be briefly written as diagonal $[x_1, x_2, x_3]$.

6. **Scalar Matrix.** A diagonal matrix in which all diagonal elements are equal is called a scalar matrix.

In symbols. The square matrix $A = [a_{ij}]_{n \times n}$ is a scalar matrix if $a_{ij} = 0$ $(i \neq j)$ and $a_{ij} = k$ for $i = j$.

e.g., $\begin{bmatrix} -2 & 0 & 0 \\ 0 & -2 & 0 \\ 0 & 0 & -2 \end{bmatrix}$ is a scalar matrix.

7. **Unit Matrix or Identity Matrix.** A *scalar* matrix *of order n* in which all diagonal elements are unity is called a **unit or identity matrix** and is generally denoted by I_n.

In Symbols. A square matrix $A = [a_{ij}]_{n \times n}$ will be a unit or identity matrix if

(*i*) $a_{ij} = 0$ for $i \neq j$ and (*ii*) $a_{ij} = 1$ for $i = j$.

8. **Tri-angular Matrix.** These are of two types :

(*a*) **Upper-triangular matrix.** It is a matrix in which all elements below the principal diagonal are zero *e.g.*, $\begin{bmatrix} 1 & -2 & i \\ 0 & 5 & -7 \\ 0 & 0 & 9 \end{bmatrix}$

(*b*) **Lower-triangular matrix.** It is a matrix in which all elements above the Principal diagonal are zero

e.g., $\begin{bmatrix} 1 & 0 & 0 \\ -5 & 7 & 0 \\ 3 & 8 & 4 \end{bmatrix}$

9. **Sub-matrix.** A matrix B obtained by deleting some rows or columns or both of a matrix A, is called a sub-matrix of A.

For example. If A = $\begin{bmatrix} 1 & 2 & 5 & 7 \\ 1 & 3 & 9 & 1 \\ 0 & 0 & 1 & 2 \end{bmatrix}$, then the matrices $\begin{bmatrix} 1 & 2 & 5 \\ 1 & 3 & 9 \end{bmatrix}, \begin{bmatrix} 2 & 5 \\ 3 & 9 \end{bmatrix}$, [0, 0, 1, 2] etc.

are sub-matrices of A.

3. EQUALITY OF MATRICES

Two matrices $A = [a_{ij}]_{m \times n}$ and $B = [b_{ij}]_{p \times q}$ are equal, if and only if

(*i*) they are of the same order *i.e.* $m = p$ and $n = q$

(*ii*) their corresponding elements are **all** equal *i.e.*, $a_{ij} = b_{ij}$ for all i and j.

If A and B are two equal matrices, then we write A = B.

4. ADDITION (SUM) OF TWO MATRICES

We can add two matrices only when they are of the **same order** and two such matrices are said to be **conformable for addition.**

Let $A = [a_{ij}]_{m \times n}$ and $B = [b_{ij}]_{m \times n}$ be two matrices of the same order $m \times n$, then **their sum A + B** *is a matrix of the same order* $m \times n$ *and is obtained by adding the corresponding elements of A and B.*

Thus, if $A = [a_{ij}]_{m \times n}$, $B = [b_{ij}]_{m \times n}$, then the sum

$$A + B = [a_{ij}]_{m \times n} + [b_{ij}]_{m \times n} = [a_{ij} + b_{ij}]_{m \times n}.$$

Remark. The elements of a matrix will be assumed to belong to some number system say of Rationals, Reals or Complex.

5. PROPERTIES OF MATRIX ADDITION (WITHOUT PROOF)

1. **Matrix Addition is Commutative.** *i.e.*, if A and B are matrices of the same order, then **A + B = B + A.**

2. **Matrix Addition is Associative.** If A, B, C be matrices of the same order, then (A + B) + C = A + (B + C).

Note. Because of associative property of addition, we write

$$(A + B) + C = A + (B + C) = A + B + C.$$

3. **Existence of Additive identity.** Given any matrix A of order $m \times n$, there exists a matrix O of order $m \times n$, each of whose element is zero such that A + O = A.

Note. The zero matrix O is called additive identity or a zero and is unique for a set of all $m \times n$ matrices.

4. **Existence of Additive Inverse.** Given a matrix A of order $m \times n$; their exists a matrix X also of the same order, so that A + X = O.

This matrix X = – $[a_{ij}]$ is **called additive inverse or Negative of A** and we shall denote it by (– A).

Thus if A = $[a_{ij}]$, then – A = $[-a_{ij}]$.

6. SUBTRACTION OF TWO MATRICES

Let A and B be two matrices of the same order (type), then subtraction of B from A is written as A – B and is defined as sum of A and – B.

Thus, A – B = A + (– B).

Hence A – B is obtained by subtracting from each element of A the corresponding element of B.

7. MULTIPLICATION OF A MATRIX BY A SCALAR

Let A = $[a_{ij}]_{m \times n}$ be any matrix and k any scalar, then the multiplication of A by the scalar k written as k A is a matrix of order $m \times n$ obtained by multiplying each element of A by the scalar k. Thus,

If $$A = [a_{ij}]_{m \times n}, \text{ then}$$
$$kA = k[a_{ij}]_{m \times n} = [k.a_{ij}]_{m \times n}.$$

For example. If $A = \begin{bmatrix} -1 & 2 & 7 & 8 \\ 3 & 4 & -2 & 7 \\ 1 & 2 & 3 & 4i \end{bmatrix}$ is a matrix of order 3 × 4 and 5 is a scalar, then

$$5A = \begin{bmatrix} -5 & 10 & 35 & 40 \\ 15 & 20 & -10 & 35 \\ 5 & 10 & 15 & 20i \end{bmatrix}.$$

8. PROPERTIES OF MULTIPLICATION OF A MATRIX BY A SCALAR

If A = $[a_{ij}]$ and B = $[b_{ij}]$ be any two matrices of the same type $m \times n$ and x and y are scalars, then

(*i*) $x(A + B) = xA + xB$ (*ii*) $(x + y) A = xA + yA$

(*iii*) $x(yA) = (xy)A$ (*iv*) There exists a scalar 1 so that 1.A = A.

9. MULTIPLICATION OF TWO MATRICES

Let A = $[a_{ij}]$ and B = $[b_{ij}]$ be two matrices, then the product AB in this order is defined if the number of columns in A (pre-factor) = the number of rows in B (post-factor), and

(*i*) Number of rows in AB = the number of rows in A.

(*ii*) Number of columns in AB = the number of cols. in B.

(*iii*) The (i, j)th element of AB = sum of products of the elements of ith row of A with the corresponding elements of the jth column of B.

In Symbols. If $A = [a_{ij}]_{m \times n}$ and $B = [b_{ij}]_{n \times p}$ be two matrices, then the product AB is defined and is a matrix $C = [c_{ij}]_{m \times p}$ of order $m \times p$.

where $c_{ij} = (i, j)$th element of C(= AB)

= Sum of the products of elements of ith row with the corresponding elements of jth column

$$= (a_{i1}\, a_{i2} \,......\, a_{in}) \begin{pmatrix} b_{1j} \\ b_{2j} \\ \cdots \\ b_{nj} \end{pmatrix}$$

$$= a_{i1}b_{1j} + a_{i2}b_{2j} + ... + a_{in}b_{nj}$$

or

$$\textbf{(i, j)th element of AB} = \sum_{k=1}^{n} a_{ik}b_{kj}.$$

Remarks. 1. If the product AB is defined, then the matrices A and B are said to be conformable for multiplication AB.

2. If AB is defined, BA may or may not be defined.

3. Method of multiplication is known as Row-by-Column method.

10. PROPERTIES OF MATRIX MULTIPLICATION

Property 1. Matrix Multiplication is associative. *If A, B, C are matrices of the order $m \times n$, $n \times p$, $p \times q$ respectively, then (AB)C = A(BC).*

Note. (AB)C and A(BC) both can be written = ABC.

Property 2. Distributive Laws :

(*i*) If A, B, C are three matrices of type $m \times n$, $n \times p$, $n \times p$ respectively, then

A(B + C) = AB + AC [*Left* Distributive Law]

(*ii*) If A, B, C are matrices of order $p \times m$, $n \times p$, $n \times p$ then

(B + C)A = BA + CA [*Right* Distributive Law]

Property 3. If A be any $n \times n$ matrix, then

$$AI_n = A = I_nA.$$

Property 4. Matrix Multiplication is not commutative.

Prove that the product of matrices is not commutative in general i.e., prove AB ≠ BA, discussing all possibilities.

Proof. Case I. AB is defined but BA is not defined.

Let A be of order 3 × 2 and B be of order 2 × 4.

∴ AB is defined and is a matrix of order 3 × 4.

But BA is not defined ∴ AB ≠ BA.

Case. II. AB and BA are both defined but are of different order.

Let A be of order 2 × 3 and matrix B of order 3 × 2.

∴ AB is defined and is a matrix of order 2 × 2.

BA is also defined and is a matrix of order 3 × 3.

∴ $AB \neq BA$.

Case III. *AB and BA are both defined and both are of the same order, yet AB ≠ BA.*

Let $A = \begin{bmatrix} 2 & 3 \\ -1 & 4 \end{bmatrix}, B = \begin{bmatrix} 2 & 1 \\ 7 & 5 \end{bmatrix}$

be two square matrices of the same order 2 × 2.

$$\therefore \quad AB = \begin{bmatrix} 2 & 3 \\ -1 & 4 \end{bmatrix}\begin{bmatrix} 2 & 1 \\ 7 & 5 \end{bmatrix} = \begin{bmatrix} 4+21 & 2+15 \\ -2+28 & -1+20 \end{bmatrix} = \begin{bmatrix} 25 & 17 \\ 26 & 19 \end{bmatrix}.$$

$$BA = \begin{bmatrix} 2 & 1 \\ 7 & 5 \end{bmatrix}\begin{bmatrix} 2 & 3 \\ -1 & 4 \end{bmatrix} = \begin{bmatrix} 4-1 & 6+4 \\ 14-5 & 21+20 \end{bmatrix} = \begin{bmatrix} 3 & 10 \\ 9 & 41 \end{bmatrix}$$

Thus, $AB \neq BA$.

Property 5. Give an example of matrices A and B such that A ≠ O, B ≠ O, but AB = O.

Or **Prove that AB = O, does not imply either A = O or B = O.**

For example, Let $A = \begin{bmatrix} 1 & 1 \\ 1 & 1 \end{bmatrix}, B = \begin{bmatrix} 1 & 0 \\ -1 & 0 \end{bmatrix}$

∴ $A \neq O, B \neq O$

$$AB = \begin{bmatrix} 1 & 1 \\ 1 & 1 \end{bmatrix}\begin{bmatrix} 1 & 0 \\ -1 & 0 \end{bmatrix} = \begin{bmatrix} 1-1 & 0+0 \\ 1-1 & 0+0 \end{bmatrix} = \begin{bmatrix} 0 & 0 \\ 0 & 0 \end{bmatrix}$$

Thus $A \neq O, B \neq O$, yet $AB = O$.

Note. If AB = O, BA may *or* may not be zero.

Property 6. Cancelation law in multiplication does not hold i.e., Give an example of matrices A, B, C such that

AB = AC, but B ≠ C.

For example, Let $A = \begin{bmatrix} 1 & 1 \\ 1 & 1 \end{bmatrix}, B = \begin{bmatrix} 3 & 1 \\ 2 & 1 \end{bmatrix}, C = \begin{bmatrix} 2 & 1 \\ 3 & 1 \end{bmatrix}$

$$\therefore \quad AB = \begin{bmatrix} 1 & 1 \\ 1 & 1 \end{bmatrix}\begin{bmatrix} 3 & 1 \\ 2 & 1 \end{bmatrix} = \begin{bmatrix} 3+2 & 1+1 \\ 3+2 & 1+1 \end{bmatrix} = \begin{bmatrix} 5 & 2 \\ 5 & 2 \end{bmatrix}$$

$$AC = \begin{bmatrix} 1 & 1 \\ 1 & 1 \end{bmatrix}\begin{bmatrix} 2 & 1 \\ 3 & 1 \end{bmatrix} = \begin{bmatrix} 2+3 & 1+1 \\ 2+3 & 1+1 \end{bmatrix} = \begin{bmatrix} 5 & 2 \\ 5 & 2 \end{bmatrix}$$

Thus, AB = AC, whereas B ≠ C.

11. POSITIVE INTEGRAL POWERS OF A MATRIX

If A is a square matrix, then A. A is also a square matrix of the same order and we write $A.A = A^2$; $A.A.A = A^3$ etc.

∴ For all positive integers m and n, the following results hold

(1) $A^m = A . A . A ...... m$ times (2) $A^m . A^n = A^{m+n}$

(3) $(A^m)^n = A^{mn}$ (4) We define that $A^0 = I$.

12. DEFINITIONS OF IDEMPOTENT, NILPOTENT, AND INVOLUTORY MATRIX

(i) A square matrix A is said to be Idempotent if $A^2 = A$.

For example, the matrix $A = \begin{bmatrix} 2 & -2 & -4 \\ -1 & 3 & 4 \\ 1 & -2 & -3 \end{bmatrix}$ is idempotent. (Verify that $A^2 = A$).

(ii) A square matrix A is called Nilpotent if there exists a positive integer m such that $A^m = O$. If m is the least positive integer such that $A^m = O$, then m is called the index of the nilpotent matrix A.

For example, the matrices $\begin{bmatrix} ab & b^2 \\ -a^2 & -ab \end{bmatrix}, \begin{bmatrix} 1 & 1 \\ -1 & -1 \end{bmatrix}$ are nilpotent (Verify that $A^2 = O$).

(iii) A square matrix A is said to be Involutory if $A^2 = I$.

For example, the matrix $A = \begin{bmatrix} \sqrt{2} & 1 \\ -1 & -\sqrt{2} \end{bmatrix}$ is involutory.

13. TRANSPOSE OF A MATRIX

Let A be any given matrix of the order $m \times n$, then a matrix obtained from A by changing its rows into columns and columns into rows is called the **transpose of a matrix** A and is denoted by A′ which will be of the type $n \times m$.

In symbols. If $A = [a_{ij}]_{m \times n}$, then, $A' = [c_{ij}]_{n \times m}$, where $c_{ij} = a_{ji}$, *i.e.*, (i, j)th element of A′ = (j, i)th element of A.

For example, Let

$$A = \begin{bmatrix} 1 & 2 & 3 & -1 \\ 2 & -3 & 4 & -5 \\ 6 & 7 & -8 & 2 \end{bmatrix}$$

then $$A' = \begin{bmatrix} 1 & 2 & 6 \\ 2 & -3 & 7 \\ 3 & 4 & -8 \\ -1 & -5 & 2 \end{bmatrix}.$$

14. THEOREM

If A′ and B′ denote transpose of A and B, then

1. $(A')' = A$

2. $(A + B)' = A' + B'$, A, B are conformable for addition

3. $(kA)' = kA'$, k is any scalar

4. $(AB)' = B'A'$, A, B are conformable for multiplication

(Reversal Law)

Generalising this Reversal Law result (4), we have $(A_1A_2 \ldots A_n)' = A_n' \ldots\ldots A_2'A_1'$.

5. $(A^n)' = (A')^n$, where A is a square matrix and n is a positive integer.

15. CONJUGATE OF A MATRIX

Let A be a given matrix of order $m \times n$ over the complex number system, then a matrix obtained from A by replacing each of its elements by their corresponding complex conjugates is called the **conjugate of A** and is denoted by $\overline{A}$, where $\overline{A}$ is also of the same order $m \times n$.

In notation we can define as

If $A = [a_{ij}]_{m \times n}$, then

$$\overline{A} = [b_{ij}]_{m \times n}, \text{ where } b_{ij} = \overline{a}_{ij}$$

For example. Let

$$A = \begin{bmatrix} 2+i & 2 & 5i \\ 5i+7 & -8 & 4i-3 \\ 2 & 5+i & 4-2i \end{bmatrix}$$

$$\therefore \quad \overline{A} = \begin{bmatrix} 2-i & 2 & -5i \\ -5i+7 & -8 & -4i-3 \\ 2 & 5-i & 4+2i \end{bmatrix}$$

It is to be noted that conjugate complex of $5i + 7$ is $-5i + 7$.

16. THEOREM

If $\overline{A}$ and $\overline{B}$ denote the conjugates of A and B, respectively, then

1. $\overline{(\overline{A})} = A$.

2. $\overline{(A+B)} = \overline{A} + \overline{B}$, where A and B are conformable for addition.

3. $\overline{(kA)} = \overline{k}\,.\,\overline{A}$, where k is any complex number.

4. $\overline{(AB)} = \overline{A}.\,\overline{B}$.

5. $(\overline{A})^n = (\overline{A^n})$.

17. TRANSPOSED CONJUGATE OF A MATRIX

The transposed of the conjugate or conjugate of the transpose of a matrix A is called *Transposed Conjugate* of A and is denoted by A^θ or by A^*. Thus

$$A^\theta = (\overline{A})' = (\overline{A'}).$$

Thus if $A = [a_{ij}]$, then $A^\theta = [\alpha_{ij}]$ where $\alpha_{ij} = \overline{a}_{ji}$

i.e., (i, j)th element of A^θ = The conjugate complex of the (j, i)th element of A. For example, if

$$A = \begin{bmatrix} 1-2i & 2+3i & 4 \\ -7 & 8i & 5 \\ 0 & 6i+5 & 4 \end{bmatrix}$$

then
$$\overline{A} = \begin{bmatrix} 1+2i & 2-3i & 4 \\ -7 & -8i & 5 \\ 0 & -6i+5 & 4 \end{bmatrix}$$

and
$$A^\theta = (\overline{A})' = \begin{bmatrix} 1+2i & -7 & 0 \\ 2-3i & -8i & -6i+5 \\ 4 & 5 & 4 \end{bmatrix}.$$

18. THEOREM

If A^θ and B^θ be the transposed conjugates of A and B respectively, then

1. $(A^\theta)^\theta = A$

2. $(A + B)^\theta = A^\theta + B^\theta$, A, B are of the same order

3. $(kA)^\theta = \bar{k} A^\theta$, k is any complex number

4. $(AB)^\theta = B^\theta A^\theta$, A, B are conformable for multiplication. (Reversal Law)

5. $(A^n)^\theta = (A^\theta)^n$ where A is a square matrix and n is a positive integer.

Generalising the Reversal Law of result (4), we have $(A_1A_2...A_n)^\theta = A_n^\theta ...A_2^\theta A_1^\theta$.

19. SYMMETRIC MATRIX

Def. A matrix A is said to be a **symmetric** matrix if A′ = A, *i.e.* if the transpose of a matrix is equal to itself.

Let $A = [a_{ij}]_{m\times n}$

$\therefore$ $A' = [\alpha_{ij}]_{n\times m}$, where $\alpha_{ij} = a_{ji}$.

The matrix A will be symmetric, if and only if, **A = A′**

i.e. if and only if $m = n$ and $a_{ij} = \alpha_{ij} = a_{ji}$. Thus we have

Definition. *A square matrix $A = [a_{ij}]$ is symmetric if $a_{ij} = a_{ji}$ for all i and j, i.e.*

A square matrix is symmetric if and only if (*i, j*)th element = (*j, i*)th element.

For example, the matrices

$$\begin{bmatrix} 1 & 2 & -4 \\ 2 & 5 & -7 \\ -4 & -7 & 8 \end{bmatrix}, \begin{bmatrix} a & h & g \\ h & b & f \\ g & f & c \end{bmatrix}, \begin{bmatrix} 1+3i & 2i \\ 2i & 5 \end{bmatrix}$$

are all symmetric matrices.

20. SKEW-SYMMETRIC MATRIX

Def. A matrix A is said to be a *skew-symmetric matrix*, if A′ = – A *i.e.* if its transpose equals to its negative.

Let $A = [a_{ij}]_{m\times n}$

$\therefore$ $A' = [\alpha_{ij}]_{m\times n}$, where $\alpha_{ij} = a_{ji}$

A is skew-symmetric, if and only if, A′ = – A

i.e. iff $m = n$ and $\alpha_{ij} = -a_{ij}$

i.e. iff $m = n$ and $a_{ji} = -a_{ij}$

i.e. iff $m = n$ and $a_{ij} = -a_{ji}$

Hence we can also define.

Definition. A square matrix $A = [a_{ij}]$ is skew-symmetric if

$a_{ij} = -a_{ji}$ for all i and j, and

when $j = i$, we get $a_{ii} = -a_{ii}$ i.e. $2a_{ii} = 0$ or $a_{ii} = 0$

i.e.* In a skew-symmetric matrix, the *diagonal elements are all zero.

For example, the matrices

$$\begin{bmatrix} 0 & 1 & 2 \\ -1 & 0 & -5 \\ -2 & 5 & 0 \end{bmatrix}, \begin{bmatrix} 0 & h & b \\ -h & 0 & f \\ -b & -f & 0 \end{bmatrix}, \begin{bmatrix} 0 & 2i & i+2 \\ -2i & 0 & 5 \\ -i-2 & -5 & 0 \end{bmatrix}$$

are all skew-symmetric matrices.

21. HERMITIAN MATRIX

A matrix 'A' is said to be *Hermitian*, if $A^\theta = A$.

That is, a square matrix $A = [a_{ij}]_{n\times n}$ is Hermitian, if $a_{ij} = \bar{a}_{ji}$, for all i and j *i.e.*, (i, j)th element is equal to conjugate of (j, i)th element.

In particular for $i = j$, $a_{ii} = \bar{a}_{ii}$.

$\Rightarrow$ a_{ii} is real.

Thus all diagonal elements of a Hermitian matrix are real.

For example, the matrices $\begin{bmatrix} 2 & 1+i \\ 1-i & 5 \end{bmatrix}, \begin{bmatrix} 1 & 2i & 5i+3 \\ -2i & 0 & 4+3i \\ -5i+3 & 4-3i & 2 \end{bmatrix}$ are Hermitian.

22. SKEW-HERMITIAN MATRIX

A matrix A is said to be *skew-Hermitian*, if $A^\theta = -A$ *i.e.*, iff A is square and $a_{ij} = -\bar{a}_{ji}$.

When $i = j$, we have $a_{ii} = -\bar{a}_{ii}$ *i.e.* diagonal elements are either purely imaginary or zero.

For example, the matrices,

$$\begin{bmatrix} 2i & 1+i \\ -1+i & 0 \end{bmatrix}, \begin{bmatrix} 3i & 1+i & 5i+7 \\ -1+i & -2i & 5 \\ 5i-7 & -5 & 0 \end{bmatrix}$$

are skew-Hermitian.

Example 1. *If A is a square matrix, then show that*

(i) A + A′ is symmetric.

(ii) A – A′ is skew-symmetric.

Sol. (*i*) $(A + A')' = A' + (A')' = A' + A = A + A'$.

Thus, the transpose of $(A + A')$ is $A + A'$ itself.

$\therefore$ $A + A'$ is symmetric.

(*ii*) $(A - A')' = A' - (A')' = A' - A = -(A - A')$.

$\therefore$ The transpose of $(A - A')$ is negative of itself.

$\therefore$ $A - A'$ is skew-symmetric.

Example 2. *If A and B are two symmetric matrices of the same order, show that AB is symmetric if and only if AB = BA.*

Sol. A and B are both symmetric.

$\therefore$ We have $A' = A$ and $B' = B$.

Let AB be symmetric.

$$\therefore \quad (AB)' = AB$$

$$\Rightarrow \quad B'A' = AB \quad i.e. \quad BA = AB.$$

Conversely. Let it be given that AB = BA.

$$\therefore \quad (AB)' = B'A' = BA = AB.$$

$\therefore$ AB is symmetric.

Example 3. *Show that every square matrix can be expressed in one and only one way as sum of a symmetric matrix and a skew-symmetric matrix.*

Sol. Let A be a square matrix.

$$\therefore \quad A = \tfrac{1}{2}(2A) = \tfrac{1}{2}(A + A' + A - A')$$

$\tfrac{1}{2}(A + A') + \tfrac{1}{2}(A - A') = P + Q$, where

$$P = \tfrac{1}{2}(A + A') \quad \text{and} \quad Q = \tfrac{1}{2}(A - A')$$

Now, $\quad P = \tfrac{1}{2}(A + A')$

$$\therefore \quad P' = [\tfrac{1}{2}(A + A')]' = \tfrac{1}{2}(A + A')' = \tfrac{1}{2}[A' + (A')'] = \tfrac{1}{2}(A' + A) = P.$$

$\therefore$ P is symmetric.

Also $\quad Q = \tfrac{1}{2}(A - A')$

$$\Rightarrow \quad Q' = \tfrac{1}{2}(A' - A) = -\tfrac{1}{2}(A - A') = -Q$$

$\therefore$ Q is skew-symmetric.

Hence A has been written as the sum of a symmetric matrix and a skew-symmetric matrix.

Uniqueness. If possible, let

$$A = L + M, \qquad \text{...}(i)$$

where L is symmetric and M is a skew-symmetric matrix.

$$\therefore \quad A' = (L + M)' = L' + M'$$

$$\therefore \quad A' = L - M \qquad \text{...}(ii)$$

$[\because \quad L' = L, M' = -M]$

$$(i) + (ii) \Rightarrow \quad L = \frac{A + A'}{2} = P$$

$$(i) - (ii) \Rightarrow \quad M = \frac{A - A'}{2} = Q.$$

Hence the representation : $A = \dfrac{1}{2}(A + A') + \dfrac{1}{2}(A - A')$ is unique.

Example 4. *Express* $A = \begin{bmatrix} 1 & 3 & 5 \\ -6 & 8 & 3 \\ -4 & 6 & 5 \end{bmatrix}$

as the sum of a symmetric matrix and a skew-symmetric matrix.

Sol. Here

$$A = \begin{bmatrix} 1 & 3 & 5 \\ -6 & 8 & 3 \\ -4 & 6 & 5 \end{bmatrix} \quad \therefore \quad A' = \begin{bmatrix} 1 & -6 & -4 \\ 3 & 8 & 6 \\ 5 & 3 & 5 \end{bmatrix}$$

Now $\quad A = \tfrac{1}{2}(A + A') + \tfrac{1}{2}(A - A'),$

where $\tfrac{1}{2}(A + A')$ is symmetric and $\tfrac{1}{2}(A - A')$ is skew-symmetric.

$$(A + A') = \begin{bmatrix} 1 & 3 & 5 \\ -6 & 8 & 3 \\ -4 & 6 & 5 \end{bmatrix} + \begin{bmatrix} 1 & -6 & -4 \\ 3 & 8 & 6 \\ 5 & 3 & 5 \end{bmatrix} = \begin{bmatrix} 2 & -3 & 1 \\ -3 & 16 & 9 \\ 1 & 9 & 10 \end{bmatrix}$$

$$\therefore \quad \tfrac{1}{2}(A + A') = \begin{bmatrix} 1 & -\frac{3}{2} & \frac{1}{2} \\ -\frac{3}{2} & 8 & \frac{9}{2} \\ \frac{1}{2} & \frac{9}{2} & 5 \end{bmatrix}, \text{ which is a symmetric matrix}$$

$$\tfrac{1}{2}(A - A') = \tfrac{1}{2}\left(\begin{bmatrix} 1 & 3 & 5 \\ -6 & 8 & 3 \\ -4 & 6 & 5 \end{bmatrix} - \begin{bmatrix} 1 & -6 & -4 \\ 3 & 8 & 6 \\ 5 & 3 & 5 \end{bmatrix}\right)$$

$$= \tfrac{1}{2}\begin{bmatrix} 0 & 9 & 9 \\ -9 & 0 & -3 \\ -9 & 3 & 0 \end{bmatrix} = \begin{bmatrix} 0 & \frac{9}{2} & \frac{9}{2} \\ -\frac{9}{2} & 0 & -\frac{3}{2} \\ -\frac{9}{2} & \frac{3}{2} & 0 \end{bmatrix}$$

which is a skew-symmetric matrix.

$$\text{Hence} \quad \begin{bmatrix} 1 & 3 & 5 \\ -6 & 8 & 3 \\ -4 & 6 & 5 \end{bmatrix} = \begin{bmatrix} 1 & -\frac{3}{2} & \frac{1}{2} \\ -\frac{3}{2} & 8 & \frac{9}{2} \\ \frac{1}{2} & \frac{9}{2} & 5 \end{bmatrix} + \begin{bmatrix} 0 & \frac{9}{2} & \frac{9}{2} \\ -\frac{9}{2} & 0 & -\frac{3}{2} \\ -\frac{9}{2} & \frac{3}{2} & 0 \end{bmatrix}.$$

Example 5. *If A is any square matrix, then prove that $A + A^\theta$, AA^θ, $A^\theta A$ are all Hermitian and $A - A^\theta$ is Skew-Hermitian.*

Sol. (*i*) Now, $(A + A^\theta)^\theta = A^\theta + (A^\theta)^\theta = A^\theta + A = A + A^\theta$

$[\because \quad (A^\theta)^\theta = A \text{ and } A^\theta + A = A + A^\theta]$

$\therefore$ $A + A^\theta$ is Hermitian.

(*ii*) $(AA^\theta)^\theta = (A^\theta)^\theta . A^\theta = AA^\theta$ $\qquad [\because \quad (A^\theta)^\theta = A]$

$\therefore$ AA^θ is Hermitian

Similarly, $A^\theta A$ is Hermitian.

(*iii*) Since $(A - A^\theta)^\theta = A^\theta - (A^\theta)^\theta = A^\theta - A = -(A - A^\theta)$

$\therefore$ $A - A^\theta$ is Skew-Hermitian.

Example 6. *Every square matrix A can be expressed in one and only one way as $P + iQ$ where P and Q are Hermitian matrices.*

Sol. We can write

$$A = \tfrac{1}{2}(2A) = \tfrac{1}{2}(A + A^\theta + A - A^\theta)$$

$$= \tfrac{1}{2}(A + A^\theta) + \tfrac{1}{2}(A - A^\theta) = \tfrac{1}{2}(A + A^\theta) + i\,\frac{1}{2i}(A - A^\theta) = P + iQ$$

where $\qquad P = \tfrac{1}{2}(A + A^\theta)$ and $Q = \dfrac{1}{2i}(A - A^\theta)$

Also, $P^\theta = \frac{1}{2}(A + A^\theta)^\theta = \frac{1}{2}(A^\theta + A) = P$

and $Q^\theta = -\frac{1}{2i}(A - A^\theta)^\theta = -\frac{1}{2i}(A^\theta - A) = \frac{1}{2i}(A - A^\theta) = Q.$ $[\because \ (kA)^\theta = \bar{k} A^\theta]$

so that P and Q are both Hermitian.

Thus, A has been expressed as P + *i*Q where P and Q are Hermitian matrices.

To prove that this representation of A is unique, let if possible, A can also be written as

$A = R + iS$, where R, S are Hermitian

$\Rightarrow \quad A^\theta = (R + iS)^\theta = R^\theta - iS^\theta = R - iS \qquad (\because \ R^\theta = R \text{ and } S^\theta = S)$

$\Rightarrow \quad R = \frac{1}{2}(A + A^\theta) \text{ and } S = \frac{1}{2i}(A - A^\theta)$

$\Rightarrow \quad R = P \quad \text{and} \quad S = Q.$

Example 7. *If $A = B + iC$ is Hermitian, show that $A^\theta A$ is real if and only if B and C anti-commute (i.e. $BC = -CB$).*

Sol. Since $A = B + iC$ is Hermitian.

$\therefore \quad A^\theta = A$

Now, $A^\theta A = A \cdot A = (B + iC)(B + iC) = B^2 + i(BC + CB) - C^2$

$\therefore$ $A^\theta A$ is real if and only if $BC + CB = O$

or $BC = -CB$, *i.e.*, B and C anti-commute.

Example 8. *Show that the value of the determinant of a Skew-Symmetric matrix of odd order is always zero.*

Sol. Let A be a Skew-Symmetric matrix of order *n*, where *n* is odd.

$\therefore \quad A' = -A$

Taking determinant of both sides

$\therefore \quad |A'| = |-A| = |(-1)A|$

$\therefore \quad |A| = (-1)^n |A| \qquad (\because \ |kA| \text{ where } k \text{ is a scalar} = k^n |A|)$

$|A| = -|A| \qquad (\because \ n \text{ is odd})$

$\Rightarrow \quad 2|A| = 0 \quad \therefore \quad |A| = 0.$

Remark. $\therefore$ every skew-symmetric matrix of odd order is singular.

Example 9. *If A, B be square matrices of same order, show that $B^\theta AB$ is Hermitian or Skew-Hermitian according as A is Hermitian or Skew-Hermitian.*

Sol. (*i*) Let A be Hermitian.

$\therefore \quad A^\theta = A \qquad ...(1)$

Now, $(B^\theta AB)^\theta = B^\theta A^\theta (B^\theta)^\theta$ [By Reversal Law]

$= B^\theta A^\theta B = B^\theta AB$ [By (1)]

$\Rightarrow$ $B^\theta AB$ is Hermitian.

(*ii*) Let A be Skew-Hermitian.

$\Rightarrow \quad A^\theta = -A \qquad ...(2)$

Now, $(B^\theta AB)^\theta = B^\theta A^\theta (B^\theta)^\theta = B^\theta A^\theta B = B^\theta(-A)B$ [By (2)]

$= -B^\theta AB \quad \therefore \quad B^\theta AB$ is a Skew-Hermitian Matrix.

EXERCISE 1

1. Define the following and give one suitable example in each case :
   (*i*) Transpose of a matrix (*ii*) Symmetric matrix
   (*iii*) Skew-symmetric matrix (*iv*) Hermitian matrix
   (*v*) Skew-Hermitian matrix.
2. If A and B are symmetric, show that (AB + BA) is symmetric and (AB – BA) is skew-symmetric.
3. If $A = \begin{bmatrix} i & 0 \\ 0 & -i \end{bmatrix}$ and $B = \begin{bmatrix} 0 & i \\ -i & 0 \end{bmatrix}$,
   verify that $(AB)^\theta = B^\theta A^\theta$.
4. Show that the matrix $\begin{bmatrix} 0 & 1+i & 2+3i \\ 1-i & 1 & -i \\ 2-3i & i & 0 \end{bmatrix}$ is Hermitian.
5. Show that (*i*) $A = \begin{bmatrix} 2 & 1+i & 2+3i \\ 1-i & 1 & -i \\ 2-3i & i & 0 \end{bmatrix}$ is Hermitian.
   (*ii*) $B = \begin{bmatrix} 2i & 1+i & 2-3i \\ -1+i & 5i & 2 \\ -2-3i & -2 & 0 \end{bmatrix}$ is skew-Hermitian.
   (*iii*) *i*B is Hermitian.
6. Show that the diagonal elements of a Hermitian matrix are all real.
   [**Hint.** See Def. 3, Art. 19.]
7. If A is a Hermitian (Skew-Hermitian) matrix, then show that $\pm iA$ is skew-Hermitian (Hermitian).
8. Show that every square matrix can be uniquely expressed as the sum of a Hermitian matrix and a Skew-Hermitian matrix.
   [**Hint.** Write $A = \frac{1}{2}(A + A^\theta) + \frac{1}{2}(A - A^\theta) = P + Q$ and show that $P^\theta = P$ and $Q^\theta = -Q$.]
9. If A and B are Hermitian, show that :
   (*i*) AB + BA is Hermitian
   (*ii*) AB – BA is skew-Hermitian.
   (*iii*) AB is Hermitian if and only if AB = BA.
   (*iv*) BAB and ABA are Hermitian.
10. Show that B′AB is symmetric or skew-symmetric according as A is symmetric or skew-symmetric.
11. Show that all positive integral powers of a symmetric matrix are symmetric.
    [**Hint.** Let A be symmetric. ∴ $A' = A$ ... (1). Now for $n \in N$, $(A^n)' = (A')^n = A^n$. [By (1)]
12. Show that all positive odd integral powers of a skew-symmetric matrix are skew-symmetric while positive even integral powers are symmetric.
13. Express $A = \begin{bmatrix} 1 & 0 & 5 & 3 \\ -2 & 1 & 6 & 1 \\ 3 & 2 & 7 & 1 \\ 4 & -4 & -2 & 0 \end{bmatrix}$ as the sum of a symmetric matrix and a skew-symmetric matrix.

23. ADJOINT OF A SQUARE MATRIX

Def. *If A = $[\bar{a}_{ij}]$ is a square matrix of order n, and A_{ij} is the cofactor of a_{ij} in | A |, then the matrix*

$$[A_{ij}]' = \begin{bmatrix} A_{11} & A_{21} \cdots\cdots A_{n1} \\ A_{12} & A_{22} \cdots\cdots A_{n2} \\ \vdots & \vdots \qquad \vdots \\ A_{1n} & A_{2n} \cdots\cdots A_{nn} \end{bmatrix}$$

is called the adjoint of A and is written as adj. A.

In other words : To find adjoint of square matrix A, replace each element of A by its co-factor in | A | and take the transpose, the matrix so obtained will be adjoint of A.

Example 1. *Calculate the adjoint of* $A = \begin{bmatrix} a & b \\ c & d \end{bmatrix}$.

Sol. $| A | = \begin{bmatrix} a & b \\ c & d \end{bmatrix}$

$\therefore$ Co-factor of a, (1, 1)th element = $(-1)^2 \; | d | = d$

Co-factor of b, (1, 2)th element = $(-1)^3 \; | c | = -c$

Co-factor of c, (2, 1)th element = $(-1)^3 \; | b | = -b$

Co-factor of d, (2, 2)th element = $(-1)^4 \; | a | = a$

$$\therefore \quad \text{Adj. of } A = \begin{bmatrix} d & -c \\ -b & a \end{bmatrix}' = \begin{bmatrix} d & -b \\ -c & a \end{bmatrix}.$$

Example 2. *Calculate the adjoint of A, where* $A = \begin{bmatrix} 1 & 2 & 3 \\ 1 & -1 & 2 \\ -2 & 1 & 1 \end{bmatrix}$.

Sol. Let C_{ij} denote cofactor of (i, j)th element in A.

$$\therefore \quad C_{11} = (-1)^2 \begin{vmatrix} -1 & 2 \\ 1 & 1 \end{vmatrix} = (-1-2) = -3$$

$$C_{12} = (-1)^3 \begin{vmatrix} 1 & 2 \\ -2 & 1 \end{vmatrix} = -(1+4) = -5$$

$$C_{13} = (-1)^4 \begin{vmatrix} 1 & -1 \\ -2 & 1 \end{vmatrix} = (1-2) = -1$$

$$C_{21} = (-1)^3 \begin{vmatrix} 2 & 3 \\ 1 & 1 \end{vmatrix} = -(2-3) = 1$$

$$C_{22} = (-1)^4 \begin{vmatrix} 1 & 3 \\ -2 & 1 \end{vmatrix} = (1+6) = 7$$

$$C_{23} = (-1)^5 \begin{vmatrix} 1 & 2 \\ -2 & 1 \end{vmatrix} = -(1+4) = -5$$

$$C_{31} = (-1)^4 \begin{vmatrix} 2 & 3 \\ -1 & 2 \end{vmatrix} = (4+3) = 7$$

$$C_{32} = (-1)^5 \begin{vmatrix} 1 & 3 \\ 1 & 2 \end{vmatrix} = -(2-3) = 1$$

$$C_{33} = (-1)^6 \begin{vmatrix} 1 & 2 \\ 1 & -1 \end{vmatrix} = (-1-2) = -3$$

$$\therefore \quad \text{adj. A} = \begin{bmatrix} -3 & -5 & -1 \\ 1 & 7 & -5 \\ 7 & 1 & -3 \end{bmatrix}' = \begin{bmatrix} -3 & 1 & 7 \\ -5 & 7 & 1 \\ -1 & -5 & -3 \end{bmatrix}.$$

Example 3. *Calculate the adjoint of the diagonal matrix*

$$A = \begin{bmatrix} d_1 & 0 & 0 \\ 0 & d_2 & 0 \\ 0 & 0 & d_3 \end{bmatrix}.$$

Sol. Co-factor of $d_1 = (-1)^2 \begin{vmatrix} d_2 & 0 \\ 0 & d_3 \end{vmatrix} = d_2 d_3$

Co-factor of $d_2 = (-1)^4 \begin{vmatrix} d_1 & 0 \\ 0 & d_3 \end{vmatrix} = d_1 d_3$

Co-factor of $d_3 = (-1)^6 \begin{vmatrix} d_1 & 0 \\ 0 & d_2 \end{vmatrix} = d_1 d_2$

Clearly cofactor of each of the six non-diagonal entries is zero.

$$\therefore \quad \text{adj. A} = \begin{bmatrix} d_2 d_3 & 0 & 0 \\ 0 & d_3 d_1 & 0 \\ 0 & 0 & d_1 d_2 \end{bmatrix}' = \begin{bmatrix} d_2 d_3 & 0 & 0 \\ 0 & d_3 d_1 & 0 \\ 0 & 0 & d_1 d_2 \end{bmatrix}$$

This shows that adjoint of a diagonal matrix is a diagonal matrix.

24. THEOREM

If A be a n-square matrix, then

$$\textbf{A(adj. A) = (adj. A) A = | A | } \mathbf{I_n}$$

where I_n denotes the unit matrix of order n.

Example 4. *Find the adjoint of matrix* $\begin{bmatrix} 2 & 1 & 3 \\ 3 & 1 & 2 \\ 1 & 2 & 3 \end{bmatrix}$ *and verify the theorem A(adj. A) = (adj. A)A = | A | I.*

Sol. $| A | = \begin{vmatrix} 2 & 1 & 3 \\ 3 & 1 & 2 \\ 1 & 2 & 3 \end{vmatrix}$

Expanding by first row

$| A | = 2(3-4) - 1(9-2) + 3(6-1) = -2 - 7 + 15 = 6$

Let C_{ij} denote the co-factor of (i, j)th element of A.

$$\therefore \quad C_{11} = \begin{vmatrix} 1 & 2 \\ 2 & 3 \end{vmatrix} = 3 - 4 = -1$$

$$C_{12} = -\begin{vmatrix} 3 & 2 \\ 1 & 3 \end{vmatrix} = -(9-2) = -7$$

$$C_{13} = \begin{vmatrix} 3 & 1 \\ 1 & 2 \end{vmatrix} = 6 - 1 = 5$$

$$C_{21} = -\begin{vmatrix} 1 & 3 \\ 2 & 3 \end{vmatrix} = -(3-6) = 3$$

$$C_{22} = \begin{vmatrix} 2 & 3 \\ 1 & 3 \end{vmatrix} = 6 - 3 = 3$$

$$C_{23} = -\begin{vmatrix} 2 & 1 \\ 1 & 2 \end{vmatrix} = -(4-1) = -3$$

$$C_{31} = \begin{vmatrix} 1 & 3 \\ 1 & 2 \end{vmatrix} = 2 - 3 = -1$$

$$C_{32} = -\begin{vmatrix} 2 & 3 \\ 3 & 2 \end{vmatrix} = -(4-9) = 5$$

$$C_{33} = \begin{vmatrix} 2 & 1 \\ 3 & 1 \end{vmatrix} = 2 - 3 = -1$$

$$\therefore \quad \text{Adj. A} = \begin{bmatrix} -1 & -7 & 5 \\ 3 & 3 & -3 \\ -1 & 5 & -1 \end{bmatrix}' = \begin{bmatrix} -1 & 3 & -1 \\ -7 & 3 & 5 \\ 5 & -3 & -1 \end{bmatrix}$$

$$\text{A (adj. A)} = \begin{bmatrix} 2 & 1 & 3 \\ 3 & 1 & 2 \\ 1 & 2 & 3 \end{bmatrix} \begin{bmatrix} -1 & 3 & -1 \\ -7 & 3 & 5 \\ 5 & -3 & -1 \end{bmatrix}$$

$$= \begin{bmatrix} 2.(-1) + 1.(-7) + 3.5 & 2.3 + 1.3 + 3.(-3) & 2.(-1) + 1.5 + 3.(-1) \\ 3.(-1) + 1.(-7) + 2.5 & 3.3 + 1.3 + 2.(-3) & 3.(-1) + 1.5 + 2.(-1) \\ 1.(-1) + 2.(-7) + 3.5 & 1.3 + 2.3 + 3.(-3) & 1.(-1) + 2.5 + 3.(-1) \end{bmatrix}$$

$$= \begin{bmatrix} 6 & 0 & 0 \\ 0 & 6 & 0 \\ 0 & 0 & 6 \end{bmatrix} = 6 \begin{bmatrix} 1 & 0 & 0 \\ 0 & 1 & 0 \\ 0 & 0 & 1 \end{bmatrix} = |A| I_3$$

$\therefore$ A (adj. A) = $|A|\, I_3$.

Similarly, it can be shown that (adj. A)A = $|A| \,.\, I_3$

Hence the verification.

25. INVERSE OF A SQUARE MATRIX

Definition. Let A be an n-square matrix. If there exists an n-square matrix B such that

$$AB = BA = I_n, \text{ then}$$

the matrix A is said to be invertible and the matrix B is called the **inverse** of the matrix A.

Note 1. From the definition given above, it is very clear that if B is the inverse of A, then A is also the inverse of B.

2. A non-square matrix does not have any inverse.

26. THEOREM. *Inverse of a square matrix, if it exists is unique.*

Proof. Let A be any n-rowed square invertible matrix.

If possible, let B and C both be inverses of A.

$\therefore$ $AB = BA = I_n$ ...(1)

[$\because$ B is inverse of A]

and $AC = CA = I_n$ ...(2)

[$\because$ C is inverse of A]

Since A, B, C are all square matrices of the same order n.

$\therefore$ The product CAB is defined and

$$CAB = C(AB) = CI_n \quad \text{[Using (1)]}$$
$$= C \quad \text{...(3)}$$

Again, $$CAB = (CA)B = I_nB \quad \text{[Using (2)]}$$
$$= B \quad \text{...(4)}$$

From (3) and (4), we get $B = C$

Thus, inverse of A is unique.

Note. The inverse of A shall, in general, be denoted by A^{-1}.

27. SINGULAR AND NON-SINGULAR MATRIX

Def. *A square matrix A is said to be singular or non-singular according as* $|A| = O$ *or* $|A| \neq O$.

For example, the matrix $\begin{bmatrix} 4 & 2 \\ 6 & 3 \end{bmatrix}$ is singular and $\begin{bmatrix} 1 & 7 \\ 9 & 2 \end{bmatrix}$ is non-singular.

28. THEOREM

The necessary and sufficient condition for a square matrix A to possess the inverse is that $|A| \neq 0$ (i.e., A is non-singular).

Proof. The condition is necessary. Given that A is invertible (*i.e.*, A possesses inverse), to show that A is non-singular.

Let A possess inverse B (say).

$$\therefore \quad AB = BA = I_n \quad \text{[By def. of inverse]}$$

Taking determinants, we get $|AB| = |BA| = |I_n| = I$

i.e., $$|A| . |B| = I \neq O$$

But $|A|$ and $|B|$ are scalars (numbers)

$$\therefore \quad |A| \neq 0.$$

$\therefore$ A is a non-singular matrix.

The condition is sufficient. Given that A is non-singular. To show that A has inverse.

$\because$ A is non-singular, $\therefore$ $|A| \neq 0$

Consider the matrix $$B = \frac{\text{adj. } A}{|A|}$$

Now, $$AB = A . \left(\frac{\text{adj. } A}{|A|}\right) = \frac{1}{|A|} (A)(\text{adj. } A) = \frac{1}{|A|} |A| I_n = I_n$$

Also, $$BA = \left(\frac{\text{adj. } A}{|A|}\right) A = \frac{1}{|A|} (\text{adj. } A)(A) = \frac{1}{|A|} |A| I_n = I_n$$

Thus, $$AB = BA = I_n$$

$\therefore$ A is invertible and B is its inverse.

$$\therefore \quad A^{-1} = B = \frac{\text{adj. } A}{|A|}.$$

Remember. If A is non-singular, then $A^{-1} = \dfrac{\text{adj. A}}{|\text{A}|}$.

Note. The above theorem gives us one of the methods to compute the inverse of a non-singular matrix. We illustrate the method by examples given below.

Example 1. *Find the inverse of matrix* $A = \begin{bmatrix} a & b \\ c & d \end{bmatrix}$ *where ad – bc ≠ 0.*

Sol. Given matrix is $A = \begin{bmatrix} a & b \\ c & d \end{bmatrix}$ $\therefore$ $|A| = \begin{vmatrix} a & b \\ c & d \end{vmatrix} = ad - bc \neq 0$

i.e., A is non-singular, $\therefore$ A possesses inverse

$$\text{Adj. A} = \begin{bmatrix} d & -c \\ -b & a \end{bmatrix}' = \begin{bmatrix} d & -b \\ -c & a \end{bmatrix}$$

$$\therefore \quad A^{-1} = \frac{\text{adj. A}}{|A|} = \begin{bmatrix} \dfrac{d}{ad-bc} & \dfrac{-b}{ad-bc} \\ \dfrac{-c}{ad-bc} & \dfrac{a}{ad-bc} \end{bmatrix} \qquad [\because \; |A| = ad - bc]$$

Example 2. *D is the diagonal matrix*

$$\begin{bmatrix} d_1 & 0 & 0 & 0 \\ 0 & d_2 & 0 & 0 \\ 0 & 0 & d_3 & 0 \\ 0 & 0 & 0 & d_4 \end{bmatrix}$$

where none of the elements d_1, d_2, d_3, d_4 *is zero. Find* D^{-1}.

Sol.
$$D = \begin{bmatrix} d_1 & 0 & 0 & 0 \\ 0 & d_2 & 0 & 0 \\ 0 & 0 & d_3 & 0 \\ 0 & 0 & 0 & d_4 \end{bmatrix}$$

$$|D| = \begin{vmatrix} d_1 & 0 & 0 & 0 \\ 0 & d_2 & 0 & 0 \\ 0 & 0 & d_3 & 0 \\ 0 & 0 & 0 & d_4 \end{vmatrix}$$

$$= d_1 d_2 d_3 d_4 \neq 0 \qquad [\because \; \text{none of } d_1, d_2, d_3, d_4 \text{ is zero}]$$

Now co-factor of $d_1 = (-1)^2 \begin{vmatrix} d_2 & 0 & 0 \\ 0 & d_3 & 0 \\ 0 & 0 & d_4 \end{vmatrix} = d_2 d_3 d_4.$

Similarly, co-factors of d_2, d_3, d_4 are respectively $d_1d_3d_4$; $d_1d_2d_4$; $d_1d_2d_3$.

Clearly co-factor of each of the remaining twelve non-diagonal entries of matrix A is zero.

$$\therefore \quad \text{adj. A} = \begin{bmatrix} d_2d_3d_4 & 0 & 0 & 0 \\ 0 & d_1d_3d_4 & 0 & 0 \\ 0 & 0 & d_1d_2d_4 & 0 \\ 0 & 0 & 0 & d_1d_2d_3 \end{bmatrix}$$

Now, $$A^{-1} = \frac{\text{Adj. A}}{|A|} = \frac{\text{Adj. A}}{d_1 d_2 d_3 d_4}$$

$$= \begin{bmatrix} \frac{1}{d_1} & 0 & 0 & 0 \\ 0 & \frac{1}{d_2} & 0 & 0 \\ 0 & 0 & \frac{1}{d_3} & 0 \\ 0 & 0 & 0 & \frac{1}{d_4} \end{bmatrix} = \begin{bmatrix} d_1^{-1} & 0 & 0 & 0 \\ 0 & d_2^{-1} & 0 & 0 \\ 0 & 0 & d_3^{-1} & 0 \\ 0 & 0 & 0 & d_4^{-1} \end{bmatrix}$$

Hence if D = diag. $[d_1, d_2, d_3, d_4]$, where $d_1 d_2 d_3 d_4 \neq 0$

then D^{-1} = diag. $[d_1^{-1}, d_2^{-1}, d_3^{-1}, d_4^{-1}]$.

Remark. The method is quite general and can be extended to a non-singular diagonal matrix of any order.

Example 3. *Find the inverse of the matrix A given by*

$$A = \begin{bmatrix} 9 & 5 & 6 \\ 7 & -1 & 8 \\ 3 & 4 & 2 \end{bmatrix}.$$

Sol. Here $$|A| = \begin{vmatrix} 9 & 5 & 6 \\ 7 & -1 & 8 \\ 3 & 4 & 2 \end{vmatrix}$$

Expanding by first row $= 9(-2-32) - 5(14-24) + 6(28+3) = -70 \neq 0.$

$\therefore$ A is non-singular and hence A^{-1} exists

$$\text{adj. A} = \begin{bmatrix} -34 & 10 & 31 \\ 14 & 0 & -21 \\ 46 & -30 & -44 \end{bmatrix}'$$

(Finding co-factors is being left as an exercise for the reader)

i.e., $$\text{adj. A} = \begin{bmatrix} -34 & 14 & 46 \\ 10 & 0 & -30 \\ 31 & -21 & -44 \end{bmatrix}$$

Hence, $$A^{-1} = \frac{\text{adj. A}}{|A|} = \begin{bmatrix} \frac{-34}{-70} & \frac{14}{-70} & \frac{46}{-70} \\ \frac{10}{-70} & \frac{0}{-70} & \frac{-30}{-70} \\ \frac{31}{-70} & \frac{-21}{-70} & \frac{-44}{-70} \end{bmatrix} \quad \textit{i.e.,} \quad A^{-1} = \begin{bmatrix} \frac{17}{35} & \frac{-1}{5} & -\frac{23}{35} \\ \frac{-1}{7} & 0 & \frac{3}{7} \\ \frac{-31}{70} & \frac{3}{10} & \frac{22}{35} \end{bmatrix}.$$

29. THEOREM

1. If A is invertible, then so is A^{-1} and $(A^{-1})^{-1}$ = A.

2. If A and B are square matrices of order n, then AB is invertible if and only if A and B are invertible and then $(AB)^{-1} = B^{-1} A^{-1}$.

Proof. 1. As A is invertible $\therefore$ A^{-1} exists

and $$AA^{-1} = A^{-1}A = I.$$

This shows (by def.) that A^{-1} is also invertible and inverse of A^{-1} is A

i.e., $$(A^{-1})^{-1} = A.$$

2. $$|AB| = |A| . |B|$$

$\therefore$ $|AB| \neq 0$, if and only if $|A| \neq 0$ and $|B| \neq 0$

i.e., AB is non-singular, if and only if A and B are both non-singular, which is the same as AB is invertible if and only if A and B are invertible.

Let A and B be invertible and their inverse be A^{-1} and B^{-1}. All these matrices A, B, A^{-1}, B^{-1} are squared matrices of same order n.

Now $$(AB)(B^{-1}A^{-1}) = A(BB^{-1})A^{-1} \quad \text{[Associative law]}$$
$$= (AI)A^{-1} = AA^{-1} = I$$

Again $$(B^{-1}A^{-1})(AB) = B^{-1}(A^{-1}A)B = B^{-1}IB = B^{-1}B = I$$

i.e., $$(AB)(B^{-1}A^{-1}) = (B^{-1}A^{-1})(AB) = I$$

This shows that inverse of AB is $B^{-1}A^{-1}$ or $\mathbf{(AB)^{-1} = B^{-1}A^{-1}}$.

Cor. 1. The result can be extended to the inverse of the product of more than two matrices *i.e.* if $A_1, A_2, \ldots\ldots A_k$ are non-singular matrices of order n, then

$$(A_1 . A_2 \ldots\ldots A_{k-1} . A_k)^{-1} = A_k^{-1} . A^{-1}_{k-1} \ldots\ldots A_2^{-1} . A_1^{-1}.$$

Cor. 2. Put $A_1 = A_2 \ldots\ldots A_k = A$ in result of Cor. 1.

$\therefore$ $$(A.A.A \ldots\ldots k \text{ times})^{-1} = A^{-1} . A^{-1} \ldots\ldots k \text{ times}$$

i.e., $$(A^k)^{-1} = (A^{-1})^k$$

and we write both of these as A^{-k}.

Hence $$A^{-k} = (A^k)^{-1} = (A^{-1})^k.$$

30. THEOREM

1. If A is non-singular, then prove that so is A′

and $$\mathbf{(A')^{-1} = (A^{-1})'.}$$

Proof. $\because$ A is non-singular $\therefore$ A is invertible and let A^{-1} be the inverse of A.

$$AA^{-1} = A^{-1}A = I \quad \text{...(1)}$$

Taking transpose of both sides

$$(AA^{-1})' = (A^{-1}A)' = (I)'$$

i.e., $$(A^{-1})'A' = A'(A^{-1})' = I \quad \text{(Reversal Law)}$$

This shows that inverse of A′ exists and it is $(A^{-1})'$.

Hence A′ is invertible and $(A')^{-1} = (A^{-1})'$.

2. If A is non-singular, then prove that A^{θ} is also non-singular, and

$$\mathbf{(A^{\theta})^{-1} = (A^{-1})^{\theta}.}$$

Proof. $\because$ A is non-singular $\therefore$ A^{-1} exists and

$$AA^{-1} = A^{-1}A = I$$

Taking conjugate transpose of both sides

$\therefore$ $$(AA^{-1})^{\theta} = (A^{-1}A)^{\theta} = (I)^{\theta}$$

or $$(A^{-1})^{\theta}A^{\theta} = A^{\theta}(A^{-1})^{\theta} = I.$$

This shows that the inverse of A^{θ} exists and it is $(A^{-1})^{\theta}$.

$\therefore$ A^θ is non-singular and

$$(A^\theta)^{-1} = (A^{-1})^\theta.$$

Example 1. *If A is a non-singular matrix, then prove that*

$$\det (A^{-1}) = (\det A)^{-1}.$$

Sol. $\because$ A is non-singular, so A^{-1} exists and

$$AA^{-1} = I$$

Taking determinants of both sides

$$\therefore \quad | AA^{-1} | = | I |$$

or $$| A | \, | A^{-1} | = I$$

Divide by $| A | \neq 0$, we get

$$| A^{-1} | = \frac{1}{|A|} = (| A |)^{-1}$$

which can also be written as $\det A^{-1} = (\det A)^{-1}$.

Example 2. *Prove that adjoint of a non-singular matrix is non-singular.*

Sol. Let A be a non-singular matrix of order *n*.

$$\Rightarrow \quad | A | \neq 0.$$

Since $$A(\text{adj. } A) = | A | I_n$$

Taking determinants of both sides

$$| A (\text{adj. } A) | = | \, | A | I_n \, |$$

$$\Rightarrow \quad | A | \, . \, | \text{adj. } A | = | A |^n \, | I_n | \qquad [\because \; | kA | = k^n \, | A |]$$

$$\Rightarrow \quad | A | \, . \, | \text{adj. } A | = | A |^n \qquad (\because \; | I_n | = 1)$$

Dividing by $| A |$, $| \text{adj. } A | = | A |^{n-1} \neq 0$

$\Rightarrow$ adj. A is non-singular.

Example 3. *If A is a non-singular matrix of order n, then prove that*

$$adj \, (adj \, A) = | A |^{n-2} A.$$

Sol. We know that $$A \, . \, \text{adj. } A = | A | I_n \qquad ...(1)$$

Replacing A by Adj. A in the above relation,

we have $$\text{adj. } A \text{ adj. } (\text{adj. } A) = | \text{adj. } A | I_n = | A |^{n-1} I_n$$

[Reproduce Ex. 2 above to prove that $| \text{adj. } A | = | A |^{n-1}$]

Pre-multiplying both sides by A,

$$A \text{ adj. } A \text{ adj. } (\text{adj. } A) = A \, | A |^{n-1} I_n$$

or $$| A | I_n \text{ adj. } (\text{adj. } A) = | A |^{n-1} A I_n \qquad [\text{By (1)}]$$

or $$| A | \text{ adj. } (\text{Adj. } A) = | A |^{n-1} A \qquad [\because \; I_n . B = B = B . I_n]$$

Dividing both sides by $| A | \neq 0$ ($\because$ A is non-singular),

we have $$\text{adj. } (\text{adj. } A) = | A |^{n-2} A.$$

Example 4. *If A and B are two non-singular matrices of same order n, then prove that adj. (AB) = (adj. B) (adj. A).*

Sol. $\because$ A and B are non-singular

$$\therefore \quad | A | \neq 0 \quad \text{and} \quad | B | \neq 0$$

and $$| AB | = | A | \, | B | \neq 0$$

$\therefore$ A^{-1}, B^{-1} and $(AB)^{-1}$ all exist.

Also we know that for every non-singular matrix P,

$$P^{-1} = \frac{\text{adj. } P}{|P|}$$

$\therefore$ adj. P = | P | P^{-1} ...(1)

Putting P = AB in (1), we have

$$\text{adj. } (AB) = |AB| (AB)^{-1}$$
$$= |A| |B| B^{-1} A^{-1} \text{ (Reversal Law)}$$
$$= |A| |B| \frac{\text{adj. } B}{|B|} \frac{\text{adj. } A}{|A|} = (\text{adj. } B)(\text{adj. } A).$$

EXERCISE 2

1. Define the inverse of a matrix, and show that whenever it exists, it is unique.
2. Prove that a square matrix A is invertible if and only if, | A | $\neq 0$, where | A | denotes the determinant of A.
3. Prove that the inverse of product of two matrices is equal to the product of their inverses but in reverse order.
4. Calculate the inverse of the following matrices whenever exists :

   (i) $\begin{bmatrix} 3 & 5 & 7 \\ 2 & -3 & 1 \\ 1 & 1 & 2 \end{bmatrix}$ (ii) $\begin{bmatrix} 1 & 1 & 3 \\ 1 & 3 & -3 \\ -2 & -4 & -4 \end{bmatrix}$

   (iii) $\begin{bmatrix} 1 & 2 & 1 \\ 3 & 1 & 2 \\ 0 & 1 & 2 \end{bmatrix}$.
5. If A = $\begin{bmatrix} 0 & 0 & 1 \\ 0 & 1 & 0 \\ 1 & 0 & 0 \end{bmatrix}$, show that $A^{-1} = A$.
6. Let A and B be invertible square matrices of order n. Does $(A + B)^{-1}$ exist ? Justify by giving example.
7. If B is non-singular, prove that | B^{-1} AB | = | A | A and B being square matrices of the same order.
8. If A is an n-square non-singular matrix, prove that | adj. A | = | A $|^{n-1}$.
   [**Hint.** Reproduce Ex. 2 Page 67]
9. If the non-singular matrix A is symmetric, prove that A^{-1} is also symmetric.
10. If the matrices A and B commute, then A^{-1} and B^{-1} also commute.

ANSWERS

4. (i) $\begin{bmatrix} 7 & 3 & -26 \\ 3 & 1 & -11 \\ -5 & -2 & 19 \end{bmatrix}$ (ii) $\begin{bmatrix} 3 & 1 & 3/2 \\ -5/4 & -1/4 & -3/4 \\ -1/4 & -1/4 & -1/4 \end{bmatrix}$ (iii) $\begin{bmatrix} 0 & \frac{1}{3} & -\frac{1}{3} \\ \frac{2}{3} & -\frac{2}{9} & -\frac{1}{9} \\ -\frac{1}{3} & \frac{1}{9} & \frac{5}{9} \end{bmatrix}$.

6. $(A + B)^{-1}$ may or may not exist. For example, let A = $\begin{bmatrix} 1 & 0 \\ 0 & 1 \end{bmatrix}$ and B = $\begin{bmatrix} -1 & 0 \\ 0 & -1 \end{bmatrix}$.

31. TO SOLVE A SYSTEM OF LINEAR EQUATIONS

$$\left.\begin{aligned} a_1x + b_1y + c_1z = d_1 \\ a_2x + b_2y + c_2z = d_2 \\ a_3x + b_3y + c_3z = d_3 \end{aligned}\right\} \quad ...(1)$$

As explained earlier these equations, can be put in the form AX = B ...(2)

where
$$A = \begin{bmatrix} a_1 & b_1 & c_1 \\ a_2 & b_2 & c_2 \\ a_3 & b_3 & c_3 \end{bmatrix}, X = \begin{bmatrix} x \\ y \\ z \end{bmatrix}, B = \begin{bmatrix} d_1 \\ d_2 \\ d_3 \end{bmatrix}$$

To solve the equations means to find the column matrix X, so as to satisfy (2). We shall give solution of these equations when $|A| \neq 0$ *i.e.,* when the **co-efficient matrix is non-singular.**

32. THEOREM

If AX = B is a system of linear equations with $|A| \neq 0$, then $X = A^{-1}B$ is the unique solution.

Proof. The system of equation is AX = B ...(1)

$\therefore$ $|A| \neq 0$ *i.e.,* A is non-singular $\therefore$ A^{-1} exists

Pre-multiply both sides of (1) by A^{-1}

$\therefore$ $A^{-1}(AX) = A^{-1}B$

or $(A^{-1}A)X = A^{-1}B$ or $IX = A^{-1}B$ or $X = A^{-1}B$

which gives the solution of equation (1).

The solution is unique. For, if possible, let X_1 and X_2 be the two solutions of (1).

$\therefore$ $AX_1 = B$

$AX_2 = B$

$\therefore$ $AX_1 = AX_2$

Pre-multiplying by A^{-1} which exists

$\therefore$ $A^{-1}(AX_1) = A^{-1}(AX_2)$ or $(A^{-1}A)X_1 = (A^{-1}A)X_2$

i.e., $X_1 = X_2$

Hence the solution is **unique.**

We illustrate the method with the help of the following examples :

Example. *Solve the following equations by matrices*

$$x + 4y + 6z + 25 = 0$$
$$6x + y + 4z + 17 = 0$$
$$4x + 6y + z + 24 = 0.$$

Sol. Writing the given equations as

$$x + 4y + 6z = -25$$
$$6x + y + 4z = -17$$
$$4x + 6y + z = -24 \quad ...(1)$$

These can be written in the matrix form as
$$\begin{bmatrix} 1 & 4 & 6 \\ 6 & 1 & 4 \\ 4 & 6 & 1 \end{bmatrix}\begin{bmatrix} x \\ y \\ z \end{bmatrix} = \begin{bmatrix} -25 \\ -17 \\ -24 \end{bmatrix}$$

i.e., $AX = B$ ...(2)

where $A = \begin{bmatrix} 1 & 4 & 6 \\ 6 & 1 & 4 \\ 4 & 6 & 1 \end{bmatrix}$

Solution of (1), *i.e.* that of (2) is $X = A^{-1} B$

$\therefore$ **To find A^{-1}** $|A| = \begin{vmatrix} 1 & 4 & 6 \\ 6 & 1 & 4 \\ 4 & 6 & 1 \end{vmatrix}$

Expanding by first row $|A| = 1(1 - 24) - 4(6 - 16) + 6(36 - 4)$

$$= -23 + 40 + 192 = 209 \neq 0$$

$$\text{Adj. } A = \begin{bmatrix} -23 & 10 & 32 \\ 32 & -23 & 10 \\ 10 & 32 & -23 \end{bmatrix}' = \begin{bmatrix} -23 & 32 & 10 \\ 10 & -23 & 32 \\ 32 & 10 & -23 \end{bmatrix}$$

(Finding co-factors is being left as an exercise for the reader)

$$\therefore \quad A^{-1} = \frac{\text{adj. } A}{|A|} = \begin{bmatrix} \frac{-23}{209} & \frac{32}{209} & \frac{10}{209} \\ \frac{10}{209} & \frac{-23}{209} & \frac{32}{209} \\ \frac{32}{209} & \frac{10}{209} & \frac{-23}{209} \end{bmatrix}$$

Solution is $X = A^{-1} B$ *i.e.,* $\begin{bmatrix} x \\ y \\ z \end{bmatrix} = \begin{bmatrix} \frac{-23}{209} & \frac{32}{209} & \frac{10}{209} \\ \frac{10}{209} & \frac{-23}{209} & \frac{32}{209} \\ \frac{32}{209} & \frac{10}{209} & \frac{-23}{209} \end{bmatrix} \begin{bmatrix} -25 \\ -17 \\ -24 \end{bmatrix}$

Multiplying the matrices on R.H.S. and equating it with matrix on L.H.S., we have

$$x = \frac{-23}{209}(-25) + \frac{32}{209}(-17) + \frac{10}{209}(-24)$$

$$= \frac{575 - 544 - 240}{209} = \frac{-209}{209} = -1$$

$$y = \frac{10}{209}(-25) + \left(\frac{-23}{209}\right)(-17) + \frac{32}{209}(-24)$$

$$= \frac{-250 + 391 - 768}{209} = \frac{-627}{209} = -3$$

$$z = \frac{32}{209}(-25) + \frac{10}{209}(-17) + \left(\frac{-23}{209}\right)(-24)$$

$$= \frac{-800 - 170 + 552}{209} = \frac{-418}{209} = -2$$

Required solution is $x = -1, y = -3, z = -2$.

Note. The students are suggested to verify the solution by putting these values in the given equations.

EXERCISE 3

1. Solve the following linear equation in x, y, z by matrices :

(a)
$$x + z = 7$$
$$2x + y = 7$$
$$3x + 2y + z = 17$$

(b)
$$2x + 3y + 5z = 15$$
$$3x + 5y + 2z = 12$$
$$5x + 2y + 3z = 13$$

(c)
$$x + y + z + 6 = 0$$
$$x + 2y + 3z + 14 = 0$$
$$x + 4y + 9z + 36 = 0.$$

2. Find the inverse of the coefficient matrix of the equations

$$x + y + z = \lambda$$
$$ax + by + cz = \mu$$
$$a^2x + b^2y + c^2z = \nu$$

and hence solve these equations. It being given that $a \neq b \neq c$.

ANSWERS

1. (a) $x = 2, y = 3, z = 5$ (b) $x = 1, y = 1, z = 2$ (c) $-1, -2, -3$

2.
$$\begin{bmatrix} \dfrac{bc}{(a-b)(a-c)} & \dfrac{-(b+c)}{(a-b)(a-c)} & \dfrac{1}{(a-b)(a-c)} \\ \dfrac{ca}{(b-a)(b-c)} & \dfrac{-(c+a)}{(b-a)(b-a)} & \dfrac{1}{(b-a)(b-c)} \\ \dfrac{ab}{(c-a)(c-b)} & \dfrac{-(a+b)}{(c-a)(c-b)} & \dfrac{1}{(c-b)(c-a)} \end{bmatrix}$$

$$x = \frac{\lambda bc - \mu(b+c) + \nu}{(a-b)(a-c)}, \; y = \frac{\lambda ca - \mu(c+a) + \nu}{(b-a)(b-c)}, \; z = \frac{\lambda ab - \mu(a+b) + \nu}{(c-b)(c-a)}.$$

4

Rank, Row Rank and Column Rank of a Matrix and Linear Equations

1. Minor of order k of a Matrix

Definition. *A minor of order k of a matrix A is the determinant of a k-square sub-matrix of A.* That is if from an $m \times n$ matrix A, any k rows and k columns are retained ($0 < k \leq \min(m, n)$) and others are deleted, then the determinant of the resulting k-square sub-matrix is a minor of order k of A. If A is a $n \times n$ square matrix, then $|A|$ is also a minor of order n of matrix A.

For example, in the matrix

$A = \begin{bmatrix} 0 & 2 & -1 & 0 \\ 2 & 4 & -3 & -1 \\ 4 & 5 & 2 & 2 \end{bmatrix}$, deleting first and fourth columns and third row, we get the sub-matrix $\begin{bmatrix} 2 & -1 \\ 4 & -3 \end{bmatrix}$

Its determinant $\begin{vmatrix} 2 & -1 \\ 4 & -3 \end{vmatrix} = -2$ is a minor of order 2 of A.

Similarly, $\begin{vmatrix} 0 & 0 \\ 2 & -1 \end{vmatrix} = 0$ is also a minor of order 2 of A.

$\begin{vmatrix} 0 & 2 & -1 \\ 2 & 4 & -3 \\ 4 & 5 & 2 \end{vmatrix} = 22$ is a minor of order 3 of A.

Remark. There can be more than one minor of order k for each integer k.

2. Rank of a Matrix

Definition. *A non-zero matrix A is said to have rank r if*

(i) A has at least one non-zero minor of order r and

(ii) every minor of order r + 1 of A, if any, is zero.

Rank of a matrix is denoted by $\rho(A)$. *If* $A = O$, *then* $\rho(A)$ is defined as *0.*

Note. From the theory of determinants, we know that the determinant of a k-square matrix A, ($k \geq 2$), can be expressed as a linear combination of determinants of sub-matrices of order $k - 1$ of A. Thus, if all minors of order $r + 1$ of a matrix A vanish, then so do all the minors of order $r + 2$ and continuing thus, we find that all minors of order larger than r of A vanish.

For example :

(*i*) Let $A = \begin{bmatrix} 1 & 2 & 3 \\ 2 & 3 & 4 \\ 3 & 5 & 7 \end{bmatrix}$

Here, $|A| = 1(21 - 20) - 2(14 - 12) + 3(10 - 9) = 0$

But, $\begin{vmatrix} 1 & 2 \\ 2 & 3 \end{vmatrix} = -1 \neq 0 \quad \therefore \quad \rho(A) = 2.$

(*ii*) Let $A = \begin{bmatrix} 1 & 2 & -1 & 0 \\ 3 & 1 & 4 & 2 \\ 1 & -3 & 6 & 2 \end{bmatrix}$

We can see that all the four minors of order 3 of A namely $\begin{vmatrix} 1 & 2 & -1 \\ 3 & 1 & 4 \\ 1 & -3 & 6 \end{vmatrix}, \begin{vmatrix} 1 & 2 & 0 \\ 3 & 1 & 2 \\ 1 & -3 & 2 \end{vmatrix},$ $\begin{vmatrix} 1 & -1 & 0 \\ 3 & 4 & 2 \\ 1 & 6 & 2 \end{vmatrix}, \begin{vmatrix} 2 & -1 & 0 \\ 1 & 4 & 2 \\ -3 & 6 & 2 \end{vmatrix}$ vanish. (*i.e.* become zero)

A has several non-zero minors of order 2

For instance, $\begin{vmatrix} 1 & 2 \\ 3 & 1 \end{vmatrix} = -5 \neq 0$

Hence, $\rho(A) = 2.$

(*iii*) Let $A = \begin{bmatrix} 1 & -1 & 2 & 0 \\ 0 & 2 & 3 & -1 \\ 0 & 0 & 0 & 8 \\ 0 & 0 & 0 & -3 \end{bmatrix}$

Only minor of order 4 of A is $|A|$ and $|A| = 0$ on expanding by first column but the minor $\begin{vmatrix} 1 & -1 & 0 \\ 0 & 2 & -1 \\ 0 & 0 & 8 \end{vmatrix} = 16$ of order 3 is non-zero.

Hence, $\rho(A) = 3.$

(*iv*) Let $A = \begin{bmatrix} 1 & a & b & 0 \\ 0 & c & d & 1 \\ 1 & a & b & 0 \\ 0 & c & d & 1 \end{bmatrix}$

Here, $|A| = 0$ and all minors of order 3 are zero because either two rows or two columns are identical.

However, the second order minor $\begin{vmatrix} 1 & 0 \\ 0 & 1 \end{vmatrix} = 1 \neq 0.$

Hence, $\rho(A) = 2.$

(*v*) $\rho(I_n) = n$, where I_n is the identity matrix of order *n*. $(\because |I_n| = 1 \neq 0)$

Remarks 1. If A is not a zero matrix, then it has at least, one non-zero entry in it so that A has a non-zero minor of order 1 at least. **Thus for a non-zero matrix A, $\rho(A) \geq 1$.** For the zero matrix A, we defined $\rho(A) = 0$.

2. If A is an $m \times n$ matrix, then $\rho(A) \leq \min.(m, n)$.

3. If A has a non-zero minor of order *k*, then $\rho(A) \geq k$.

4. If every minor of order $k + 1$ of A vanishes, then $\rho(A) \leq k$.

5. If A is a diagonal matrix, $\rho(A)$ is the number of non-zero entries in the diagonal.

3. Working Procedure for Finding Rank of a Matrix

Working procedure for finding rank of a matrix A (as is clear from the above five examples) is : We first consider minors of the largest order (say *m*) which can be formed. If there exists a non-zero minor of order *m*, then $\rho(A) = m$. If each minor of order *m* is zero, consider minors of order $m - 1$. If there is one non-zero minor of order $m - 1$, then $\rho(A) = m - 1$. In case, each minor of order $m - 1$ is zero, we consider all minors of order $m - 2$ and repeat the procedure. However, in practice, this procedure is inconvenient and tedious. In practice, we generally first reduce the matrix to Echelon or triangular form—the procedure to be explained later.

Example 1. *If A is a non-zero column matrix and B is a non-zero row matrix, show that $\rho(AB) = 1$.*

Sol. Let $A = \begin{bmatrix} a_{11} \\ a_{21} \\ \vdots \\ a_{m1} \end{bmatrix}$ be a non-zero column matrix and $B = [b_{11}\ b_{12} \ldots\ldots b_{1n}]$ be a non-zero row matrix.

$$\therefore \quad AB = \begin{bmatrix} a_{11}b_{11} & a_{11}b_{12} & \cdots & a_{11}b_{1n} \\ a_{21}b_{11} & a_{21}b_{12} & \cdots & a_{21}b_{1n} \\ \cdots & \cdots & \cdots & \cdots \\ a_{m1}b_{11} & a_{m1}b_{12} & \cdots & a_{m1}b_{1n} \end{bmatrix}$$

Because $A \neq 0$, therefore at least one element of A say $a_{i1} \neq 0$.

Again because $B \neq 0$, therefore at least one element of B say $b_{1k} \neq 0$.

$\therefore$ (*i, k*) the element of $AB = a_{i1}\, b_{1k} \neq 0$

$\therefore$ Matrix $AB \neq O$ $(\because$ At least its one element $a_{i1}\, b_{1k} \neq 0)$

$\therefore$ $\rho(AB) \geq 1$ (By Remark 1, Art. 2) ...(1)

But $\rho(AB) \leq \rho(A)$ ...(2) (By Art. Proved later in this chapter)

Again A is a **non-zero** column matrix (*i.e.* having one column only)

$\therefore$ $\rho(A) = 1$ (See Remark 2, Art. 2) ...(3)

From (2) and (3), we have $\rho(AB) \leq 1$ ...(4)

From (1) and (4), $\rho(AB) = 1$.

Example 2. *If A is an n-square matrix of rank n – 1, then show that adj. A ≠ O.*

Sol. Since A is an *n*-square matrix of rank $n - 1$, therefore A has at least one non-zero $n - 1$ square minor. But, every $n - 1$ square minor of matrix A is equal in magnitude to the cofactor of some element in | A |. Therefore, at least one element in | A | has its co-factor non- zero.

So, at last one element of adj. A is non-zero.

Hence, adj. $A \neq O$.

Example 3. *Show that no skew-symmetric matrix is of rank 1.*

Sol. Let us prove that result for a 4 × 4 skew-symmetric matrix. The method of proof can however be used for a skew-symmetric matrix of any order.

Let $A = \begin{bmatrix} 0 & h & g & l \\ -h & 0 & f & m \\ -g & -f & 0 & n \\ -l & -m & -n & 0 \end{bmatrix}$ be a skew-symmetric matrix.

[Here $a_{ii} = 0$ and $a_{ij} = -a_{ji}$ $\forall$ i and j]

If f, g, h, l, m, n are all zero, then $\rho(A) = 0$. If at least one of these elements, say, g is non-zero, then at least one second order minor $\begin{vmatrix} 0 & g \\ -g & 0 \end{vmatrix}$ of the matrix A is g^2 ($\neq 0$).

$\therefore$ $\rho(A) \geq 2$ (By Remark 3, Art. 2)

Hence, $\rho(A) \neq 1$.

Note. $\rho(A') = \rho(A)$, $\rho(\overline{A}) = \rho(A)$, $\rho(A^\theta) = \rho(A)$.

4. Singular and Non-singular Matrices

Definition. *An n-square matrix A is called a* **non-singular matrix** *if* $\rho(A) = n$, *i.e., iff at least one minor of order n of n × n matrix A is non-zero i.e. iff* $|A| \neq 0$ ($\because$ $|A|$ *is the only minor of order n of matrix A). Otherwise A is called* **singular matrix.**

Since for two n-square matrices, A and B,

$|AB| = |A| . |B|$, we have :

(i) the product of two or more non-singular n-square matrices is non-singular.

(ii) the product of two or more n-square matrices is singular if at least one of the matrices is singular.

EXERCISE 1

1. Find the rank of the following matrices :

(i) $\begin{bmatrix} 1 & 2 & 3 \\ 3 & 2 & 1 \\ 1 & 1 & 1 \end{bmatrix}$ (ii) $\begin{bmatrix} 0 & -1 & 2 \\ 4 & 3 & 1 \\ 4 & 2 & 3 \end{bmatrix}$ (iii) $\begin{bmatrix} 1 & 2 & 3 \\ -2 & 4 & -9 \\ 1 & 2 & 3 \end{bmatrix}$.

2. Find the rank of the following matrices :

(i) $\begin{bmatrix} 2 & -2 & 0 & 6 \\ 4 & 2 & 0 & 2 \\ 1 & -1 & 0 & 3 \\ 1 & -2 & 1 & 2 \end{bmatrix}$ (ii) $\begin{bmatrix} 2 & -1 & 3 & 4 \\ 0 & 3 & 4 & 1 \\ 2 & 3 & 7 & 5 \\ 2 & 5 & 11 & 6 \end{bmatrix}$ (iii) $\begin{bmatrix} 1 & -1 & 3 & 6 \\ 1 & 3 & -3 & -4 \\ 5 & 3 & 3 & 11 \end{bmatrix}$

(iv) $\begin{bmatrix} 2 & 0 & 0 & 0 \\ 0 & 0 & 0 & 0 \\ 0 & 5 & 0 & 0 \\ 0 & 0 & 7 & 0 \end{bmatrix}$ (v) $\begin{bmatrix} 3 & 4 & 1 & 2 \\ 3 & 2 & 1 & 4 \\ 7 & 6 & 2 & 5 \end{bmatrix}$ (vi) $\begin{bmatrix} 1 & 2 & -1 & 3 \\ -2 & -4 & 4 & -7 \\ 1 & 2 & 1 & 2 \end{bmatrix}$.

3. If $A = \begin{bmatrix} 1 & 1 & -1 \\ 2 & -3 & 4 \\ 3 & -2 & 3 \end{bmatrix}$, $B = \begin{bmatrix} -1 & -2 & -1 \\ 6 & 12 & 6 \\ 5 & 10 & 5 \end{bmatrix}$, find ranks of A, B, A + B, AB and BA by the help of minors of the corresponding matrices.

4. If $A = \begin{bmatrix} 0 & 1 & 0 & 0 \\ 0 & 0 & 1 & 0 \\ 0 & 0 & 0 & 1 \\ 0 & 0 & 0 & 0 \end{bmatrix}$, find $\rho(A)$ and $\rho(A^2)$.

5. Show that rank of a matrix is greater than or equal to the rank of every sub-matrix thereof.

ANSWERS

1. (*i*) 2 (*ii*) 2 (*iii*) 2

2. (*i*) 3 (*ii*) 3 (*iii*) 3 (*iv*) 3 (*v*) 3 (*vi*) 2

3. $\rho(A) = 2, \rho(B) = 1, \rho(A + B) = 2, \rho(AB) = 0, \rho(BA) = 1$

4. $\rho(A) = 3, \rho(A^2) = 2.$

ELEMENTARY OPERATIONS (OR TRANSFORMATIONS)

5. Elementary Row Operations

The following three types of operations on the rows of a martix are called ***elementary row operations :***

1. The interchange of any two distinct rows of a matrix.

2. The multiplication of any row (each element of the row) by a fixed non-zero scalar.

3. The addition to the elements of a row, a fixed scalar multiple of the corresponding elements of another row.

6. Elementary Column Operations

The following three types of operations on the columns of a matrix are called ***elementary column operations :***

1. The interchange of any two distinct columns of a matrix.

2. The multiplication of any column (each element of the column) by a fixed non-zero scalar.

3. The addition to the elements of a column, a fixed scalar multiple of the corresponding elements of another column.

7. Elementary Operations

Definition. *The above six operations—three elementary row operations and three column operations, are called* ***elementary operations.***

Following symbols will be used to denote the six elementary operations :

I. $R_{i,j}$ for the interchange of the *i*th and *j*th rows.

II. $R_i(\lambda)$ for multiplication of the *i*th row by $\lambda \neq 0$.

III. $R_{i,j}(\lambda)$ for addition to the *i*th row, the product of the *j*th row by λ.

The corresponding column operations will be denoted by $C_{i,j}$, $C_i(\lambda)$, $C_{i,j}(\lambda)$ respectively.

Illustration 1. *Let* $A = \begin{bmatrix} 2 & 2 & -2 & 3 \\ 0 & -5 & 4 & 1 \\ 1 & -1 & 1 & 5 \end{bmatrix}.$

We have $A \xrightarrow{R_{1,2}} \begin{bmatrix} 0 & -5 & 4 & 1 \\ 2 & 2 & -2 & 3 \\ 1 & -1 & 1 & 5 \end{bmatrix},$

$$A \xrightarrow{R_2(3)} \begin{bmatrix} 2 & 2 & -2 & 3 \\ 0 & -15 & 12 & 3 \\ 1 & -1 & 1 & 5 \end{bmatrix}$$

and $A \xrightarrow{R_{2,3}(3)} \begin{bmatrix} 2 & 2 & -2 & 3 \\ 3 & -8 & 7 & 16 \\ 1 & -1 & 1 & 5 \end{bmatrix}.$

Illustration. 2. *Let* $A = \begin{bmatrix} 0 & 2 & 1 \\ 1 & 4 & -5 \\ -1 & 0 & 3 \\ 3 & 1 & 1 \end{bmatrix}.$

We have $A \xrightarrow{C_{2,3}} \begin{bmatrix} 0 & 1 & 2 \\ 1 & -5 & 4 \\ -1 & 3 & 0 \\ 3 & 1 & 1 \end{bmatrix},$

$$A \xrightarrow{C_3(2)} \begin{bmatrix} 0 & 2 & 2 \\ 1 & 4 & -10 \\ -1 & 0 & 6 \\ 3 & 1 & 2 \end{bmatrix}$$

and $A \xrightarrow{C_{1,3}(2)} \begin{bmatrix} 2 & 2 & 1 \\ -9 & 4 & -5 \\ 5 & 0 & 3 \\ 5 & 1 & 1 \end{bmatrix}.$

Remark. Any column operation on a matrix is equivalent to applying the corresponding elementary row operation on the transpose A′ of A and then taking the transpose. Thus, there is a 1 – 1 correspondence between the elementary column operations on A and the elementary row operations of the same type on A′.

Also the elementary operations do not alter the order of a matrix.

8. Theorem

The elementary operations do not alter the rank of a matrix.

Proof. We consider only elementary row operations.

Let A be an $m \times n$ matrix and $\rho(A) = r$.

Then, every minor of order $r + 1$, if any, of A is zero. Let B be the matrix obtained from A by a row operation. Denote by $|S|$ any minor of order $r + 1$ of A and by $|T|$ the minor of order $r + 1$ of B having the same position as $|S|$.

Case I. Let the row operation be $R_{i,j}$

Its effect on $|S|$ is *(i)* to leave it unchanged or *(ii)* to interchange two of its rows or *(iii)* to interchange one of its rows with a row not of $|S|$. In case *(i)*, $|T| = |S| = 0$.

In case *(ii)*, $|T| = -|S| = 0$.

In case *(iii)*, $|T|$ is, except possibly for sign, another minor of order $r + 1$ of A and hence is 0.

Case II. Let the row operation be $R_i(\lambda)$

Its effect on $|S|$ is either *(i)* to leave it unchanged or *(ii)* to multiply one of its rows by λ. Then, $|T| = |S| = 0$ or $|T| = \lambda |S| = 0$.

Case III. Let the row operation be $R_{i,j}(\lambda)$

Its effect on $|S|$ is *(i)* to leave it unchanged or *(ii)* to increase one of its rows by λ times another of its rows or *(iii)* to increase one of its rows by λ times a row not of $|S|$.

In case *(i)* and *(ii)*, $|T| = |S| = 0$.

In case *(iii)*, $|T| = |S| \pm \lambda$ (another minor of order $r + 1$ of A) $= 0 + \lambda \,.\, 0 = 0$.

$\therefore$ Every minor $|T|$ of matrix B of order $(r + 1)$ is zero.

$\therefore$ By Art. 1 (Def. of Rank), $\rho(B) \leq r\ (= \rho(A))$ ...(1)

Again because A can be obtained back from B by an elementary operation of the same type (which got us B from A),

therefore we have $\rho(A) \leq \rho(B)$ ...(2)

From (1) and (2), we have

$$\rho(A) = \rho(B)$$

Hence elementary row operations do not alter the rank of a matrix. Similarly elementary column operations also don't alter the rank of a matrix.

Remark. *We also conclude that the rank of a matrix remains unaltered by a finite chain of elementary operations.*

9. The Inverse of an Elementary Operation

The inverse of an elementary operation is an operation which nullifies the effect of the elementary operation. That is, after A has been subjected to one of the elementary operations and then the resulting matrix has been subjected to the inverse of that elementary operation, the final result is the matrix A.

For example, the effect of the elementary operation $R_{1,2}(3)$ on

$$A = \begin{bmatrix} 1 & 3 & 5 \\ 2 & -1 & 7 \\ 0 & 2 & 3 \end{bmatrix} \text{ is to produce}$$

$$B = \begin{bmatrix} 7 & 0 & 26 \\ 2 & -1 & 7 \\ 0 & 2 & 3 \end{bmatrix}.$$

The effect of the elementary operation $R_{1,2}(-3)$ on B is to produce A again.

Thus, $R_{1,2}(3)$ and $R_{1,2}(-3)$ are inverse elementary row operations.

The inverse elementary operations are :

(i) $R_{i,j}^{-1} = R_{i,j}$, $C_{i,j}^{-1} = C_{i,j}$

(ii) $R_i^{-1}(\lambda) = R_i\left(\frac{1}{\lambda}\right)$, $C_i^{-1}(\lambda) = C_i\left(\frac{1}{\lambda}\right)$

(iii) $R_{i,j}^{-1}(\lambda) = R_{i,j}(-\lambda)$, $C_{i,j}^{-1}(\lambda) = C_{i,j}(-\lambda)$.

Hence, the inverse of an elementary operation is an elementary operation of the same type.

EQUIVALENT MATRICES

10. Row Equivalent

Definition. *A matrix A is said to be row equivalent to a matrix B if B can be obtained from A by applying in succession a finite number of elementary row operations on A* and we write $A \overset{R}{\sim} B$.

Row-equivalent matrices have the same order and same rank.(By Art. 8)

For example, let $$A = \begin{bmatrix} 1 & 2 & -1 & 4 \\ 2 & 4 & 3 & 5 \\ -1 & -2 & 6 & -7 \end{bmatrix}.$$

Applying successively the elementary row operations $R_{2,1}(-2)$, $R_{3,1}(1)$, $R_{3,2}(-1)$, we have

$$A \xrightarrow{R_{2,1}(-2)} \begin{bmatrix} 1 & 2 & -1 & 4 \\ 0 & 0 & 5 & 3 \\ -1 & -2 & 6 & -7 \end{bmatrix}$$

$$\xrightarrow{R_{3,1}(1)} \begin{bmatrix} 1 & 2 & -1 & 4 \\ 0 & 0 & 5 & -3 \\ 0 & 0 & 5 & -3 \end{bmatrix}$$

$$\xrightarrow{R_{3,2}(-1)} \begin{bmatrix} 1 & 2 & -1 & 4 \\ 0 & 0 & 5 & -3 \\ 0 & 0 & 0 & 0 \end{bmatrix} = B.$$

Thus, $$A \overset{R}{\sim} B.$$

Since all the four minors of B of order 3 are zero and a minor of order 2 namely $\begin{bmatrix} -1 & 4 \\ 5 & -3 \end{bmatrix}$ $= 3 - 20 = -17 \neq 0$, therefore $\rho(B) = 2$.

Hence $\rho(A) = 2$. (By Art. 8)

11. Column Equivalent

Definition. *A matrix A is said to be column-equivalent to a matrix B if B can be obtained from A by applying in succession a finite number of elementary column operations on A and we write* $A \overset{C}{\sim} B$.

Column-equivalent matrices have the same order and the same rank.

For example, let $$A = \begin{bmatrix} 0 & 1 & -1 & 3 \\ 2 & 4 & 0 & 1 \\ 1 & 3 & 1 & -5 \end{bmatrix}$$

Then, $$A \xrightarrow{C_{1,2}} \begin{bmatrix} 1 & 0 & -1 & 3 \\ 4 & 2 & 0 & 1 \\ 3 & 1 & 1 & -5 \end{bmatrix}$$

$$\xrightarrow{C_{3,1}(1)} \begin{bmatrix} 1 & 0 & 0 & 3 \\ 4 & 2 & 4 & 1 \\ 3 & 1 & 4 & -5 \end{bmatrix}$$

$$\xrightarrow{C_{4,1}(-3)} \begin{bmatrix} 1 & 0 & 0 & 0 \\ 4 & 2 & 4 & -11 \\ 3 & 1 & 4 & -14 \end{bmatrix} = B$$

Then, $A \overset{C}{\sim} B.$

12. Equivalent Matrices

Definition. *Two matrices A and B are called equivalent if one can be obtained from the other by applying in succession a finite number of elementary operations and we write* $A \sim B$.

The matrices A and B in the above example are also equivalent.

EQUIVALENT MATRICES HAVE SAME RANK

13. Working procedure for finding rank using elementary operations

Since rank of a matrix remains unchanged by elementary operations, the elements of a given matrix are reduced by the repeated application of elementary operations till, we get as many zeros as possible in the resultant equivalent matrix. At this stage the largest non-zero minor is easily picked out for obtaining the rank.

For example, to find the rank of the matrix, let.

$$A = \begin{bmatrix} 1 & 2 & 3 & 0 \\ 2 & 4 & 3 & 2 \\ 3 & 2 & 1 & 3 \\ 6 & 8 & 7 & 5 \end{bmatrix}$$

$$\therefore\ A \xrightarrow{R_4 \to R_4 - R_1 - R_2 - R_3} \begin{bmatrix} 1 & 2 & 3 & 0 \\ 2 & 4 & 3 & 2 \\ 3 & 2 & 1 & 3 \\ 0 & 0 & 0 & 0 \end{bmatrix}$$

$$\xrightarrow{R_{2,1}(-2),\ R_{3,1}(-3)} \begin{bmatrix} 1 & 2 & 3 & 0 \\ 0 & 0 & -3 & 2 \\ 0 & -4 & -8 & 3 \\ 0 & 0 & 0 & 0 \end{bmatrix}$$

$$\xrightarrow{C_{2,1}(-2),\, C_{3,1}(-3)} \begin{bmatrix} 1 & 0 & 0 & 0 \\ 0 & 0 & -3 & 2 \\ 0 & -4 & -8 & 3 \\ 0 & 0 & 0 & 0 \end{bmatrix}$$

Since a third order minor $\begin{vmatrix} 1 & 0 & 0 \\ 0 & 0 & -3 \\ 0 & -4 & -8 \end{vmatrix} \neq 0$

and the only minor of order 4 is zero (Because the 4th row contains only zeros).

$\therefore$ $\rho(A) = 3.$

EXERCISE 2

1. Find the ranks of the matrices of Exercise 1, using elementary operations.
2. Show that the relation '~' in the set of all $m \times n$ matrices is an equivalence relation. (*i.e.*, show that it is reflexive, symmetric and transitive).
3. Show that if two matrices A and B have same size and same rank, they are equivalent.
4. Using elementary operations reduce the following matrix A to triangular form and show that rank of A is 3.

$$A = \begin{bmatrix} 5 & 3 & 14 & 4 \\ 0 & 1 & 2 & 1 \\ 1 & -1 & 2 & 0 \end{bmatrix}.$$

5. Show that interchange of a pair of rows does not change the rank of the matrix.
   [**Hint.** Reproduce Art. 8 case I.]
6. Show that multiplication of the elements of a row, by a non-zero number does not change the rank of the matrix.
   [**Hint.** Reproduce Art. 8 case II.]

14. Row-Echelon Matrix

Definition. *An $m \times n$ matrix $A = [a_{ij}]$ is said to be a row-echelon matrix if the following conditions are satisfied :*

(*i*) The non-zero rows of A are at the top *i.e.*, if A has r $(0 \le r \le m)$ non-zero rows then the rows $R_1, R_2, ..., R_r$ are non-zero while $R_{r+1}, R_{r+2}, ..., R_m$ are rows of zeros.

(*ii*) The first non-zero entry in any non-zero row is 1. It is called the **leading entry** of the row.

(*iii*) The positions (column indices) of the leading entries are strictly increasing with the rows (For instance, in the first matrix given below, the leading entries of R_1, R_2 and R_3 occur in C_1, C_3 and C_4 respectively).

For example, the matrices

$$\begin{bmatrix} 1 & 2 & 3 & 2 & 5 \\ 0 & 0 & 1 & 0 & 3 \\ 0 & 0 & 0 & 1 & 7 \\ 0 & 0 & 0 & 0 & 0 \end{bmatrix}, \begin{bmatrix} 1 & -2 & -1 & 4 \\ 0 & 0 & 1 & 3 \\ 0 & 0 & 0 & 0 \end{bmatrix} \text{ and } \begin{bmatrix} 1 & 2 & 0 & 1 & 5 \\ 0 & 0 & 1 & 2 & 1 \\ 0 & 0 & 0 & 1 & 1 \\ 0 & 0 & 0 & 0 & 0 \end{bmatrix}$$

are matrices in row-echelon form.

It is clear from the examples that if a matrix is in row-echelon form, then any column that contains a leading entry 1 must contain zeros below the leading entry 1. The entries above the leading entry 1, however are arbitrary.

A matrix in row-echelon form satisfying the following additional condition is said to be in row reduced echelon form.

(iv) Each column which contains a leading entry 1 of a row has all the other entries zero.

For example, the matrices

$$\begin{bmatrix} 1 & 2 & 0 & 0 & 6 \\ 0 & 0 & 1 & 1 & -1 \\ 0 & 0 & 0 & 0 & 0 \\ 0 & 0 & 0 & 0 & 0 \end{bmatrix}, \begin{bmatrix} 1 & a & 0 & b & 0 \\ 0 & 0 & 1 & c & 0 \\ 0 & 0 & 0 & 0 & 1 \\ 0 & 0 & 0 & 0 & 0 \end{bmatrix}$$

are in *row reduced echelon form.*

A zero matrix and an identity matrix are always in row reduced echelon form.

Note. *(i)* Row reduced echelon form is preferred to a row-echelon form in the solution of a system of linear equations as from the former a solution can be readily obtained while to get a solution from the latter, some more algebraic manipulations will be required.

(ii) The rank of a matrix when it is reduced to Row Echelon form is equal to **number of non-zero** rows of the row-echelon matrix.

ROW RANK AND COLUMN RANK OF A MATRIX

15. Definition

Row Rank. *The number of non-zero rows in the row-echelon form of a matrix A, is called the row rank of A and is denoted by* $\rho_R(A)$.

Since all row-echelon forms of a square matrix (if there are more than one) have the same number of non-zero rows, row rank of a matrix is independent of the row-echelon form of the matrix.

We have the general result :

Every matrix is row equivalent to a row echelon matrix.

Example 1. *Reduce to row echelon form the matrix*

$$A = \begin{bmatrix} 1 & -2 & -1 & 4 \\ 2 & -4 & 3 & 5 \\ -1 & 2 & 6 & -7 \end{bmatrix}.$$

Also find the row-rank of A.

Sol. Since the top entry of the first column of A is already 1, we subtract suitable multiples of the first row from the other rows to make the other entries in the first column zero. This is done on performing the operations $R_{2,1}(-2)$ and $R_{3,1}(1)$. Thus

$$A \overset{R}{\sim} \begin{bmatrix} 1 & -2 & -1 & 4 \\ 0 & 0 & 5 & -3 \\ 0 & 0 & 5 & -3 \end{bmatrix}$$

On removing the first row, the remaining submatrix has its first two columns zero and the top entry in its third column is 5. We now, multiply second row by $\frac{1}{5}$ to make its leading entry 1.

$$\therefore \qquad A \overset{R}{\sim} \begin{bmatrix} 1 & -2 & -1 & 4 \\ 0 & 0 & 1 & -\frac{3}{5} \\ 0 & 0 & 5 & -3 \end{bmatrix}$$

Operating $R_{3,\,2}(-5)$, we have

$$A \overset{R}{\sim} \begin{bmatrix} 1 & -2 & -1 & 4 \\ 0 & 0 & 1 & -\frac{3}{5} \\ 0 & 0 & 0 & 0 \end{bmatrix}$$

which is in row-echelon form. Since the row-echelon form of A has two non-zero rows, therefore row rank of A is 2.

Remark. It may be noted that row-echelon form of a matrix is not **unique** whereas row-rank is unique.

Example 2. *Find a row echelon matrix which is row equivalent to*

$$A = \begin{bmatrix} 0 & 0 & -2 & 3 & 1 \\ 2 & 4 & 1 & 4 & 3 \\ 1 & 2 & -3 & 1 & 2 \\ 4 & 8 & 2 & 3 & 5 \end{bmatrix} \quad \textit{and find } \rho_R(A).$$

Sol. Note that top entry in first column is zero. To make the top entry in the first column as non-zero (and in particular 1), we operate $R_{1,\,3}$ and get

$$A \overset{R}{\sim} \begin{bmatrix} 1 & 2 & -3 & 1 & 2 \\ 2 & 4 & 1 & 4 & 3 \\ 0 & 0 & -2 & 3 & 1 \\ 4 & 8 & 2 & 3 & 5 \end{bmatrix}$$

Here leading entry in first column is 1. Performing $R_{2,\,1}(-2)$ and $R_{4,\,1}(-4)$, we make other entries of first column zero. Thus,

$$A \overset{R}{\sim} \begin{bmatrix} 1 & 2 & -3 & 1 & 2 \\ 0 & 0 & 7 & 2 & -1 \\ 0 & 0 & -2 & 3 & 1 \\ 0 & 0 & 14 & -1 & -3 \end{bmatrix}$$

After removing the first row, we get a sub-matrix in which first two columns are zero. In the third column, the top entry itself is not zero, (if it were zero, we would interchange some two rows to make the top entry non-zero). We make it 1 by performing $R_2\left(\frac{1}{7}\right)$ and get

$$A \overset{R}{\sim} \begin{bmatrix} 1 & 2 & -3 & 1 & 2 \\ 0 & 0 & 1 & \frac{2}{7} & -\frac{1}{7} \\ 0 & 0 & -2 & 3 & 1 \\ 0 & 0 & 14 & -1 & -3 \end{bmatrix}$$

Performing $R_{3,2}(2)$ and $R_{4,1}(-14)$, we have [since the entries preceeding 1 in the second row are zero, these row operations will not affect the first two columns of A]

$$A \overset{R}{\sim} \begin{bmatrix} 1 & 2 & -3 & 1 & 2 \\ 0 & 0 & 1 & \frac{2}{7} & -\frac{1}{7} \\ 0 & 0 & 0 & \frac{25}{7} & \frac{5}{7} \\ 0 & 0 & 0 & -5 & -1 \end{bmatrix}$$

Deleting the first two rows, we see that the first non-zero column is the fourth column of the submatrix.

Further, in that submatrix, the entry of the first row is non-zero. To make it 1, we perform $R_3\left(\frac{7}{25}\right)$ and get

$$A \overset{R}{\sim} \begin{bmatrix} 1 & 2 & -3 & 1 & 2 \\ 0 & 0 & 1 & \frac{2}{7} & -\frac{1}{7} \\ 0 & 0 & 0 & 1 & \frac{1}{5} \\ 0 & 0 & 0 & -5 & -1 \end{bmatrix}$$

Operating $R_{4,3}(5)$, we have

$$A \overset{R}{\sim} \begin{bmatrix} 1 & 2 & -3 & 1 & 2 \\ 0 & 0 & 1 & \frac{2}{7} & -\frac{1}{7} \\ 0 & 0 & 0 & 1 & \frac{1}{5} \\ 0 & 0 & 0 & 0 & 0 \end{bmatrix}$$

which is the required row echelon matrix row equivalent to A.

Also $\rho_R(A) = 3$.

Example 3. *Reduce to row reduced echelon form the matrix*

$$A = \begin{bmatrix} 0 & 1 & 3 & -1 & 4 \\ 2 & 0 & -4 & 1 & 2 \\ 1 & 4 & 2 & 0 & -1 \\ 3 & 4 & -2 & 1 & 1 \\ 6 & 9 & -1 & 1 & 6 \end{bmatrix}$$

and find $\rho_R(A)$.

Sol. To make the top entry of the first column of A non-zero and in particular 1, we operate $R_{1,3}$ and get

$$A \overset{R}{\sim} \begin{bmatrix} 1 & 4 & 2 & 0 & -1 \\ 2 & 0 & -4 & 1 & 2 \\ 0 & 1 & 3 & -1 & 4 \\ 3 & 4 & -2 & 1 & 1 \\ 6 & 9 & -1 & 1 & 6 \end{bmatrix}$$

Operating $R_{2,1}(-2)$, $R_{4,1}(-3)$ and $R_{5,1}(-6)$, we have

$$A \overset{R}{\sim} \begin{bmatrix} 1 & 4 & 2 & 0 & -1 \\ 0 & -8 & -8 & 1 & 4 \\ 0 & 1 & 3 & -1 & 4 \\ 0 & -8 & -8 & 1 & 4 \\ 0 & -15 & -13 & 1 & 12 \end{bmatrix}$$

To make the leading entry of second row 1, we operate $R_{2,3}$ and get

$$A \overset{R}{\sim} \begin{bmatrix} 1 & 4 & 2 & 0 & -1 \\ 0 & 1 & 3 & -1 & 4 \\ 0 & -8 & -8 & 1 & 4 \\ 0 & -8 & -8 & 1 & 4 \\ 0 & -15 & -13 & 1 & 12 \end{bmatrix}$$

Operating $R_{1,2}(-4)$, $R_{3,2}(8)$, $R_{4,2}(8)$, $R_{5,2}(15)$, we have

$$A \overset{R}{\sim} \begin{bmatrix} 1 & 0 & -10 & 4 & -17 \\ 0 & 1 & 3 & -1 & 4 \\ 0 & 0 & 16 & -7 & 36 \\ 0 & 0 & 16 & -7 & 36 \\ 0 & 0 & 32 & -14 & 72 \end{bmatrix}$$

Operating $R_3\left(\frac{1}{16}\right)$, we have

$$A \overset{R}{\sim} \begin{bmatrix} 1 & 0 & -10 & 4 & -17 \\ 0 & 1 & 3 & -1 & 4 \\ 0 & 0 & 1 & -\frac{7}{16} & \frac{9}{4} \\ 0 & 0 & 16 & -7 & 36 \\ 0 & 0 & 32 & -14 & 72 \end{bmatrix}$$

Operating $R_{1,3}(10)$, $R_{2,3}(-3)$, $R_{4,3}(-16)$, $R_{5,3}(-32)$, we have

$$A \overset{R}{\sim} \begin{bmatrix} 1 & 0 & 0 & -\frac{3}{8} & \frac{11}{2} \\ 0 & 1 & 0 & \frac{5}{16} & -\frac{11}{4} \\ 0 & 0 & 1 & -\frac{7}{16} & \frac{9}{4} \\ 0 & 0 & 0 & 0 & 0 \\ 0 & 0 & 0 & 0 & 0 \end{bmatrix}$$

which is the row reduced echelon form.

$$\rho_R(A) = 3.$$

Example 4. *Reduce to a row reduced echelon form, the matrix*

$$A = \begin{bmatrix} 0 & 1 & 3 & -1 & 3 & 1 \\ 0 & 1 & 3 & 0 & 2 & 3 \\ 0 & 2 & 6 & 1 & 3 & 9 \\ 0 & 4 & 12 & -2 & 10 & 7 \end{bmatrix}$$

and find $\rho_R(A)$.

Sol. First non-zero column in A is the second column with 1 as its top entry. Operating $R_{2,1}(-1)$, $R_{3,1}(-2)$, $R_{4,1}(-4)$, we have

$$A \overset{R}{\sim} \begin{bmatrix} 0 & 1 & 3 & -1 & 3 & 1 \\ 0 & 0 & 0 & 1 & -1 & 2 \\ 0 & 0 & 0 & 3 & -3 & 7 \\ 0 & 0 & 0 & 2 & -2 & 3 \end{bmatrix}$$

Operating $R_{1,2}(1)$, $R_{3,2}(-3)$, $R_{4,2}(-2)$, we have

$$A \overset{R}{\sim} \begin{bmatrix} 0 & 1 & 3 & 0 & 2 & 3 \\ 0 & 0 & 0 & 1 & -1 & 2 \\ 0 & 0 & 0 & 0 & 0 & 1 \\ 0 & 0 & 0 & 0 & 0 & -1 \end{bmatrix}$$

Operating $R_{1,3}(-3)$, $R_{2,3}(-2)$, $R_{4,3}(1)$, we have

$$A \overset{R}{\sim} \begin{bmatrix} 0 & 1 & 3 & 0 & 2 & 0 \\ 0 & 0 & 0 & 1 & -1 & 0 \\ 0 & 0 & 0 & 0 & 0 & 1 \\ 0 & 0 & 0 & 0 & 0 & 0 \end{bmatrix}$$

which is in row reduced echelon form

$$\rho_R(A) = 3.$$

Remark. (*i*) The above method also gives a method to row reduce a square matrix to an **upper triangular matrix.** However in that case the procedure can be further simplified.

(*ii*) The row reduced echelon matrix equivalent to a given matrix is **unique.**

16. Column Rank

Similar to the concept of row echelon form, we can define the concept of the column echelon form and have results analogous to row echelon matrices.

The number of non-zero rows in the column-echelon form of matrix A, is called the column rank of A and is denoted by $\rho_C(A)$.

For example, the matrices $\begin{bmatrix} 0 & 0 & 0 & 0 \\ 1 & 0 & 0 & 0 \\ 3 & 0 & 0 & 0 \\ 0 & 1 & 0 & 0 \\ 2 & -1 & 0 & 0 \\ 0 & 0 & 1 & 0 \end{bmatrix}$ and $\begin{bmatrix} 1 & 0 & 0 & 0 & 0 \\ 0 & 1 & 0 & 0 & 0 \\ 0 & 0 & 1 & 0 & 0 \\ -3 & 5 & -7 & 0 & 0 \\ 11 & -11 & 9 & 0 & 0 \\ 9 & 9 & 29 & 0 & 0 \end{bmatrix}$

are in column echelon form. Their column ranks are 3 each.

Every matrix is column equivalent to a column echelon matrix.

In fact, if A is in row echelon form, then A′ is the column echelon form.

Example 5. *Reduce the matrix A to column-echelon form and find the column rank of A*

where
$$A = \begin{bmatrix} 1 & 2 & 0 & -1 \\ 2 & 6 & -3 & -3 \\ 3 & 10 & -6 & -5 \end{bmatrix}$$

Sol.
$$A = \begin{bmatrix} 1 & 2 & 0 & -1 \\ 2 & 6 & -3 & -3 \\ 3 & 10 & -6 & -5 \end{bmatrix}$$

Operating $C_{2,1}(-2)$ and $C_{4,1}(1)$, we have $A \overset{C}{\sim} \begin{bmatrix} 1 & 0 & 0 & 0 \\ 2 & 2 & -3 & -1 \\ 3 & 4 & -6 & -2 \end{bmatrix}$

Operating $C_{2,4}(2)$, $C_{3,4}(-3)$, we have $A \overset{C}{\sim} \begin{bmatrix} 1 & 0 & 0 & 0 \\ 2 & 0 & 0 & -1 \\ 3 & 0 & 0 & -2 \end{bmatrix}$

Operating $C_{2,4}$, we have $A \overset{C}{\sim} \begin{bmatrix} 1 & 0 & 0 & 0 \\ 2 & -1 & 0 & 0 \\ 3 & -2 & 0 & 0 \end{bmatrix}$

which is column-echelon form.

Since it has only two non-zero columns, therefore column rank of A is 2.

EXERCISE 3

1. Find the row rank of the matrices :

(*i*) $\begin{bmatrix} 0 & -1 & 2 \\ 4 & 3 & 1 \\ 4 & 2 & 3 \end{bmatrix}$ (*ii*) $\begin{bmatrix} 3 & 4 & 1 & 2 \\ 3 & 2 & 1 & 4 \\ 7 & 6 & 2 & 5 \end{bmatrix}$

(*iii*) $\begin{bmatrix} 1 & 2 & 3 \\ -2 & 4 & -9 \\ 1 & 2 & 3 \end{bmatrix}$ (*iv*) $\begin{bmatrix} 6 & 1 & 3 & 8 \\ 4 & 2 & 6 & -1 \\ 10 & 3 & 9 & 7 \\ 16 & 4 & 12 & 15 \end{bmatrix}$

(*v*) $\begin{bmatrix} -1 & 2 & 0 & 0 \\ 0 & -1 & 8 & 0 \\ 1 & 0 & 0 & 0 \\ 0 & 0 & 0 & 1 \end{bmatrix}$ (*vi*) $\begin{bmatrix} 2 & 3 & -1 & -1 \\ 1 & -1 & -2 & -4 \\ 3 & 1 & 3 & -2 \\ 6 & 3 & 0 & -7 \end{bmatrix}$

(*vii*) $\begin{bmatrix} 1 & 2 & -1 & 0 \\ 3 & 1 & 4 & 2 \\ 1 & -3 & 6 & 2 \end{bmatrix}$

2. Reduce the following matrices to the *row reduced echelon* form. Also find their row ranks :

(*i*) $\begin{bmatrix} 1 & 2 & -3 \\ 2 & 5 & -4 \end{bmatrix}$ (*ii*) $\begin{bmatrix} 1 & 2 & 3 & 4 \\ 3 & 4 & 1 & 2 \\ 4 & 3 & 1 & 2 \end{bmatrix}$

(*iii*) $\begin{bmatrix} 1 & 1 & 1 & 2 \\ 2 & 1 & -3 & -6 \\ 3 & -1 & 1 & 2 \end{bmatrix}$ (*iv*) $\begin{bmatrix} 1 & -1 & 1 & 1 & 1 \\ 1 & -1 & 2 & 3 & 1 \\ 2 & -2 & 1 & 0 & 2 \\ 1 & 1 & -1 & -3 & 3 \end{bmatrix}$

3. **Using elementary operations, reduce the following matrices to triangular form and hence find the rank of each matrix.**

(*i*) $\begin{bmatrix} 1 & 1 & 1 & 2 \\ 2 & 1 & -3 & -6 \\ 3 & -1 & 1 & 2 \end{bmatrix}$ (*ii*) $\begin{bmatrix} 1 & 1 & 2 & 3 \\ 1 & 3 & 0 & 3 \\ 1 & -2 & -3 & -3 \\ 1 & 1 & 2 & 3 \end{bmatrix}$.

4. **Reduce the matrices given in the above question number 3 to column echelon form and hence find the column rank of each matrix.**

ANSWERS

1. (*i*) 2 (*ii*) 3 (*iii*) 2 (*iv*) 2 (*v*) 4 (*vi*) 3 (*vii*) 2

2. (*i*) $\begin{bmatrix} 1 & 0 & -7 \\ 0 & 1 & 2 \end{bmatrix}$ and row rank is 2 (*ii*) $\begin{bmatrix} 1 & 0 & 0 & \frac{1}{9} \\ 0 & 1 & 0 & \frac{1}{9} \\ 0 & 0 & 1 & \frac{11}{9} \end{bmatrix}$ and row rank is 3

(*iii*) $\begin{bmatrix} 1 & 0 & 0 & 0 \\ 0 & 1 & 0 & 0 \\ 0 & 0 & 1 & 2 \end{bmatrix}$ and row rank is 3 (*iv*) $\begin{bmatrix} 1 & 0 & 0 & -1 & 2 \\ 0 & 1 & 0 & 0 & 1 \\ 0 & 0 & 1 & 2 & 0 \\ 0 & 0 & 0 & 0 & 0 \end{bmatrix}$ and row rank is 3

3. (*i*) 3 (*ii*) 3 **4.** (*i*) 3 (*ii*) 3.

17. Normal Form of a Matrix

We have seen that successive application of elementary row operations on a matrix reduce it to the row echelon form. We can naturally expect that if we apply both elementary row and column operations, we may get a still simpler matrix—called the *normal* (*or canonical*) *form* of the matrix.

By applying successively elementary operations, we can reduce any matrix A of rank $r > 0$ in one of the forms :

$$I_r, \begin{bmatrix} I_r & O \\ O & O \end{bmatrix}, [I_r \ \ O], \begin{bmatrix} I_r \\ O \end{bmatrix}$$

(where I_r is the identity matrix of order r and O represents zero matrices) *called its normal form.*

If $r = 0$, then A is in normal form if and only if A = O. If, A is a normal form as above, then $\rho(A) = r$. If A is an n-square matrix in the normal form, then A is non-singular, if and only if $A = I_n$.

18. Theorem

If A is an $m \times n$ matrix of rank r, then A is equivalent to the matrix $\begin{bmatrix} I_r & O \\ O & O \end{bmatrix}$ *in the normal form.*

Proof. If A is a null matrix, then it is in the normal form.

So, let $A \neq O$.

Let $a_{ij} = \lambda$ be a non-zero entry of A. We can bring this to (1, 1) position (top left position) by interchanging the first row with the ith row and then the first column with jth column. Then, A is equivalent to a matrix A_1 with $\lambda \neq 0$ as its (1, 1) entry. Multiplying the first row by $\frac{1}{\lambda}$, we make the (1, 1) entry of A_1 as 1. Subtracting suitable multiples of first row from the remaining rows, we get 0 in all but first entry of first column. Similarly, subtracting suitable multiples of the first column from the remaining columns of A_1, we get all but the first entry in the first row zero. If the resulting matrix is A_2, then

$$A \sim A_2 = \begin{bmatrix} 1 & O \\ O & B_1 \end{bmatrix} = \begin{bmatrix} I_1 & O \\ O & B_1 \end{bmatrix}$$

If $B_1 = O$, we stop, otherwise we repeat the above argument with B_1. The row and column operations do not affect the entries in the first row and first column. Thus,

$$A \sim \begin{bmatrix} 1 & 0 & O \\ 0 & 1 & O \\ O & O & B_2 \end{bmatrix} = \begin{bmatrix} I_2 & O \\ O & B_2 \end{bmatrix}$$

Repeating the argument, after a finite number of row and column operations, we have $A \sim \begin{bmatrix} I_k & O \\ O & O \end{bmatrix}$, which is in the normal form. Since $\rho(A) = r$ and rank of $\begin{bmatrix} I_k & O \\ O & O \end{bmatrix}$ is k.

$\therefore \quad r = k$ and $A \sim \begin{bmatrix} I_r & O \\ O & O \end{bmatrix}$.

We, now, illustrate by examples how to reduce a matrix to the normal form.

Example 1. *Reduce to the normal form the matrix* $A = \begin{bmatrix} 2 & 2 & 2 \\ 1 & 2 & 1 \\ 3 & 4 & 3 \end{bmatrix}$

Hence, find $\rho(A)$.

Sol.
$$A = \begin{bmatrix} 2 & 2 & 2 \\ 1 & 2 & 1 \\ 3 & 4 & 3 \end{bmatrix}$$

Operating $R_{1,2}$
$$A \sim \begin{bmatrix} 1 & 2 & 1 \\ 2 & 2 & 2 \\ 3 & 4 & 3 \end{bmatrix}$$

Operating $R_{2,1}(-2)$, $R_{3,1}(-3)$, we have
$$A \sim \begin{bmatrix} 1 & 2 & 1 \\ 0 & -2 & 0 \\ 0 & -2 & 0 \end{bmatrix}$$

Operating $C_{2,1}(-2)$, $C_{3,1}(-1)$, we have

$$A \sim \begin{bmatrix} 1 & 0 & 0 \\ 0 & -2 & 0 \\ 0 & -2 & 0 \end{bmatrix}$$

Operating $R_2\left(-\frac{1}{2}\right)$, $$A \sim \begin{bmatrix} 1 & 0 & 0 \\ 0 & 1 & 0 \\ 0 & -2 & 0 \end{bmatrix}$$

Finally, operating $R_{3,2}(2)$, we have $A \sim \begin{bmatrix} 1 & 0 & 0 \\ 0 & 1 & 0 \\ 0 & 0 & 0 \end{bmatrix} = \begin{bmatrix} I_2 & O \\ O & O \end{bmatrix}$

which is in normal form $\rho(A)$ is 2.

Example 2. *Reduce the matrix* $A = \begin{bmatrix} 1 & -1 & 2 & -1 \\ 4 & 2 & -1 & 2 \\ 2 & 2 & -2 & 0 \end{bmatrix}$ *to* $[I_3 \;\; O]$. *Hence, find* $\rho(A)$.

Sol. Operating $R_{2,1}(-4)$ and $R_{3,1}(-2)$, we have

$$A \sim \begin{bmatrix} 1 & -1 & 2 & -1 \\ 0 & 6 & -9 & 6 \\ 0 & 4 & -6 & 2 \end{bmatrix}$$

Operating $C_{2,1}(1)$, $C_{3,1}(-2)$, $C_{4,1}(1)$, we have

$$A \sim \begin{bmatrix} 1 & 0 & 0 & 0 \\ 0 & 6 & -9 & 6 \\ 0 & 4 & -6 & 2 \end{bmatrix}$$

Operating $R_2\left(\frac{1}{6}\right)$, $$A \sim \begin{bmatrix} 1 & 0 & 0 & 0 \\ 0 & 1 & -3/2 & 1 \\ 0 & 4 & -6 & 2 \end{bmatrix}$$

Operating $R_{3,2}(-4)$, $$A \sim \begin{bmatrix} 1 & 0 & 0 & 0 \\ 0 & 1 & -3/2 & 1 \\ 0 & 0 & 0 & -2 \end{bmatrix}$$

Operating $C_{3,2}\left(\frac{3}{2}\right)$, $C_{4,2}(-1)$, $$A \sim \begin{bmatrix} 1 & 0 & 0 & 0 \\ 0 & 1 & 0 & 0 \\ 0 & 0 & 0 & -2 \end{bmatrix}$$

Operating $R_3\left(-\frac{1}{2}\right)$, $$A \sim \begin{bmatrix} 1 & 0 & 0 & 0 \\ 0 & 1 & 0 & 0 \\ 0 & 0 & 0 & 1 \end{bmatrix}$$

Operating $C_{3,4}$ $$A \sim \begin{bmatrix} 1 & 0 & 0 & 0 \\ 0 & 1 & 0 & 0 \\ 0 & 0 & 1 & 0 \end{bmatrix} = [I_3 \; O]$$

$\therefore \qquad \rho(A) = \text{rank of } [I_3 \; O] = 3.$

ELEMENTARY MATRICES

19. Elementary Matrix

Definition. *A matrix obtained from an identity matrix I_n, by performing an elementary operation on it is called an elementary matrix of order n.*

We denote by $E_{i,j}$, $E_i(\lambda)$ and $E_{i,j}(\lambda)$, the elementary matrices obtained from the identity matrix by applying the elementary row operations $R_{i,j}$, $R_i(\lambda)$ and $R_{i,j}(\lambda)$ on I respectively. $F_{i,j}$, $F_i(\lambda)$ and $F_{i,j}(\lambda)$ denote the elementary matrices obtained from the identity matrix by applying the elementary column operations $C_{i,j}$, $C_i(\lambda)$ and $C_{i,j}(\lambda)$ on I.

For example, (*i*) for $I = I_3$,

$$E_{1,2} = \begin{bmatrix} 0 & 1 & 0 \\ 1 & 0 & 0 \\ 0 & 0 & 1 \end{bmatrix} = F_{1,2},\ E_2(\lambda) = \begin{bmatrix} 1 & 0 & 0 \\ 0 & \lambda & 0 \\ 0 & 0 & 1 \end{bmatrix} = F_2(\lambda)$$

and

$$E_{2,3}(\lambda) = \begin{bmatrix} 1 & 0 & 0 \\ 0 & 1 & \lambda \\ 0 & 0 & 1 \end{bmatrix} = F_{3,2}(\lambda)$$

(*ii*) For $I = I_4$

$$E_{2,3} = \begin{bmatrix} 1 & 0 & 0 & 0 \\ 0 & 0 & 1 & 0 \\ 0 & 1 & 0 & 0 \\ 0 & 0 & 0 & 1 \end{bmatrix} = F_{2,3},\ E_3(5) = \begin{bmatrix} 1 & 0 & 0 & 0 \\ 0 & 1 & 0 & 0 \\ 0 & 0 & 5 & 0 \\ 0 & 0 & 0 & 1 \end{bmatrix} = F_3(5)$$

and

$$E_{2,1}(3) = \begin{bmatrix} 1 & 0 & 0 & 0 \\ 3 & 1 & 0 & 0 \\ 0 & 0 & 1 & 0 \\ 0 & 0 & 0 & 1 \end{bmatrix} = F_{1,2}(3).$$

Note. $E_{i,j} = F_{i,j}$, $E_i(\lambda) = F_i(\lambda)$ and $E_{i,j}(\lambda) = F_{j,i}(\lambda)$.

We, now examine the relationship between elementary operations on a matrix and the elementary matrices.

20. Theorem

If B = PA, then the effect of an elementary row operation on the pre-multiplier P is the same as applying this operation on the rows of B and conversely.

Proof. Let A be $m \times n$ and P be $l \times m$ so that B(= PA) is $l \times n$. Writing P as a column matrix of rows, we have

$$P = \begin{bmatrix} R_1 \\ \vdots \\ R_i \\ \vdots \\ R_j \\ \vdots \\ R_l \end{bmatrix} \qquad \text{Then,} \quad B = PA = \begin{bmatrix} R_1A \\ \vdots \\ R_iA \\ \vdots \\ R_jA \\ \vdots \\ R_lA \end{bmatrix}$$

Operating $R_{i,j}$ on P, the transformed matrix (assuming $i < j$) is

$$R_{i,j}\,P = \begin{bmatrix} R_1 \\ \vdots \\ R_j \\ \vdots \\ R_i \\ \vdots \\ R_l \end{bmatrix} \quad \therefore \quad (R_{i,j}P)A = \begin{bmatrix} R_1A \\ \vdots \\ R_jA \\ \vdots \\ R_iA \\ \vdots \\ R_lA \end{bmatrix}$$

which is obtained on applying R_{ij}, on B.

Similar is the case when the row operation $R_i(\lambda)$ is to be considered. Now, consider the operation $R_{i,j}(\lambda)$. On applying this operation to P (assuming $i < j$).

$$R_{i,j}(\lambda)\,P = \begin{bmatrix} R_1 \\ \vdots \\ R_i + \lambda R_j \\ \vdots \\ R_j \\ \vdots \\ R_l \end{bmatrix}$$

$$\therefore \qquad (R_{i,j}(\lambda)P)A = \begin{bmatrix} R_1A \\ \vdots \\ (R_i + \lambda R_j)A \\ \vdots \\ R_jA \\ \vdots \\ R_lA \end{bmatrix} = \begin{bmatrix} R_1A \\ \vdots \\ R_iA + \lambda R_jA \\ \vdots \\ R_jA \\ \vdots \\ R_lA \end{bmatrix}$$

which is obtained by adding to the ith row λ times the jth row of B.

The converse part is proved by retracing the steps.

Cor. *If A is an $m \times n$ matrix and E an elementary row matrix of order m, then EA is the same as the matrix obtained from A by applying the elementary row operation to which E corresponds and conversely.*

Proof. Proof follows from the above theorem by taking $P = I_m$, because $A = IA$.

Similar to the above results, we have results corresponding to elementary column operations which we state without proof.

21. Theorem

If $B = AQ$, then the effect of an elementary column operation on the post-multiplier Q is the same as applying this operation on the columns of B and conversely.

Cor. *If A is an $m \times n$ matrix and F is an elementary column matrix of order n, then AF is the same as the matrix obtained from A by applying the elementary column operations to which F corresponds and conversely.*

Combining the above corollaries and the theorem 8, we conclude the following important result :

The pre-multiplication or post-multiplication by any elementary matrix and as such by any series of elementary matrices, does not alter the rank of a matrix.

Example. (*a*) *Perform the operations (i) $R_{1,3}$, (ii) $R_2(3)$ and (iii) $R_{2,3}(-1)$ on the matrix*

$$A = \begin{bmatrix} 1 & 2 & 0 & 5 \\ 2 & -3 & 1 & 0 \\ 1 & -1 & 4 & 2 \end{bmatrix}$$

and express the result as EA.

(*b*) *Perform the operations (i) $C_{1,3}$, (ii) $C_2(3)$ and (iii) $C_{2,3}(-1)$ on the matrix A and express the result as AF.*

Sol. (*a*)
$$I = \begin{bmatrix} 1 & 0 & 0 \\ 0 & 1 & 0 \\ 0 & 0 & 1 \end{bmatrix}$$

$$\therefore \quad E_{1,3} = \begin{bmatrix} 0 & 0 & 1 \\ 0 & 1 & 0 \\ 1 & 0 & 0 \end{bmatrix}, E_2(3) = \begin{bmatrix} 1 & 0 & 0 \\ 0 & 3 & 0 \\ 0 & 0 & 1 \end{bmatrix}$$

$$E_{2,3}(-1) = \begin{bmatrix} 1 & 0 & 0 \\ 0 & 1 & -1 \\ 0 & 0 & 1 \end{bmatrix}$$

$\therefore$ (*i*) $R_{1,3} \cdot A = E_{1,3} \cdot A$

$$= \begin{bmatrix} 0 & 0 & 1 \\ 0 & 1 & 0 \\ 1 & 0 & 0 \end{bmatrix} \begin{bmatrix} 1 & 2 & 0 & 5 \\ 2 & -3 & 1 & 0 \\ 1 & -1 & 4 & 2 \end{bmatrix} = \begin{bmatrix} 1 & -1 & 4 & 2 \\ 2 & -3 & 1 & 0 \\ 1 & 2 & 0 & 5 \end{bmatrix}.$$

(*ii*) $R_2(3)A = E_2(3)A$

$$= \begin{bmatrix} 1 & 0 & 0 \\ 0 & 3 & 0 \\ 0 & 0 & 1 \end{bmatrix} \begin{bmatrix} 1 & 2 & 0 & 5 \\ 2 & -3 & 1 & 0 \\ 1 & -1 & 4 & 2 \end{bmatrix} = \begin{bmatrix} 1 & 2 & 0 & 5 \\ 6 & -9 & 3 & 0 \\ 1 & -1 & 4 & 2 \end{bmatrix}.$$

(*iii*) $R_{2,3}(-1)A = E_{2,3}(-1) \cdot A$

$$= \begin{bmatrix} 1 & 0 & 0 \\ 0 & 1 & -1 \\ 0 & 0 & 1 \end{bmatrix} \begin{bmatrix} 1 & 2 & 0 & 5 \\ 2 & -3 & 1 & 0 \\ 1 & -1 & 4 & 2 \end{bmatrix} = \begin{bmatrix} 1 & 2 & 0 & 5 \\ 1 & -2 & -3 & -2 \\ 1 & -1 & 4 & 2 \end{bmatrix}$$

(*b*)
$$I = \begin{bmatrix} 1 & 0 & 0 & 0 \\ 0 & 1 & 0 & 0 \\ 0 & 0 & 1 & 0 \\ 0 & 0 & 0 & 1 \end{bmatrix}$$

$$\therefore \quad F_{1,3} = \begin{bmatrix} 0 & 0 & 1 & 0 \\ 0 & 1 & 0 & 0 \\ 1 & 0 & 0 & 0 \\ 0 & 0 & 0 & 1 \end{bmatrix}, F_2(3) = \begin{bmatrix} 1 & 0 & 0 & 0 \\ 0 & 3 & 0 & 0 \\ 0 & 0 & 1 & 0 \\ 0 & 0 & 0 & 1 \end{bmatrix}$$

and
$$F_{2,3}(-1) = \begin{bmatrix} 1 & 0 & 0 & 0 \\ 0 & 1 & 0 & 0 \\ 0 & -1 & 1 & 0 \\ 0 & 0 & 0 & 1 \end{bmatrix}.$$

$\therefore$ (*i*) $C_{1,3}A = A \cdot F_{1,3}$

$$= \begin{bmatrix} 1 & 2 & 0 & 5 \\ 2 & -3 & 1 & 0 \\ 1 & -1 & 4 & 2 \end{bmatrix} \begin{bmatrix} 0 & 0 & 1 & 0 \\ 0 & 1 & 0 & 0 \\ 1 & 0 & 0 & 0 \\ 0 & 0 & 0 & 1 \end{bmatrix} = \begin{bmatrix} 0 & 2 & 1 & 5 \\ 1 & -3 & 2 & 0 \\ 4 & -1 & 1 & 2 \end{bmatrix}.$$

(*ii*) $C_2(3) . A = A . (F_2(3))$

$$= \begin{bmatrix} 1 & 2 & 0 & 5 \\ 2 & -3 & 1 & 0 \\ 1 & -1 & 4 & 2 \end{bmatrix} \begin{bmatrix} 1 & 0 & 0 & 0 \\ 0 & 3 & 0 & 0 \\ 0 & 0 & 1 & 0 \\ 0 & 0 & 0 & 1 \end{bmatrix} = \begin{bmatrix} 1 & 6 & 0 & 5 \\ 2 & -9 & 1 & 0 \\ 1 & -3 & 4 & 2 \end{bmatrix}.$$

(*iii*) $C_{2,3}(-1)A = A . F_{2,3}(-1)$

$$= \begin{bmatrix} 1 & 2 & 0 & 5 \\ 2 & -3 & 1 & 0 \\ 1 & -1 & 4 & 2 \end{bmatrix} \begin{bmatrix} 1 & 0 & 0 & 0 \\ 0 & 1 & 0 & 0 \\ 0 & -1 & 1 & 0 \\ 0 & 0 & 0 & 1 \end{bmatrix} = \begin{bmatrix} 1 & 2 & 0 & 5 \\ 2 & -4 & 1 & 0 \\ 1 & -5 & 4 & 2 \end{bmatrix}.$$

22. Theorem

(*i*) *The elementary matrices are non-singular.*

(*ii*) *The inverse of an elementary matrix is an elementary matrix of the same type.*

Proof. Since $E_{i,j}$ is the matrix obtained from the identity matrix I by applying the operation $R_{i,j}$, therefore, if we apply the operation $R_{j,i}$ on $E_{i,j}$, we get back I. The same operation can be done by pre-multiplying $E_{i,j}$ by $E_{i,j}$.

$\therefore$ $E_{i,j} . E_{i,j} = I$

$\Rightarrow$ $(E_{i,j})^{-1} = E_{i,j}$ (By def. of inverse of a matrix)

$\Rightarrow$ **$E_{i,j}$ is invertible (and hence non-singular) and is its own inverse.**

Similarly, $(E_i(\lambda))^{-1} = E_i\left(\frac{1}{\lambda}\right)$

Finally, $E_{i,j}(\lambda)$ is the matrix obtained from the identity matrix I by applying the operation $R_{i,j}(\lambda)$. Therefore, if we apply the operation $R_{i,j}(-\lambda)$ to $E_{i,j}(\lambda)$, we get back I.

Thus, $E_{i,j}(-\lambda) . E_{i,j}(\lambda) = I.$

$\Rightarrow$ **$E_{i,j}(\lambda)$ is invertible and $(E_{i,j}(\lambda))^{-1} = E_{i,j}(-\lambda)$**

In an analogous manner, we can prove that

$$(F_{i,j})^{-1} = F_{i,j},\ (F_i(\lambda))^{-1} = F_i\left(\frac{1}{\lambda}\right),\ F(_{i,j}(\lambda))^{-1} = F_{i,j}(-\lambda).$$

Remark. One can prove part (*a*) even by observing that the determinant of the elementary matrices $E_{i,j}$, $E_i(\lambda)$ and $E_{i,j}(\lambda)$ are -1, λ and 1 respectively and each one of the three values is non-zero but in that case we shall not get inverse matrix of the elementary matrix.

23. Theorem

Two matrices A and B are equivalent if and only if there exist non-singular matrices P and Q such that PAQ= B.

Proof. Let A and B be equivalent. Then, B can be obtained from A by applying in succession a finite number of row and column operations. Let $E_1, E_2, ..., E_r$; $F_1, F_2, ..., F_s$ be the elementary matrices corresponding to the row and column operations which reduce A to B. Since each elementary row (column) operation of a matrix can be brought about by pre-multiplication (post-multiplication) with the corresponding elementary matrix, therefore, we have

$$E_r...E_2E_1AF_1F_2...F_s = B$$

$\Rightarrow$ $PAQ = B$

where $P = E_r...E_2E_1$ and $Q = F_1F_2...F_s$

P and Q are non-singular matrices since the elementary matrices are non-singular and product of non-singular matrices is non-singular.

Cor. 1. *If A is an $m \times n$ matrix of rank r, then there exist non-singular matrices P and Q of order m and n respectively, such that*

$$PAQ = \begin{bmatrix} I_r & O \\ O & O \end{bmatrix}.$$

Proof. Since A is of rank r, A is equivalent to the matrix $\begin{bmatrix} I_r & O \\ O & O \end{bmatrix}$ in the normal form.

$\therefore$ there exist non-singular matrices P and Q of order m and n respectively, such that

$$PAQ = \begin{bmatrix} I_r & O \\ O & O \end{bmatrix}. \qquad \textbf{(Normal Form)}$$

Remark. The non-singular matrices P and Q obtained in arriving at the above **Normal form** are **not unique**.

Cor. 2. *If A is an n-square non-singular matrix, there exist non-singular matrices, P and Q such that $PAQ = I_n$.*

Proof. Since every matrix is equivalent to its normal form

$$\therefore \qquad A \sim I_n.$$

$\therefore$ there exist non-singular matrices P and Q (as in the above theorem) such that

$$PAQ = I_n.$$

Cor. 3. *Every non-singular matrix can be expressed as a product of elementary matrices.*

Proof. Let A be an n-square non-singular matrix.

$\therefore$ there exist non-singular matrices P, Q which are products of elementary matrices such that

$$PAQ = I_n$$

Let $\qquad P = E_1E_2...E_r$ and $Q = F_1F_2...F_s$

where $E_1, E_2, ..., E_r$ and $F_1, F_2, ... F_s$ are elementary matrices.

Thus, $\qquad E_1E_2...E_r\, A\, F_1\, F_2...F_s = I_n$

$\Rightarrow$ $\mathbf{A = E_r^{-1} ... E_2^{-1}\, E_1^{-1}\, F_s^{-1} ... F_2^{-1}\, F_1^{-1}}$ ...(1)

$\Rightarrow$ A is the product of elementary matrices since the inverse of an elementary matrix is an elementary matrix.

Cor. 4. (*i*) *If A is non-singular, then $\rho(AB) = \rho(B)$ and $\rho(BA) = \rho(B)$.*

(*ii*) *If P and Q are non-singular, then $\rho(PAQ) = \rho(A)$.*

Proof. Every non-singular matrix can be expressed as the product of elementary matrices. Also, elementary row (column) operations are equivalent to pre-(post) multiplication with the corresponding elementary matrices and elementary matrices do not alter the rank of a matrix. Hence, the result.

Example 1. *For the following matrices, find non-singular matrices P and Q such that PAQ is in the normal form and hence find out rank of matrix A.*

$$(i)\ A = \begin{bmatrix} 1 & 1 & 2 \\ 1 & 2 & 3 \\ 0 & -1 & -1 \end{bmatrix} \qquad (ii)\ A = \begin{bmatrix} 1 & 2 & -1 & 2 \\ 2 & 5 & -2 & 3 \\ 1 & 2 & 1 & 2 \end{bmatrix}.$$

Sol. (*i*) We write A = IAI, *i.e.*,

$$\begin{bmatrix}1&1&2\\1&2&3\\0&-1&-1\end{bmatrix}=\begin{bmatrix}1&0&0\\0&1&0\\0&0&1\end{bmatrix}A\begin{bmatrix}1&0&0\\0&1&0\\0&0&1\end{bmatrix}$$

Each elementary row (column) operation of the product can be effected by subjecting the prefactor (post-factor) of A to the same.

Operating $R_{2,1}(-1)$, we have

$$\begin{bmatrix}1&1&2\\0&1&1\\0&-1&-1\end{bmatrix}=\begin{bmatrix}1&0&0\\-1&1&0\\0&0&1\end{bmatrix}A\begin{bmatrix}1&0&0\\0&1&0\\0&0&1\end{bmatrix}$$

Operating $C_{2,1}(-1)$, $C_{3,1}(-2)$, we have

$$\begin{bmatrix}1&0&0\\0&1&1\\0&-1&-1\end{bmatrix}=\begin{bmatrix}1&0&0\\-1&1&0\\0&0&1\end{bmatrix}A\begin{bmatrix}1&-1&-2\\0&1&0\\0&0&1\end{bmatrix}$$

Operating $R_{3,2}(1)$,

$$\begin{bmatrix}1&0&0\\0&1&1\\0&0&0\end{bmatrix}=\begin{bmatrix}1&0&0\\-1&1&0\\-1&1&1\end{bmatrix}A\begin{bmatrix}1&-1&-2\\0&1&0\\0&0&1\end{bmatrix}$$

Operating $C_{3,2}(-1)$,

$$\begin{bmatrix}1&0&0\\0&1&0\\0&0&0\end{bmatrix}=\begin{bmatrix}1&0&0\\-1&1&0\\-1&1&1\end{bmatrix}A\begin{bmatrix}1&-1&-1\\0&1&-1\\0&0&1\end{bmatrix}$$

Thus, we have the normal form with

$$P=\begin{bmatrix}1&0&0\\-1&1&0\\-1&1&1\end{bmatrix},\ Q=\begin{bmatrix}1&-1&-1\\0&1&-1\\0&0&1\end{bmatrix}$$

(*ii*)
$$\begin{bmatrix}1&2&-1&2\\2&5&-2&3\\1&2&1&2\end{bmatrix}=\begin{bmatrix}1&0&0\\0&1&0\\0&0&1\end{bmatrix}A\begin{bmatrix}1&0&0&0\\0&1&0&0\\0&0&1&0\\0&0&0&1\end{bmatrix}$$

Operating $R_{2,1}(-2)$, $R_{3,1}(-1)$

$$\begin{bmatrix}1&2&-1&2\\0&1&0&-1\\0&0&2&0\end{bmatrix}=\begin{bmatrix}1&0&0\\-2&1&0\\-1&0&1\end{bmatrix}A\begin{bmatrix}1&0&0&0\\0&1&0&0\\0&0&1&0\\0&0&0&1\end{bmatrix}$$

Operating $C_{2,1}(-2)$, $C_{3,1}(1)$, $C_{4,1}(-2)$

$$\begin{bmatrix}1&0&0&0\\0&1&0&-1\\0&0&2&0\end{bmatrix}=\begin{bmatrix}1&0&0\\-2&1&0\\-1&0&1\end{bmatrix}A\begin{bmatrix}1&-2&1&2\\0&1&0&0\\0&0&1&0\\0&0&0&1\end{bmatrix}$$

Operating $C_{4,2}(1)$

$$\begin{bmatrix}1&0&0&0\\0&1&0&0\\0&0&2&0\end{bmatrix}=\begin{bmatrix}1&0&0\\-2&1&0\\-1&0&1\end{bmatrix}A\begin{bmatrix}1&-2&1&-4\\0&1&0&1\\0&0&1&0\\0&0&0&1\end{bmatrix}$$

Operating $R_3\left(\frac{1}{2}\right)$, $\begin{bmatrix} 1 & 0 & 0 & 0 \\ 0 & 1 & 0 & 0 \\ 0 & 0 & 1 & 0 \end{bmatrix} = \begin{bmatrix} 1 & 0 & 0 \\ -2 & 1 & 0 \\ -\frac{1}{2} & 0 & \frac{1}{2} \end{bmatrix} A \begin{bmatrix} 1 & -2 & 1 & -4 \\ 0 & 1 & 0 & 1 \\ 0 & 0 & 1 & 0 \\ 0 & 0 & 0 & 1 \end{bmatrix}$

which is in normal form with $P = \begin{bmatrix} 1 & 0 & 0 \\ -2 & 1 & 0 \\ -\frac{1}{2} & 0 & \frac{1}{2} \end{bmatrix}, Q = \begin{bmatrix} 1 & -2 & 1 & -4 \\ 0 & 1 & 0 & 1 \\ 0 & 0 & 1 & 0 \\ 0 & 0 & 0 & 1 \end{bmatrix}$.

Example 2. *Express* $A = \begin{bmatrix} 1 & 2 & -1 \\ 2 & 5 & -2 \\ 1 & 2 & 1 \end{bmatrix}$ *as a product of elementary matrices.*

Sol. $A = \begin{bmatrix} 1 & 2 & -1 \\ 2 & 5 & -2 \\ 1 & 2 & 1 \end{bmatrix}$

operating $R_{2,1}(-2)$ and $R_{3,1}(-1)$ on Matrix A, to make entries below $a_{11} = 1$ as zeros.

$$\therefore \quad A \sim \begin{bmatrix} 1 & 2 & -1 \\ 0 & 1 & 0 \\ 0 & 0 & 2 \end{bmatrix} \quad ...(1)$$

We know that Elementary matrices corresponding to operations $R_{2,1}(-2)$ and $R_{3,1}(-1)$ are obtained by operating

$R_{2,1}(-2)$ and $R_{3,1}(-1)$ on $I_3 = \begin{bmatrix} 1 & 0 & 0 \\ 0 & 1 & 0 \\ 0 & 0 & 1 \end{bmatrix}$.

Let these elementary matrices be E_1 and E_2 respectively.

$\therefore$ E_1^{-1} = matrix obtained by operating $R_{2,1}(2)$ on I_3 (By Art. 22)

$$= \begin{bmatrix} 1 & 0 & 0 \\ 2 & 1 & 0 \\ 0 & 0 & 1 \end{bmatrix}$$

and E_2^{-1} = Matrix obtained by operating

$R_{3,1}(1)$ on I_3 $= \begin{bmatrix} 1 & 0 & 0 \\ 0 & 1 & 0 \\ 1 & 0 & 1 \end{bmatrix}$

Now operating $C_{2,1}(-2)$ and $C_{3,1}(1)$ on (1) to make entries adjacent to $a_{11} = 1$ as zeros

$$\therefore \quad \text{From (1)} \quad A \sim \begin{bmatrix} 1 & 0 & 0 \\ 0 & 1 & 0 \\ 0 & 0 & 2 \end{bmatrix} \quad ...(2)$$

We know that elementary matrices corresponding to operations $C_{2,1}(-2)$ and $C_{3,1}(1)$ are obtained by operating $C_{2,1}(-2)$ and $C_{3,1}(1)$ on I_3.

Let these elementary matrices be F_1 and F_2 respectively.

$\therefore$ F_1^{-1} = Matrix obtained by operating

$C_{2,1}(2)$ on I_3 (By Art. 22)

$$= \begin{bmatrix} 1 & 2 & 0 \\ 0 & 1 & 0 \\ 0 & 0 & 1 \end{bmatrix}$$

and F_2^{-1} = Matrix obtained by operating $C_{3,1}(-1)$ on I_3

$$= \begin{bmatrix} 1 & 0 & -1 \\ 0 & 1 & 0 \\ 0 & 0 & 1 \end{bmatrix}$$

operating $R_3\left(\frac{1}{2}\right)$ on (2) to make $a_{33} = 1$; we have

$$A \sim \begin{bmatrix} 1 & 0 & 0 \\ 0 & 1 & 0 \\ 0 & 0 & 1 \end{bmatrix} = I_3$$

(It may be remarked that we could operate $C_3\left(\frac{1}{2}\right)$ on (2) instead of $R_3\left(\frac{1}{2}\right)$.

Hence we have reduced A to I_3.

Let E_3 be the elementary matrix corresponding to the operation $R_3\left(\frac{1}{2}\right)$.

∴ E_3^{-1} = Matrix obtained by operating

$R_3(2)$ on I_3 $= \begin{bmatrix} 1 & 0 & 0 \\ 0 & 1 & 0 \\ 0 & 0 & 2 \end{bmatrix}$

∴ $E_3E_2E_1AF_1F_2 = I$

⇒ $A = (E_3^{-1}E_2^{-1}E_1^{-1})(F_2^{-1}F_1^{-1})$.

Putting values of $E_1^{-1}, E_2^{-1}, E_3^{-1}, F_2^{-1}, F_1^{-1}$ we have

$$A = \begin{bmatrix} 1 & 0 & 0 \\ 0 & 1 & 0 \\ 0 & 0 & 2 \end{bmatrix}\begin{bmatrix} 1 & 0 & 0 \\ 0 & 1 & 0 \\ 1 & 0 & 1 \end{bmatrix} . \begin{bmatrix} 1 & 0 & 0 \\ 2 & 1 & 0 \\ 0 & 0 & 1 \end{bmatrix}\begin{bmatrix} 1 & 0 & -1 \\ 0 & 1 & 0 \\ 0 & 0 & 1 \end{bmatrix}\begin{bmatrix} 1 & 2 & 0 \\ 0 & 1 & 0 \\ 0 & 0 & 1 \end{bmatrix}.$$

24. Inverse of Matrix by Elementary Transformations

Since each non-singular matrix A is the product of elementary matrices,

∴ A^{-1} being non-singular can be written as

$$A^{-1} = P_1P_2......P_s,$$

where $P_1, P_2, ..., P_s$ are elementary matrices.

Now ∵ $A^{-1}A = I$, therefore $P_1 . P_2 ...... P_s A = I$ ...(1)

Since the pre-multiplication with an elementary matrix means the corresponding elementary row transformation in the matrix.

∴ from (1), we conclude that each non-singular matrix can be reduced to the unit matrix by a series of elementary row transformations.

How to compute an inverse ?

(I) *Write A = IA.*

(II) *Perform suitable elementary row transformations on the matrix A on the left and the prefactor I on the right until, we get I = BA.*

(III) *Then B = A^{-1} is the reqd. inverse.*

Example 3. *Find the inverse of* $A = \begin{bmatrix} 1 & 2 & 1 \\ 3 & 1 & 2 \\ 0 & 1 & 2 \end{bmatrix}$ *by using elementary row operations.*

Sol. We write $\qquad A = IA$

i.e.,
$$\begin{bmatrix} 1 & 2 & 1 \\ 3 & 1 & 2 \\ 0 & 1 & 2 \end{bmatrix} = \begin{bmatrix} 1 & 0 & 0 \\ 0 & 1 & 0 \\ 0 & 0 & 1 \end{bmatrix} A$$

Operating $R_{2,1}(-3)$

$$\begin{bmatrix} 1 & 2 & 1 \\ 0 & -5 & -1 \\ 0 & 1 & 2 \end{bmatrix} = \begin{bmatrix} 1 & 0 & 0 \\ -3 & 1 & 0 \\ 0 & 0 & 1 \end{bmatrix} A$$

Operating $R_{2,3}(6)$
$$\begin{bmatrix} 1 & 2 & 1 \\ 0 & 1 & 11 \\ 0 & 1 & 2 \end{bmatrix} = \begin{bmatrix} 1 & 0 & 0 \\ -3 & 1 & 6 \\ 0 & 0 & 1 \end{bmatrix} A$$

Operating $R_{3,2}(-1)$ and $R_{1,2}(-2)$,

$$\begin{bmatrix} 1 & 0 & -21 \\ 0 & 1 & 11 \\ 0 & 0 & -9 \end{bmatrix} = \begin{bmatrix} 7 & -2 & -12 \\ -3 & 1 & 6 \\ 3 & -1 & -5 \end{bmatrix} A$$

Operating $R_3\left(-\frac{1}{9}\right)$,
$$\begin{bmatrix} 1 & 0 & -21 \\ 0 & 1 & 11 \\ 0 & 0 & 1 \end{bmatrix} = \begin{bmatrix} 7 & -2 & -12 \\ -3 & 1 & 6 \\ -\frac{1}{3} & \frac{1}{9} & \frac{5}{9} \end{bmatrix} A$$

Operating $R_{1,3}(21)$ and $R_{2,3}(-11)$

$$\begin{bmatrix} 1 & 0 & 0 \\ 0 & 1 & 0 \\ 0 & 0 & 1 \end{bmatrix} = \begin{bmatrix} 0 & \frac{1}{3} & -\frac{1}{3} \\ \frac{2}{3} & -\frac{2}{9} & -\frac{1}{9} \\ -\frac{1}{3} & \frac{1}{9} & \frac{5}{9} \end{bmatrix} A$$

Hence,
$$A^{-1} = \begin{bmatrix} 0 & \frac{1}{3} & -\frac{1}{3} \\ \frac{2}{3} & -\frac{2}{9} & -\frac{1}{9} \\ -\frac{1}{3} & \frac{1}{9} & \frac{5}{9} \end{bmatrix}.$$

Note. To find A^{-1} by elementary **column** operations, we write $A = AI$ and go on performing suitable elementary column operations on the matrix A on the left hand side and the post-factor I of A on the right hand side till we reach the result $I = AB$.

Then by def. of inverse $A^{-1} = B$. (See example 5)

Remark. We can find A^{-1} either by using **only** row operations or using **only** column operations.

If we are asked to find A^{-1} by elementary operations, then we may start either with $A = IA$ as in Example 3 above or with $A = AI$ as in the above note.

Example 4. *Find the inverse of* $A = \begin{bmatrix} -1 & -3 & 3 & -1 \\ 1 & 1 & -1 & 0 \\ 2 & -5 & 2 & -3 \\ -1 & 1 & 0 & 1 \end{bmatrix}$.

Sol. We write $\quad A = IA$

i.e.,

$$\begin{bmatrix} -1 & -3 & 3 & -1 \\ 1 & 1 & -1 & 0 \\ 2 & -5 & 2 & -3 \\ -1 & 1 & 0 & 1 \end{bmatrix} = \begin{bmatrix} 1 & 0 & 0 & 0 \\ 0 & 1 & 0 & 0 \\ 0 & 0 & 1 & 0 \\ 0 & 0 & 0 & 1 \end{bmatrix} A$$

Operating $R_{2,1}(1)$, $R_{3,1}(2)$, $R_{4,1}(-1)$, we have

$$\begin{bmatrix} -1 & -3 & 3 & -1 \\ 0 & -2 & 2 & -1 \\ 0 & -11 & 8 & -5 \\ 0 & 4 & -3 & 2 \end{bmatrix} = \begin{bmatrix} 1 & 0 & 0 & 0 \\ 1 & 1 & 0 & 0 \\ 2 & 0 & 1 & 0 \\ -1 & 0 & 0 & 1 \end{bmatrix} A$$

Operating $R_{3,2}\left(-\frac{11}{2}\right)$, $R_{4,2}(2)$, we have

$$\begin{bmatrix} -1 & -3 & 3 & -1 \\ 0 & -2 & 2 & -1 \\ 0 & 0 & -3 & \frac{1}{2} \\ 0 & 0 & 1 & 0 \end{bmatrix} = \begin{bmatrix} 1 & 0 & 0 & 0 \\ 1 & 1 & 0 & 0 \\ -\frac{7}{2} & -\frac{11}{2} & 1 & 0 \\ 1 & 2 & 0 & 1 \end{bmatrix} A$$

Operating $R_{4,3}\left(\frac{1}{3}\right)$, we have

$$\begin{bmatrix} -1 & -3 & 3 & -1 \\ 0 & -2 & 2 & -1 \\ 0 & 0 & -3 & \frac{1}{2} \\ 0 & 0 & 0 & \frac{1}{6} \end{bmatrix} = \begin{bmatrix} 1 & 0 & 0 & 0 \\ 1 & 1 & 0 & 0 \\ -\frac{7}{2} & -\frac{11}{2} & 1 & 0 \\ -\frac{1}{6} & \frac{1}{6} & \frac{1}{3} & 1 \end{bmatrix} A$$

Operating $R_{3,4}(-3)$, $R_{2,4}(6)$, $R_{1,4}(6)$, we have

$$\begin{bmatrix} -1 & -3 & 3 & 0 \\ 0 & -2 & 2 & 0 \\ 0 & 0 & -3 & 0 \\ 0 & 0 & 0 & \frac{1}{6} \end{bmatrix} = \begin{bmatrix} 0 & 1 & 2 & 6 \\ 0 & 2 & 2 & 6 \\ -3 & -6 & 0 & -3 \\ -\frac{1}{6} & \frac{1}{6} & \frac{1}{3} & 1 \end{bmatrix} A$$

Operating $R_{2,3}\left(\frac{2}{3}\right)$, $R_{1,3}(1)$, we have

$$\begin{bmatrix} -1 & -3 & 0 & 0 \\ 0 & -2 & 0 & 0 \\ 0 & 0 & -3 & 0 \\ 0 & 0 & 0 & \frac{1}{6} \end{bmatrix} = \begin{bmatrix} -3 & -5 & 2 & 3 \\ -2 & -2 & 2 & 4 \\ -3 & -6 & 0 & -3 \\ -\frac{1}{6} & \frac{1}{6} & \frac{1}{3} & 1 \end{bmatrix} A$$

Operating $R_{1,2}\left(-\frac{3}{2}\right)$, we have

$$\begin{bmatrix} -1 & 0 & 0 & 0 \\ 0 & -2 & 0 & 0 \\ 0 & 0 & -3 & 0 \\ 0 & 0 & 0 & \frac{1}{6} \end{bmatrix} = \begin{bmatrix} 0 & -2 & -1 & -3 \\ -2 & -2 & 2 & 4 \\ -3 & -6 & 0 & -3 \\ -\frac{1}{6} & \frac{1}{6} & \frac{1}{3} & 1 \end{bmatrix} A$$

Operating $R_1(-1)$, $R_2\left(-\frac{1}{2}\right)$, $R_3\left(-\frac{1}{3}\right)$, $R_4(6)$, we have

$$\begin{bmatrix} 1 & 0 & 0 & 0 \\ 0 & 1 & 0 & 0 \\ 0 & 0 & 1 & 0 \\ 0 & 0 & 0 & 1 \end{bmatrix} = \begin{bmatrix} 0 & 2 & 1 & 3 \\ 1 & 1 & -1 & -2 \\ 1 & 2 & 0 & 1 \\ -1 & 1 & 2 & 6 \end{bmatrix} A$$

Hence
$$A^{-1} = \begin{bmatrix} 0 & 2 & 1 & 3 \\ 1 & 1 & -1 & -2 \\ 1 & 2 & 0 & 1 \\ -1 & 1 & 2 & 6 \end{bmatrix}.$$

Example 5. *Find the inverse of* $A = \begin{bmatrix} 1 & 3 & 2 \\ 0 & 4 & 1 \\ 5 & 2 & 3 \end{bmatrix}$

Using only elementary column operations.

Sol. $A = AI$

i.e.,
$$\begin{bmatrix} 1 & 3 & 2 \\ 0 & 4 & 1 \\ 5 & 2 & 3 \end{bmatrix} = A \begin{bmatrix} 1 & 0 & 0 \\ 0 & 1 & 0 \\ 0 & 0 & 1 \end{bmatrix}$$

Applying $C_{2,1}(-3)$, $C_{3,1}(-2)$

$$\begin{bmatrix} 1 & 0 & 0 \\ 0 & 4 & 1 \\ 5 & -13 & -7 \end{bmatrix} = A \begin{bmatrix} 1 & -3 & -2 \\ 0 & 1 & 0 \\ 0 & 0 & 1 \end{bmatrix}$$

Applying $C_{2,3}$,

$$\begin{bmatrix} 1 & 0 & 0 \\ 0 & 1 & 4 \\ 5 & -7 & -13 \end{bmatrix} = A \begin{bmatrix} 1 & -2 & -3 \\ 0 & 0 & 1 \\ 0 & 1 & 0 \end{bmatrix}$$

Applying $C_{3,2}(-4)$

$$\begin{bmatrix} 1 & 0 & 0 \\ 0 & 1 & 0 \\ 5 & -7 & 15 \end{bmatrix} = A \begin{bmatrix} 1 & -2 & 5 \\ 0 & 0 & 1 \\ 0 & 1 & -4 \end{bmatrix}$$

Applying $C_3(1/15)$,

$$\begin{bmatrix} 1 & 0 & 0 \\ 0 & 1 & 0 \\ 5 & -7 & 1 \end{bmatrix} = A \begin{bmatrix} 1 & -2 & 1/3 \\ 0 & 0 & 1/15 \\ 0 & 1 & -4/15 \end{bmatrix}$$

Applying $C_{1,3}(-5)$, $C_{2,3}(7)$,

$$\begin{bmatrix} 1 & 0 & 0 \\ 0 & 1 & 0 \\ 0 & 0 & 1 \end{bmatrix} = A \begin{bmatrix} -2/3 & 1/3 & 1/3 \\ -1/3 & 7/15 & 1/15 \\ 4/3 & -13/15 & -4/15 \end{bmatrix}$$

$$\therefore \quad A^{-1} = \begin{bmatrix} -2/3 & 1/3 & 1/3 \\ -1/3 & 7/15 & 1/15 \\ 4/3 & -13/15 & -4/15 \end{bmatrix}.$$

EXERCISE 4

1. Reduce each of the following matrices to the normal form

(i) $\begin{bmatrix} 1 & 2 & 0 & -1 \\ 3 & 4 & 1 & 2 \\ -2 & 3 & 2 & 5 \end{bmatrix}$ (ii) $\begin{bmatrix} 0 & 2 & 3 & 4 \\ 2 & 3 & 5 & 4 \\ 4 & 8 & 13 & 12 \end{bmatrix}$

(iii) $\begin{bmatrix} 1 & 2 & -3 \\ 2 & 5 & -4 \end{bmatrix}$ (iv) $\begin{bmatrix} 9 & 0 & 2 & 3 \\ 0 & 1 & 5 & 6 \\ 4 & 5 & 3 & 0 \end{bmatrix}$

(v) $\begin{bmatrix} 1 & 2 & 1 & 0 \\ 3 & 2 & 1 & 2 \\ 2 & -1 & 2 & 5 \\ 5 & 6 & 3 & 2 \\ 1 & 3 & -1 & -3 \end{bmatrix}$

Also find rank in each case.

2. Reduce the matrix $A = \begin{bmatrix} 1 & 2 & 1 \\ -1 & 0 & 2 \\ 2 & 1 & 3 \end{bmatrix}$ to I_3 by elementary transformations.

3. Let $A = \begin{bmatrix} 2 & 1 & 4 & -2 \\ 3 & 2 & 1 & 3 \\ 4 & 3 & 2 & -4 \end{bmatrix}$

(i) From I_3, form the elementary matrices $E_{1,2}$, $E_2(3)$ and $E_{1,3}(-2)$ and check that each EA effects the corresponding row operation.

(ii) From I_4, form the elementary matrices $F_{2,4}$, $F_2(-1)$ and $F_{3,2}(3)$ and show that each AF effects the corresponding column operation.

4. For the following matrices A, find non-singular matrices P and Q such that PAQ is in the normal form and hence determine the rank of A.

(i) $\begin{bmatrix} 1 & 1 & 1 \\ 1 & -1 & -1 \\ 3 & 1 & 1 \end{bmatrix}$ (ii) $\begin{bmatrix} 1 & 2 & -1 \\ 3 & 4 & 1 \end{bmatrix}$

(iii) $\begin{bmatrix} 1 & 2 & 3 & -2 \\ 2 & -2 & 1 & 3 \\ 3 & 0 & 4 & 1 \end{bmatrix}$ *(iv)* $\begin{bmatrix} 1 & 2 & 3 \\ 3 & 2 & 1 \\ 1 & 3 & 2 \\ 2 & 1 & 3 \end{bmatrix}$

(v) $\begin{bmatrix} 1 & 1 & 2 & 3 \\ 1 & 3 & 0 & 3 \\ 1 & -2 & -3 & -3 \\ 1 & 1 & 2 & 3 \end{bmatrix}$.

5. **Express** $A = \begin{bmatrix} 1 & 3 & 3 \\ 1 & 4 & 3 \\ 1 & 3 & 4 \end{bmatrix}$ **as a product of elementary matrices.**

6. **Using elementary transformations, find the inverses of the following matrices :**

(i) $\begin{bmatrix} 1 & 2 & -1 \\ -1 & 1 & 2 \\ 2 & -1 & 1 \end{bmatrix}$ *(ii)* $\begin{bmatrix} 1 & 3 & 2 \\ 0 & 4 & 1 \\ 5 & 2 & 3 \end{bmatrix}$

(iii) $\begin{bmatrix} 1 & 1 & 3 \\ 1 & 3 & -3 \\ -2 & -4 & -4 \end{bmatrix}$ *(iv)* $\begin{bmatrix} 1 & 1 & 1 & 1 \\ 0 & 1 & 0 & 1 \\ 0 & 0 & 2 & 0 \\ 0 & 0 & 0 & 2 \end{bmatrix}$.

7. **Using only elementary column operations, find** A^{-1} **where** $A = \begin{bmatrix} 1 & 2 & 1 \\ -1 & 0 & 2 \\ 2 & 1 & 3 \end{bmatrix}$.

ANSWERS

1. *(i)* $[I_3 \;\; O]$, 3 *(ii)* $\begin{bmatrix} I_2 & O \\ O & O \end{bmatrix}$, 2 *(iii)* $[I_3 \;\; O]$, 2 *(iv)* $[I_3 \;\; O]$, 3

(v) $\begin{bmatrix} I_3 & O \\ O & O \end{bmatrix}$, 3

4. *(i)* $P = \begin{bmatrix} 1 & 0 & 0 \\ \frac{1}{2} & -\frac{1}{2} & 0 \\ -2 & -1 & 1 \end{bmatrix}$, $Q = \begin{bmatrix} 1 & -1 & 0 \\ 0 & 1 & -1 \\ 0 & 0 & 1 \end{bmatrix}$, $\rho(A) = 2$

(ii) $P = \begin{bmatrix} 1 & 0 \\ \frac{3}{2} & -\frac{1}{2} \end{bmatrix}$, $Q = \begin{bmatrix} 1 & -2 & -3 \\ 0 & 1 & 2 \\ 0 & 0 & 1 \end{bmatrix}$; $\rho(A) = 2$

$\left[\right.$ **Hint.** Operate $R_{2,1}(-3)$, $C_{2,1}(-2)$, $C_{3,1}(1)$, $R_2\left(-\frac{1}{2}\right)$, $C_{3,2}(2)$ $\left.\right]$

(*iii*) $P = \begin{bmatrix} 1 & 0 & 0 \\ -2 & 1 & 0 \\ -1 & -1 & 1 \end{bmatrix}$, $Q = \begin{bmatrix} 1 & \frac{1}{3} & -\frac{4}{3} & -\frac{1}{3} \\ 0 & -\frac{1}{6} & -\frac{5}{6} & \frac{7}{6} \\ 0 & 0 & 1 & 0 \\ 0 & 0 & 0 & 1 \end{bmatrix}$, $\rho(A) = 2$

(*iv*) $P = \begin{bmatrix} 1 & 0 & 0 & 0 \\ \frac{3}{4} & -\frac{1}{4} & 0 & 0 \\ \frac{7}{12} & -\frac{1}{12} & -\frac{4}{12} & 0 \\ -\frac{3}{2} & -\frac{1}{2} & 1 & 1 \end{bmatrix}$, $Q = \begin{bmatrix} 1 & -2 & 1 \\ 0 & 1 & -2 \\ 0 & 0 & 1 \end{bmatrix}$, $\rho(A) = 3$

(*v*) $P = \begin{bmatrix} 1 & 0 & 0 & 0 \\ -\frac{1}{2} & \frac{1}{2} & 0 & 0 \\ \frac{5}{16} & -\frac{3}{16} & -\frac{2}{16} & 0 \\ 1 & 0 & 0 & -1 \end{bmatrix}$, $Q = \begin{bmatrix} 1 & -1 & -3 & -\frac{3}{4} \\ 0 & 1 & 1 & -\frac{3}{4} \\ 0 & 0 & 1 & -\frac{3}{4} \\ 0 & 0 & 0 & 1 \end{bmatrix}$, $\rho(A) = 3$

5. $A = \begin{bmatrix} 1 & 0 & 0 \\ 1 & 1 & 0 \\ 0 & 0 & 1 \end{bmatrix} \begin{bmatrix} 1 & 0 & 0 \\ 0 & 1 & 0 \\ 1 & 0 & 1 \end{bmatrix} \begin{bmatrix} 1 & 0 & 3 \\ 0 & 1 & 0 \\ 0 & 0 & 1 \end{bmatrix} \begin{bmatrix} 1 & 3 & 0 \\ 0 & 1 & 0 \\ 0 & 0 & 1 \end{bmatrix}$

6. (*i*) $\frac{1}{14}\begin{bmatrix} 3 & -1 & 5 \\ 5 & 3 & -1 \\ -1 & 5 & 3 \end{bmatrix}$ (*ii*) $-\frac{1}{15}\begin{bmatrix} 10 & -5 & -5 \\ 5 & -7 & -1 \\ -20 & 13 & 4 \end{bmatrix}$ (*iii*) $\frac{1}{4}\begin{bmatrix} 12 & 4 & 6 \\ -5 & -1 & -3 \\ -1 & -1 & -1 \end{bmatrix}$

(*iv*) $\begin{bmatrix} 1 & -1 & -\frac{1}{2} & 0 \\ 0 & 1 & 0 & -\frac{1}{2} \\ 0 & 0 & \frac{1}{2} & 0 \\ 0 & 0 & 0 & \frac{1}{2} \end{bmatrix}$ **7.** $\frac{1}{11}\begin{bmatrix} -2 & -5 & 4 \\ 7 & 1 & -3 \\ -1 & 3 & 2 \end{bmatrix}$.

---

25. Theorem

The rank of the transpose of a matrix is the same as that of the original matrix

i.e., $$\rho(A') = \rho(A).$$

Proof. Let A be an $m \times n$ matrix of rank r. Then there is at least one square sub-matrix of A of order r whose determinant is not equal to zero. Let S be a square sub-matrix of A of order r such that $|S| \neq 0$. If S′ is the transpose of S, then obviously S′ is a sub-matrix of A′.

Since the value of a determinant remains unchanged when rows and columns are interchanged, we have,
$$|S'| = |S| \neq 0$$

Hence $$\rho(A') \geq r \qquad ...(1)$$

On the other hand, if A′ contains a square sub-matrix T of order $r + 1$, then T′ is a sub-matrix of A of order $r + 1$. Since $\rho(A) = r$, so $| T | = | T' | = 0$. Thus, A′ cannot have a square sub-matrix of order $r + 1$ whose determinant is non-zero.

Hence $\rho(A') \leq r$ ...(2)

∴ From (1) and (2), $\rho(A') = r$.

26. Theorem

The rank of the product of two matrices cannot exceed the rank of either matrix.

i.e., $\rho(AB) \leq \rho(A)$ *and* $\rho(AB) \leq \rho(B)$.

Proof. Let $\rho(A) = r$.

Using elementary transformations, $A \sim \begin{bmatrix} M \\ O \end{bmatrix}$,

where M is a matrix having first r non-zero rows.

Post multiplying by the matrix B, $AB \sim \begin{bmatrix} M \\ O \end{bmatrix} . B$.

The matrix $\begin{bmatrix} M \\ O \end{bmatrix}$ B has at the most first r non-zero rows obtained on multiplying first r non-zero rows of M with the columns of B.

Hence, $\rho(AB) = \rho\left(\begin{bmatrix} M \\ O \end{bmatrix} B\right) \leq r$ (By Def. Art. 1)

$\Rightarrow$ $\rho(AB) \leq \rho(A)$ ...(1)

This shows that the rank of product AB ≤ rank of prefactor A.

Again, $\rho(AB) = \rho(AB)'$ (By Theorem 25)

$= \rho(B'A')$

$\leq \rho(B')$ [∵ rank of product ≤ rank prefactor, proved in (1)]

$= \rho(B)$ (By Theorem 25)

$\Rightarrow$ $\rho(AB) \leq (B)$.

27. Theorem

If A is an $m \times n$ matrix of rank r, then

(i) there exists a non-singular matrix P such that $PA = \begin{bmatrix} B \\ O \end{bmatrix}$, *where B is an $r \times n$ matrix of rank r.*

(ii) there exists a non-singular matrix Q such that AQ = [C O], where C is an $m \times r$ matrix of rank r.

Proof. (*i*) Since $\rho(A) = r$, there exist non-singular matrices P and Q such that

$$PAQ = \begin{bmatrix} I_r & O \\ O & O \end{bmatrix}$$

Since Q is non-singular, Q^{-1} exists and can be written as product of elementary matrices, $Q^{-1} = F_1F_2 \ldots\ldots F_s$ (say).

$$\therefore \qquad PA = \begin{bmatrix} I_r & O \\ O & O \end{bmatrix} Q^{-1} = \begin{bmatrix} I_r & O \\ O & O \end{bmatrix} F_1F_2......F_s.$$

Since post-multiplication by elementary matrices amounts to elementary column operations, the last $m - r$ rows being zero rows in $\begin{bmatrix} I_r & O \\ O & O \end{bmatrix}$, they continue to be zero rows on applying $F_1F_2......F_s$.

Thus, $\qquad PA = \begin{bmatrix} B \\ O \end{bmatrix}$, where B is an $r \times n$ matrix

and $\qquad \rho(B) = \rho \begin{bmatrix} B \\ O \end{bmatrix} = \rho(PA) = \rho(A) = r.$

(*ii*) Similarly, on considering AQ, we can find that AQ = [C O], where C is an $m \times r$ matrix of rank r.

Let us now investigate the relationship between the rank, row rank and column rank of a matrix.

28. Theorem

For any matrix A, $\qquad \rho_R(A) = \rho(A) = \rho_C(A).$

Proof. Let $\qquad \rho(A) = r,\ \rho_R(A) = s,\ \rho_C(A) = t.$

We know that rank of a matrix is not altered on pre-or post multiplication by non-singular matrices. Likewise the row (column) rank is not altered on pre-multiplication (post multiplication) by non-singular matrices.

Since $\rho_R(A) = s$, there exists a non-singular matrix P such that PA = $\begin{bmatrix} B \\ O \end{bmatrix}$, where B consists of s non-zero rows of A. Clearly, every square sub-matrix of $\begin{bmatrix} B \\ O \end{bmatrix}$ of order $s + 1$ contains a row of zeros. Therefore, every minor of $\begin{bmatrix} B \\ O \end{bmatrix}$ of order $s + 1$ vanishes.

Hence, $\qquad \rho \begin{bmatrix} B \\ O \end{bmatrix} \leq s.$

Thus, $\qquad \rho(A) = \rho(PA) = \rho \begin{bmatrix} B \\ O \end{bmatrix} \leq s \quad \Rightarrow \quad r \leq s \qquad ...(i)$

Again, since $\rho(A) = r$, there exists a non-singular matrix Q such that QA = $\begin{bmatrix} C \\ O \end{bmatrix}$, where C is $r \times n$ matrix of rank r.

$$\therefore \qquad \rho_R(A) = \rho_R(QA) = \rho_R \begin{bmatrix} C \\ O \end{bmatrix} = \rho_R(C) \leq r. \quad \Rightarrow \quad s \leq r \qquad ...(ii)$$

$\therefore$ From (*i*) and (*ii*), $\qquad r = s.$

Similarly, $\qquad r = t.$

Hence, $\qquad r = s = t$

i.e., $\qquad \rho(A) = \rho_R(A) = \rho_C(A).$

LINEAR EQUATIONS

29. Definitions

Consider a system of m linear equations in n unknowns $x_1, x_2, \ldots\ldots, x_n$

i.e.,

$$\left.\begin{aligned} a_{11}x_1 + a_{12}x_2 + \ldots\ldots + a_{1n}x_n &= b_1 \\ a_{21}x_1 + a_{22}x_2 + \ldots\ldots + a_{2n}x_n &= b_2 \\ \ldots\ldots\ \ldots\ldots\ \ldots\ldots\ \ldots\ldots\ &\ldots\ldots \\ a_{m1}x_1 + a_{m2}x_2 + \ldots\ldots + a_{mn}x_n &= b_m \end{aligned}\right\} \quad \ldots(1)$$

in which the coefficients a_{ij} and the constant terms b_i are from a fixed number field F. (This F will be the field of rationals, reals or complex numbers). *By a solution of the system (1), we mean an ordered n—tuple $\alpha = (\alpha_1, \alpha_2, \ldots\ldots, \alpha_n)$ of elements of F if on substituting $x_1 = \alpha_1, x_2 = \alpha_2$, $x_n = \alpha_n$, all the m equations in (1) are simultaneously satisfied. When the system has a solution, it is said to be* ***consistent*** *otherwise the system is said to be* ***inconsistent.*** *The set of all solutions of a consistant system is called the solution set of the system.*

If b_i's are zero, then the system (1) is called ***homogeneous*** *system, otherwise (i.e. Even if at least one b_i is non-zero) a* ***non-homogeneous*** *system. Observe that a homogeneous system always has $x_1 = 0, x_2 = 0, \ldots\ldots, x_n = 0$ as a solution. So, a homogeneous system is always consistent. The solution $x_1 = x_2 = \ldots\ldots = x_n = 0$ is called the trivial solution of the system.*

We can write the system (1) in the matrix notation as

$$\begin{bmatrix} a_{11} & a_{12} & \ldots\ldots & a_{1n} \\ a_{21} & a_{22} & \ldots\ldots & a_{2n} \\ \ldots & \ldots & \ldots\ldots & \ldots \\ a_{m1} & a_{m2} & \ldots\ldots & a_{mn} \end{bmatrix} \begin{bmatrix} x_1 \\ x_2 \\ \ldots \\ x_n \end{bmatrix} = \begin{bmatrix} b_1 \\ b_2 \\ \ldots \\ b_m \end{bmatrix}$$

or

$$AX = B$$

where

$$A = \begin{bmatrix} a_{11} & a_{12} & \ldots & a_{1n} \\ a_{21} & a_{22} & \ldots & a_{2n} \\ \ldots & \ldots & \ldots & \ldots \\ a_{m1} & a_{m2} & \ldots & a_{mn} \end{bmatrix} \text{ is the } \textbf{coefficient matrix,}$$

$$X = \begin{bmatrix} x_1 \\ x_2 \\ \vdots \\ x_n \end{bmatrix} \text{ and } B = \begin{bmatrix} b_1 \\ b_2 \\ \vdots \\ b_m \end{bmatrix}$$

For the system (1), we may consider the **augmented matrix**

$$[A : B] = \left[\begin{array}{cccc:c} a_{11} & a_{12} & \ldots & a_{1n} & b_1 \\ a_{21} & a_{22} & \ldots & a_{2n} & b_2 \\ \ldots & \ldots & \ldots & \ldots & \ldots \\ a_{m1} & a_{m2} & \ldots & a_{mn} & b_m \end{array}\right]$$

each row of which corresponds to an equation of (1).

Following examples show that a system of linear equations may have no solution, unique solution or has infinitely many solutions.

Example 1. *Consider the system $2x + y = 1$, $x - 2y = -7$, $x + 3y = 6$ of three equations in two unknowns. This system has no solution.*

For, if $x = x_1$ and $y = y_1$ satisfy the first two equations, then

$$2x_1 + y_1 = 1$$
$$x_1 - 2y_1 = -7$$

Putting $x = x_1$ and $y = y_1$ in L.H.S. of third equation

$$x_1 + 3y_1 = (2x_1 + y_1) - (x_1 - 2y_1) = 1 - (-7) = 8 \neq 6$$

$\therefore$ $x = x_1, y = y_1$ does not satisfy the third equation. Hence, the system has no solution.

Example 2. *Consider the system $2x - 3y = 1$, $x + y = 3$.*

On solving for x and y

This system has (2, 1) as the **only solution.**

Example 3. *Consider the system*

$$x + y = 3$$
$$2x + 2y = 6.$$

In this case, any pair (x_1, y_1) which satisfies the first equation *i.e.* $x_1 + y_1 = 3$, automatically satisfies the second equation $\because$ $(2x_1 + 2y_1 = 2(x_1 + y_1) = 2 \times 3 = 6)$. Therefore, it is sufficient to solve the first equation. To solve the first equation take y arbitrarily, say $y = y_1$. Then, $x = 3 - y_1$, $y = y_1$ is a solution of the system for every value of y_1. Hence, the system has infinitely many solutions.

We now, observe some important results about the solution set of a homogeneous and a non-homogeneous system.

30. Theorem

If $AX = O$ is a homogeneous system of equations in n unknowns and $X_1 = (x_1, x_2, \ldots\ldots, x_n)$ and $X_2 = (y_1, y_2, \ldots\ldots, y_n)$ are two solutions of this system, then $X_1 + X_2 = (x_1 + y_1, x_2 + y_2, \ldots\ldots, x_n + y_n)$ is also a solution. Also, if λ is a scalar, then $\lambda X_1 = (\lambda x_1, \lambda x_2, \ldots\ldots, \lambda x_n)$ is also a solution.

Proof. Since X_1 and X_2 are solutions of $AX = O$, then

$$AX_1 = O \text{ and } AX_2 = O.$$

Now $A(X_1 + X_2) = AX_1 + AX_2 = O + O = O$

$\Rightarrow$ $X_1 + X_2$ is a solution.

Again, X_1 is also a solution.

$\therefore$ $$AX_1 = O$$

Now $A(\lambda X_1) = \lambda(AX_1) = \lambda O = O$

$\Rightarrow$ λX_1 is also a solution.

Cor. 1. *Any linear combination of solutions of $AX = O$ is also a solution of $AX = O$.*

Cor. 2. ***A system of homogeneous equations either has only the trivial solution or infinitely many solutions.***

For, if $AX = O$ has a non-trivial solution X_1 then λX_1, for $\lambda \in F$ are different solutions of $AX = O$.

Now, consider the non-homogeneous system of equations $AX = B$ and associate with the system of homogeneous equations $AX = O$. The following theorem gives the connections between the solutions of the system $AX = B$ and the solutions of $AX = O$.

31. Theorem

If X_1 is a solution of $AX = B$ and X_2 is any solution of the associated system $AX = O$, then $X_1 + X_2$ is a solution of $AX = B$. Further if Y is a solution of $AX = B$, then $Y - X_2$ is a solution of $AX = O$.

Proof. Since X_1 is a solution of AX = B and X_2 is a solution of AX = O.

$\therefore$ $$AX_1 = B \text{ and } AX_2 = O$$

Now $$A(X_1 + X_2) = AX_1 + AX_2 = B + O = B$$

$\Rightarrow$ $X_1 + X_2$ is a solution of AX = B.

Again, since Y is a solution of AX = B, therefore AY = B

Also, $$AX_1 = B$$

Now $$A(Y - X_1) = AY - AX_1 = B - B = O$$

Hence, $Y - X_1$ is a solution of AX = O.

Cor. *A system of non-homogeneous equations AX = B either has no solution, a unique solution or infinitely many solutions.*

Proof. Let AX = B be a non-homogeneous system. If it has no solution, then we have nothing to prove.

Let X_1 be a solution of this system. If AX = O has only the trivial solution, then X_1 is the unique solution of AX = B ; otherwise corresponding to every solution Y of AX = O, we have a solution $X_1 + Y$ of AX = B and hence an infinity of solutions.

32. Equivalent Systems

Two systems of linear equations over the field F and in the same number of unknowns are said to be equivalent if every solution of either system is a solution of the other, i.e., they have the same solution set.

A system of equations equivalent to (1) may be obtained from it by applying one or more transformations :

(*i*) interchanging any two equations, (*ii*) multiplying any equation by some non-zero constant in F or (*iii*) adding to any equation a constant multiple of another equation.

To solve a given system of equations, it is sufficient to solve any system equivalent to it. So, we replace the given system by an equivalent one which is easily solvable. That is, we try to reduce the system to such a system from which the solution can be easily determined. For instance, if after successive reduction, we arrive at a system which has its co-efficient matrix in the diagonal form, then the solution if exists, can be immediately written down.

Similarly, if the reduced system has its co-efficient matrix in the triangular form with non-zero diagonal entries, then the solution can be determined by successively solving these equations.

Consider an example of the system of **non-homogeneous** linear equations :

$$\left.\begin{aligned} x + 2y - z &= 3 \quad &...(i)\\ 3x - y + 2z &= 1 \quad &...(ii)\\ 2x - 2y + 3z &= 2 \quad &...(iii)\end{aligned}\right\} \qquad ...(1)$$

How do we solve it ? We first try to eliminate one of the variables, say x. To do this, we add to equation (*ii*), (– 3) times equation (*i*) and add to (*iii*), (– 2) times (*i*). The resultant system is

$$\left.\begin{aligned} x + 2y - z &= 3\\ -7y + 5z &= -8\\ -6y + 5z &= -4\end{aligned}\right\} \qquad ...(2)$$

Clearly, the systems (1) and (2) are equivalent. Let us eliminate y between the last two equations of (2). To do this, we add to the third equation. $\left(\frac{-6}{7}\right)$ times the second equation. Then, the system (2) reduces to

$$\left.\begin{aligned} x + 2y - z &= 3 \\ -7y + 5z &= -8 \\ \frac{5}{7}z &= \frac{20}{7} \end{aligned}\right\} \quad ...(3)$$

To avoid fraction, we multiply the third equation by 7 and get the equivalent system

$$\left.\begin{aligned} x + 2y - z &= 3 \\ -7y + 5z &= -8 \\ 5z &= 20 \end{aligned}\right\} \quad ...(4)$$

All the above systems (1) to (4) have the same solution set. We obtain the solution set easily from the system (5) by solving for z, y and x successively ; namely $z = 4$, $y = 4$, $x = -1$. Thus, the solution is $x = -1$, $y = 4$, $z = 4$.

In matrix notation the system (1) is

$$\begin{bmatrix} 1 & 2 & -1 \\ 3 & -1 & 2 \\ 2 & -2 & 3 \end{bmatrix}\begin{bmatrix} x \\ y \\ z \end{bmatrix} = \begin{bmatrix} 3 \\ 1 \\ 2 \end{bmatrix}$$

and the system (4) is $\begin{bmatrix} 1 & 2 & -1 \\ 0 & -7 & 5 \\ 0 & 0 & 5 \end{bmatrix}\begin{bmatrix} x \\ y \\ z \end{bmatrix} = \begin{bmatrix} 3 \\ -8 \\ 20 \end{bmatrix}$.

What we have done is that we have reduced the **augmented matrix**

$$\begin{bmatrix} 1 & 2 & -1 & : & 3 \\ 3 & -1 & 2 & : & 1 \\ 2 & -2 & 3 & : & 2 \end{bmatrix}$$

of system (1) to a row equivalent matrix $\begin{bmatrix} 1 & 2 & -1 & : & 3 \\ 0 & -7 & 5 & : & -8 \\ 0 & 0 & 5 & : & 20 \end{bmatrix}$ of system (4).

Thus, we arrive at the general method : *We successively perform suitable elementary row operations on the augmented matrix [A : B] and obtain equivalent systems of linear equations. We manipulate the row operations so that the co-efficient matrix is transformed either to a triangular matrix or row-echelon form, from which the consistency or otherwise of the system can be easily tested and solution can be easily written.*

Let us consider some examples.

Example 1. *Consider the system of equations* $x_1 + x_2 + x_3 = 4$, $2x_1 + 5x_2 - 2x_3 = 3$, $x_1 + 7x_2 - 7x_3 = 5$.

Sol. The given system can be written as AX = B where

$$A = \begin{bmatrix} 1 & 1 & 1 \\ 2 & 5 & -2 \\ 1 & 7 & -7 \end{bmatrix}, X = \begin{bmatrix} x_1 \\ x_2 \\ x_3 \end{bmatrix}, B = \begin{bmatrix} 4 \\ 3 \\ 5 \end{bmatrix}$$

We apply row operations to the augmented matrix [A : B]

$$[A : B] = \begin{bmatrix} 1 & 1 & 1 & : & 4 \\ 2 & 5 & -2 & : & 3 \\ 1 & 7 & -7 & : & 5 \end{bmatrix}$$

$$\underset{R_{3,1}(-1)}{\overset{R_{2,1}(-2)}{\sim}} \begin{bmatrix} 1 & 1 & 1 & : & 4 \\ 0 & 3 & -4 & : & -5 \\ 0 & 6 & -8 & : & 1 \end{bmatrix}$$

$$\overset{R_{3,2}(-2)}{\sim} \begin{bmatrix} 1 & 1 & 1 & : & 4 \\ 0 & 3 & -4 & : & -5 \\ 0 & 0 & 0 & : & 11 \end{bmatrix}$$

$$\underset{R_3\left(\frac{1}{11}\right)}{\overset{R_2\left(\frac{1}{3}\right)}{\sim}} \begin{bmatrix} 1 & 1 & 1 & : & 4 \\ 0 & 1 & -\frac{4}{3} & : & -\frac{5}{3} \\ 0 & 0 & 0 & : & 1 \end{bmatrix}$$

Thus, the system AX = B is equivalent to

$$x_1 + x_2 + x_3 = 4$$

$$x_2 - \frac{4}{3} x_3 = -\frac{5}{3}$$

$$0x_1 + 0x_2 + 0x_3 = 1 \quad \text{or} \quad 0 = 1 \text{ which is impossible.}$$

Thus, the system is inconsistent as the third equation is never satisfied.

Here, $\rho(A) = \rho_R(A) = 2$ (the number of non-zero rows) and

$\rho([A : B]) = 3$. (the number of non-zero rows]

So, $\rho(A) \neq \rho[A : B]$.

Example 2. *Consider the system*

$$5x_1 + 3x_2 + 14x_3 = 4,\ x_2 + 2x_3 = 1,\ x_1 - x_2 + 2x_3 = 0,\ 2x_1 + x_2 + 6x_3 = 2.$$

Sol. The augmented matrix [A : B] is $\begin{bmatrix} 5 & 3 & 14 & : & 4 \\ 0 & 1 & 2 & : & 1 \\ 1 & -1 & 2 & : & 0 \\ 2 & 1 & 6 & : & 2 \end{bmatrix}$

$$\overset{R_{1,3}}{\sim} \begin{bmatrix} 1 & -1 & 2 & : & 0 \\ 0 & 1 & 2 & : & 1 \\ 5 & 3 & 14 & : & 4 \\ 2 & 1 & 6 & : & 2 \end{bmatrix}$$

$$\underset{R_{4,1}(-2)}{\overset{R_{3,1}(-5)}{\sim}} \begin{bmatrix} 1 & -1 & 2 & : & 0 \\ 0 & 1 & 2 & : & 1 \\ 0 & 8 & 4 & : & 4 \\ 0 & 3 & 2 & : & 2 \end{bmatrix}$$

$$\underset{R_{4,2}(-3)}{\overset{R_{3,2}(-8)}{\sim}} \begin{bmatrix} 1 & -1 & 2 & : & 0 \\ 0 & 1 & 2 & : & 1 \\ 0 & 0 & -12 & : & -4 \\ 0 & 0 & -4 & : & -1 \end{bmatrix}$$

$$\overset{R_3\left(-\frac{1}{12}\right)}{\sim} \begin{bmatrix} 1 & -1 & 2 & : & 0 \\ 0 & 1 & 2 & : & 1 \\ 0 & 0 & 1 & : & \frac{1}{3} \\ 0 & 0 & -4 & : & -1 \end{bmatrix}$$

$$\overset{R_{4,3}(4)}{\sim} \begin{bmatrix} 1 & -1 & 2 & : & 0 \\ 0 & 1 & 2 & : & 1 \\ 0 & 0 & 1 & : & \frac{1}{3} \\ 0 & 0 & 0 & : & \frac{1}{3} \end{bmatrix}, \quad \overset{R_4(3)}{\sim} \begin{bmatrix} 1 & -1 & 2 & : & 0 \\ 0 & 1 & 2 & : & 1 \\ 0 & 0 & 1 & : & \frac{1}{3} \\ 0 & 0 & 0 & : & 1 \end{bmatrix}$$

The given system is equivalent to

$$x_1 - x_2 + 2x_3 = 0$$
$$x_2 + 2x_3 = 1$$
$$x_3 = \frac{1}{3}$$
$$0x_1 + 0x_2 + 0x_3 = 1 \quad \text{or } 0 = 1 \text{ which is impossible.}$$

Thus the system is inconsistent as the fourth equation is never satisfied.
Here $\rho(A) = 3$, $\rho([A : B]) = 4$ so $\rho(A) \neq \rho([A : B])$.

Example 3. *Consider the equations* $x_1 + 2x_2 - x_3 = 6,\ 3x_1 - x_2 - 2x_3 = 3,\ 4x_1 + 3x_2 + x_3 = 9.$

Sol. The augmented matrix is

$$[A : B] = \begin{bmatrix} 1 & 2 & -1 & : & 6 \\ 3 & -1 & -2 & : & 3 \\ 4 & 3 & 1 & : & 9 \end{bmatrix}$$

$$\underset{R_{3,1}(-4)}{\overset{R_{2,1}(-3)}{\sim}} \begin{bmatrix} 1 & 2 & -1 & : & 6 \\ 0 & -7 & 1 & : & -15 \\ 0 & -5 & 5 & : & -15 \end{bmatrix}$$

$$\overset{R_2(2)}{\sim} \begin{bmatrix} 1 & 2 & -1 & : & 6 \\ 0 & -14 & 2 & : & -30 \\ 0 & -5 & 5 & : & -15 \end{bmatrix}$$

$$\overset{R_{2,3}(-3)}{\sim} \begin{bmatrix} 1 & 2 & -1 & : & 6 \\ 0 & 1 & -13 & : & 15 \\ 0 & -5 & 5 & : & -15 \end{bmatrix}$$

$$\underset{\sim}{R_{3,2}(5)}\begin{bmatrix} 1 & 2 & -1 & : & 6 \\ 0 & 1 & -13 & : & 15 \\ 0 & 0 & -60 & : & 60 \end{bmatrix}$$

$$\underset{\sim}{R_3\left(-\frac{1}{60}\right)}\begin{bmatrix} 1 & 2 & -1 & : & 6 \\ 0 & 1 & -13 & : & 15 \\ 0 & 0 & 1 & : & -1 \end{bmatrix}$$

The given system is equivalent to

$$x_1 + 2x_2 - x_3 = 6$$
$$x_2 - 13x_3 = 15$$
$$x_3 = -1.$$

Solving these equations successively, we have

$$x_3 = -1, x_2 = 15 + 13x_3 = 2$$

and $\quad x_1 = 6 - 2x_2 + x_3 = 6 - 4 - 1 = 1$

Thus, $\quad x_1 = 1, x_2 = 2, x_3 = -1$ is the only (unique) solution.

Here, $\quad \rho(A) = 3 = \rho([A : B])$. (number of non-zero rows)

Example 4. *Consider the system of equations $x_1 + 2x_2 + 2x_3 = 1$, $2x_1 + x_2 + x_3 = 2$, $3x_1 + 2x_2 + 2x_3 = 3$, $x_2 + x_3 = 0$.*

Sol. The augmented matrix is

$$[A : B] = \begin{bmatrix} 1 & 2 & 2 & : & 1 \\ 2 & 1 & 1 & : & 2 \\ 3 & 2 & 2 & : & 3 \\ 0 & 1 & 1 & : & 0 \end{bmatrix}$$

$$\underset{R_{3,1}(-3)}{\overset{R_{2,1}(-2)}{\sim}}\begin{bmatrix} 1 & 2 & 2 & : & 1 \\ 0 & -3 & -3 & : & 0 \\ 0 & -4 & -4 & : & 0 \\ 0 & 1 & 1 & : & 0 \end{bmatrix}$$

$$\underset{R_3\left(-\frac{1}{4}\right)}{\overset{R_2\left(-\frac{1}{3}\right)}{\sim}}\begin{bmatrix} 1 & 2 & 2 & : & 1 \\ 0 & 1 & 1 & : & 0 \\ 0 & 1 & 1 & : & 0 \\ 0 & 1 & 1 & : & 0 \end{bmatrix}$$

$$\underset{R_{4,2}(-1)}{\overset{R_{1,2}(-2),\, R_{3,2}(-1)}{\sim}}\begin{bmatrix} 1 & 0 & 0 & : & 1 \\ 0 & 1 & 1 & : & 0 \\ 0 & 0 & 0 & : & 0 \\ 0 & 0 & 0 & : & 0 \end{bmatrix}$$

which is in row-echelon form.

Thus, the system is equivalent to $\quad x_1 = 1$

$$x_2 + x_3 = 0$$

Let $\quad x_2 = \lambda \quad \therefore \lambda + x_3 = 0 \text{ or } x_3 = -\lambda$

The solution set of this system is $x_1 = 1$, $x_2 = \lambda$, $x_3 = -\lambda$, where λ is any real number. Hence the system has infinitely many solutions for different values of λ.

Here, $\rho(A) = \rho([A : B]) = 2$(< number of unknowns).

The above examples clearly show that a given system of equations AX = B is consistent if $\rho(A) = \rho([A : B])$. The result is true, in general and is given in the following theorem due to Kronecker and Capalli.

33. Theorem

A system AX = B of m linear equations in n unknowns is consistent if and only if the coefficient matrix A and the augmented matrix [A : B] of the system have the same rank.

Proof. Since A is a submatrix of the augmented matrix

$$[A : B],\ \rho(A) \le \rho([A : B]).$$

Now, by applying elementary row operations to the augmented matrix [A : B], we reduce it to row reduced echelon matrix

$$\begin{array}{r} \\ \\ \\ \\ r\text{th row} \\ \\ \\ \end{array}\begin{bmatrix} 1 & 0 & 0 & 0 & : & \alpha_1 \\ 0 & 1 & & 0 & : & \alpha_2 \\ 0 & 0 & & 0 & : & \alpha_3 \\ \vdots & \vdots & & & & \vdots \\ 0 & 0 & & 1 & : & \alpha_r \\ \vdots & \vdots & & \vdots & \vdots & \vdots \\ 0 & 0 & & 0 & : & \alpha_m \end{bmatrix}$$

If at least one of $\alpha_{r+1}, \alpha_{r+2}, \ldots\ldots, \alpha_m$ is not zero, then the number of non-zero rows in the coefficient matrix is not equal to the number of non-zero rows in the augmented matrix.

Hence A and [A : B] do not have the same rank,

i.e., $$\rho(A) < \rho([A : B]).$$

But, this shows that at least one of the equations is $\alpha_j = 0$, $(j = r + 1, r + 2, \ldots, m)$ which is false. Hence, the system is inconsistent.

If $\alpha_{r+1} = \alpha_{r+2} = \ldots = \alpha_m = 0$, then $\rho(A) = \rho([A : B])$ and the system has one or **more solutions.**

34. Theorem

In a consistent system AX = B of rank r < n (number of unknowns) n – r of the unknowns can be assigned values arbitrarily and the other r unknowns can be expressed uniquely in terms of the n – r arbitrarily chosen unknowns.

Proof. Proof follows from the fact that when [A : B] is reduced to the row-echelon form $[A_R : C]$, then $[A_R : C]$ has exactly r non-zero rows.

35. Theorem

A system AX = B of n non-homogeneous equations in n unknowns has a unique solution provided A is non-singular, i.e. $\rho(A) = n$.

Proof. Let A be non-singular.

Then, using elementary row operations, [A : B] can be reduced to [I : C], where I is the identity matrix of order n. Then, X = C is a solution of the system.

Let, if possible, X = D be a second solution of the system.

Then, $AD = B$

Also, $AC = B$

$\Rightarrow \quad AD = AC \quad \ldots(1)$

Since A is non-singular, A^{-1} exists.

Pre-multiplying both sides of (1) by A^{-1}, we have

$$A^{-1}(AD) = A^{-1}(AC)$$

$$\Rightarrow \quad (A^{-1}A)D = (A^{-1}A)C$$

$$\Rightarrow \quad ID = IC \Rightarrow D = C$$

So, the solution is unique.

36. Finally, we **summarise :**

The system AX = B where B ≠ 0 of m linear non-homogeneous equations in n unknowns has

(*i*) **no solution if $\rho(A) \neq \rho([A : B])$**

(*ii*) **a unique solution if $\rho(A) = \rho([A : B]) = n$**

(*iii*) **an infinite number of solutions**

if $\rho(A) = \rho([A : B]) = r < n$

In this case, the solution set involves *n—r* parameters.

Example 1. *Examine the consistency of the following equations and if consistent, find the complete solution.* $x + y + z = 6$, $x + 2y + 3z = 14$, $x + 4y + 7z = 30$.

Sol. Augmented matrix is

$$[A : B] = \begin{bmatrix} 1 & 1 & 1 & : & 6 \\ 1 & 2 & 3 & : & 14 \\ 1 & 4 & 7 & : & 30 \end{bmatrix}$$

$$\underset{R_{3,1}(-1)}{\overset{R_{2,1}(-1)}{\sim}} \begin{bmatrix} 1 & 1 & 1 & : & 6 \\ 0 & 1 & 2 & : & 8 \\ 0 & 3 & 6 & : & 24 \end{bmatrix}$$

$$\overset{R_{3,2}(-3)}{\sim} \begin{bmatrix} 1 & 1 & 1 & : & 6 \\ 0 & 1 & 2 & : & 8 \\ 0 & 0 & 0 & : & 0 \end{bmatrix} \text{ which is row-Echelon form.}$$

Clearly, $\rho(A) = \rho([A : B]) = 2.$

Hence, the system is consistent. (By Art. 36)

Since $\rho(A) < 3$ (number of unknowns), the system has an **infinite** number of solutions. (By Art. 36)

Now the system reduces to

$$x + y + z = 6$$
$$y + 2z = 8$$

$$\Rightarrow \quad y = 8 - 2z, x = 6 - y - z = 6 - (8 - 2z) - z = z - 2$$

Taking $z = \lambda$, we have

$x = \lambda - 2, y = 8 - 2\lambda, z = \lambda$ (where λ is arbitrary) as the general solution.

Example 2. *Discuss the consistency of the system of equations*

$$x - y + z + 1 = 0$$
$$x - y + z - 1 = 0$$
$$x - y - z + 1 = 0.$$

Sol. The equations can be written as

$$x - y + z = -1$$
$$x - y + z = 1$$
$$x - y - z = -1.$$

The augmented matrix is

$$[A : B] = \begin{bmatrix} 1 & -1 & 1 & : & -1 \\ 1 & -1 & 1 & : & 1 \\ 1 & -1 & -1 & : & -1 \end{bmatrix}$$

$$\underset{R_{3,1}(-1)}{\overset{R_{2,1}(-1)}{\sim}} \begin{bmatrix} 1 & -1 & 1 & : & -1 \\ 0 & 0 & 0 & : & 2 \\ 0 & 0 & -2 & : & 0 \end{bmatrix}$$

$$\overset{R_{3,2}}{\sim} \begin{bmatrix} 1 & -1 & 1 & : & -1 \\ 0 & 0 & -2 & : & 0 \\ 0 & 0 & 0 & : & 2 \end{bmatrix}$$

$$\overset{R_2(-1/2)}{\sim} \begin{bmatrix} 1 & -1 & 1 & : & -1 \\ 0 & 0 & 1 & : & 0 \\ 0 & 0 & 0 & : & 2 \end{bmatrix} \text{ which is row-echelon form.}$$

Clearly $\rho(A) = 2$ and $\rho([A : B]) = 3$. (number of non-zero rows)

Since $\rho(A) \neq \rho([A : B])$, so the system is not consistent. (By Art. 36)

Example 3. *Solve the equations* $2x_1 - x_2 + 3x_3 = 3$, $x_1 + 2x_2 - x_3 - 5x_4 = 4$, $x_1 + 3x_2 - 2x_3 - 7x_4 = 5$.

Sol. The augmented matrix is

$$[A : B] = \begin{bmatrix} 2 & -1 & 3 & 0 & : & 3 \\ 1 & 2 & -1 & -5 & : & 4 \\ 1 & 3 & -2 & -7 & : & 5 \end{bmatrix}$$

$$\overset{R_{1,2}}{\sim} \begin{bmatrix} 1 & 2 & -1 & -5 & : & 4 \\ 2 & -1 & 3 & 0 & : & 3 \\ 1 & 3 & -2 & -7 & : & 5 \end{bmatrix}$$

$$\underset{R_{3,1}(-1)}{\overset{R_{2,1}(-2)}{\sim}} \begin{bmatrix} 1 & 2 & -1 & -5 & : & 4 \\ 0 & -5 & 5 & 10 & : & -5 \\ 0 & 1 & -1 & -2 & : & 1 \end{bmatrix}$$

$$\overset{R_2(-1/5)}{\sim} \begin{bmatrix} 1 & 2 & -1 & -5 & : & 4 \\ 0 & 1 & -1 & -2 & : & 1 \\ 0 & 1 & -1 & -2 & : & 1 \end{bmatrix}$$

$$\underset{\sim}{R_{3,2}(-1)} \begin{bmatrix} 1 & 2 & -1 & -5 & : & 4 \\ 0 & 1 & -1 & -2 & : & 1 \\ 0 & 0 & 0 & 0 & : & 0 \end{bmatrix}$$ which is row-echelon form.

$\therefore$ $\rho(A) = \rho([A : B]) = 2$ (number of non-zero rows)

$\therefore$ the system is consistent. (By Art. 36)

Since $\rho(A) = 2 < 4$ (number of unknowns), the number of solutions is infinite (By Art 36) and the solutions will contain $4 - 2 = 2$ arbitrary constants.

The system reduces to $x_1 + 2x_2 - x_3 - 5x_4 = 4$

and $x_2 - x_3 - 2x_4 = 1$

Taking $x_3 = a$ and $x_4 = b$, we have

$x_2 = 1 + x_3 + 2x_4 = 1 + a + 2b$

and $x_1 = 4 - 2x_2 + x_3 + 5x_4 = 4 - 2(1 + a + 2b) + a + 5b = 2 - a + b$

Here, $x_1 = 2 - a + b$, $x_2 = 1 + a + 2b$, $x_3 = a$, $x_4 = b$ (where a, b are arbitrary constants) is the general solution.

Example 4. *For what value of* λ, *the system* $\begin{bmatrix} 1 & 2 \\ 3 & \lambda \end{bmatrix} \begin{bmatrix} x \\ y \end{bmatrix} = \begin{bmatrix} 1 \\ 3 \end{bmatrix}$ *has (i) no solution (ii) unique solution (iii) more than one solution ?*

Sol. The augmented matrix is

$$[A : B] = \begin{bmatrix} 1 & 2 & : & 1 \\ 3 & \lambda & : & 3 \end{bmatrix}$$

$$\underset{\sim}{R_{2,1(-3)}} \begin{bmatrix} 1 & 2 & : & 1 \\ 0 & \lambda - 6 & : & 0 \end{bmatrix} \qquad ...(1)$$

Case I. $\lambda - 6 \neq 0$ **i.e.** $\lambda \neq 6$

Operating $R_2\left(\dfrac{1}{\lambda - 6}\right)$ in (1), from $[A : B] \sim \begin{bmatrix} 1 & 2 & : & 1 \\ 0 & 1 & : & 0 \end{bmatrix}$ which is row-echelon form.

Here $\rho(A) = \rho[A : B] = 2$ (= number of unknowns).

$\therefore$ The given system has a unique solution. (By Art. 36)

The system reduces to $x + 2y = 1$ and $y = 0$

Putting $y = 0$, we have $x = 1$

$\therefore$ The unique solution is $x = 1, y = 0$.

Case II. $\lambda - 6 = 0$ **i.e.** $\lambda = 6$

$\therefore$ From (1) $[A : B] \sim \begin{bmatrix} 1 & 2 & : & 1 \\ 0 & 0 & : & 0 \end{bmatrix}$ which is row-echelon form.

Here $\rho(A) = \rho[A : B] = 1$ (Number of non-zero rows) < 2 (number of unknowns).

Therefore the given system of equations is consistent and has infinitely many solutions.

The reduced equations are $x + 2y = 1$.

Let $y = \lambda$ $\therefore$ $x + 2\lambda = 1$ $\therefore$ $x = 1 - 2\lambda$

Hence the general solution is $x = 1 - 2\lambda$, $y = \lambda$ where λ is an arbitary constant.

Hence the system is consistent for all values of λ.

Example 5. *Discuss the solution of the system* $\begin{bmatrix} 1 & 2 \\ 3 & \lambda \end{bmatrix}\begin{bmatrix} x \\ y \end{bmatrix} = \begin{bmatrix} 1 \\ 1 \end{bmatrix}$ *for different values of* λ.

Sol. Augmented matrix is $[A : B] = \begin{bmatrix} 1 & 2 & \vdots & 1 \\ 3 & \lambda & \vdots & 1 \end{bmatrix}$

$$\overset{R_{2,1}(-3)}{\sim} \begin{bmatrix} 1 & 2 & \vdots & 1 \\ 0 & \lambda - 6 & \vdots & -2 \end{bmatrix} \quad ...(1)$$

Case I. $\lambda - 6 \neq 0$ *i.e.* $\lambda \neq 6$

Operating $R_2\left(\dfrac{1}{\lambda - 6}\right)$ in (1),

$$[A : B] \sim \begin{bmatrix} 1 & 2 & : & 1 \\ 0 & 1 & : & -\dfrac{2}{\lambda - 6} \end{bmatrix}, \text{ which is row-echelon form.}$$

Here $\rho(A) = \rho(A : B) = 2$ (= number of unknowns).

$\therefore$ The given system has a unique solution. (By Art. 36)

The given system of equations reduces to $x + 2y = 1$ and $y = -\dfrac{2}{\lambda - 6}$

$$\therefore \quad x = 1 - 2y = 1 + \frac{4}{\lambda - 6} = \frac{\lambda - 6 + 4}{\lambda - 6} = \frac{\lambda - 2}{\lambda - 6}$$

Hence $x = \dfrac{\lambda - 2}{\lambda - 6}$ and $y = -\dfrac{2}{\lambda - 6}$ is the unique solution for each $\lambda \neq 6$.

Case II. $\lambda - 6 = 0$ *i.e.* $\lambda = 6$

$\therefore$ From (1) $[A : B] \sim \begin{bmatrix} 1 & 2 & : & 1 \\ 0 & 0 & : & -2 \end{bmatrix}$

Operating $R_3\left(\dfrac{-1}{2}\right)$, $[A : B] \sim \begin{bmatrix} 1 & 2 & : & 1 \\ 0 & 0 & : & 1 \end{bmatrix}$, which is row-echelon form.

Here $\rho(A) (= 1) \neq \rho(A : B) (= 2)$

$\therefore$ The system is inconsistent if $\lambda = 6$.

$\therefore$ For no λ, the system has infinitely many of solutions.

Example 6. *For what* λ, *the equations* $x + y + z = 1$, $x + 2y + 4z = \lambda$, $x + 4y + 10z = \lambda^2$ *have a solution and solve completely in each case.*

Sol. The augmented matrix is

$$[A : B] = \begin{bmatrix} 1 & 1 & 1 & : & 1 \\ 1 & 2 & 4 & : & \lambda \\ 1 & 4 & 10 & : & \lambda^2 \end{bmatrix}$$

$$\underset{R_{3,1}(-1)}{\overset{R_{2,1}(-1)}{\sim}} \begin{bmatrix} 1 & 1 & 1 & : & 1 \\ 0 & 1 & 3 & : & \lambda - 1 \\ 0 & 3 & 9 & : & \lambda^2 - 1 \end{bmatrix}$$

$$\underset{\sim}{R_{3,2}(-3)} \begin{bmatrix} 1 & 1 & 1 & : & 1 \\ 0 & 1 & 3 & : & \lambda-1 \\ 0 & 0 & 0 & : & \lambda^2-3\lambda+2 \end{bmatrix} \quad ...(1)$$

Thus, $\rho(A) = 2.$

$\therefore$ The system is consistent if $\rho([A : B]) = \rho(A) = 2.$

i.e., if $\lambda^2 - 3\lambda + 2 = 0$ which implies $\lambda = 1, 2.$

Case (*i*) $\lambda = 1$, then from (1), the system of equations reduces to $x + y + z = 1, y + 3z = 0.$

Taking $z = a$(some real number), $y = -3a$, $x = 1 - y - z = 1 + 3a - a = 1 + 2a$

Thus, the general solution is given by $x = 1 + 2a, y = -3a, z = a$, where a is arbitrary.

Case (*ii*) $\lambda = 2$, then the system reduces to $x + y + z = 1, y + 3z = 2 - 1 = 1$

Taking $z = a, y = 1 - 3a, x = 1 - y - z = 1 - (1 - 3a) - a = 2a.$

Thus, the general solution is given by $x = 2a, y = 1 - 3a, z = a$, where a is arbitrary.

Example 7. *For what values of* λ, μ *the system of equations :*

$$x + y + z = 6$$
$$x + 2y + 3z = 10$$
$$x + 2y + \lambda z = \mu$$

has (i) no solution, (ii) a unique solution, (iii) an infinite number of solutions.

Sol. The augmented matrix of the system is

$$[A : B] = \begin{bmatrix} 1 & 1 & 1 & : & 6 \\ 1 & 2 & 3 & : & 10 \\ 1 & 2 & \lambda & : & \mu \end{bmatrix}$$

$$\underset{R_{3,1}(-1)}{\overset{R_{2,1}(-1)}{\sim}} \begin{bmatrix} 1 & 1 & 1 & : & 6 \\ 0 & 1 & 2 & : & 4 \\ 0 & 1 & \lambda-1 & : & \mu-6 \end{bmatrix}$$

$$\underset{\sim}{R_{3,2}(-1)} \begin{bmatrix} 1 & 1 & 1 & : & 6 \\ 0 & 1 & 2 & : & 4 \\ 0 & 0 & \lambda-3 & : & \mu-10 \end{bmatrix} \quad ...(1)$$

Case I. $\lambda - 3 \neq 0$ ***i.e.*** $\lambda \neq 3$

operating $R_3\left(\frac{1}{\lambda-3}\right)$ in (1), $[A : B] \sim \begin{bmatrix} 1 & 1 & 1 & : & 6 \\ 0 & 1 & 2 & : & 4 \\ 0 & 0 & 1 & : & \frac{\mu-10}{\lambda-3} \end{bmatrix}$ which is row-echelon form.

Here $\rho(A) = \rho[A : B] = 3$ for all values of μ ($\because$ Number of non-zero rows is 3)

= Number of unknowns.

$\therefore$ The given system is consistent and has unique solution for $\lambda \neq 3$ (and for all values of μ)

Case II. $\lambda - 3 = 0$ **i.e.** $\lambda = 3$

$\therefore$ From (1) $$[A : B] \sim \begin{bmatrix} 1 & 1 & 1 & : & 6 \\ 0 & 1 & 2 & : & 4 \\ 0 & 0 & 0 & : & \mu-10 \end{bmatrix} \quad ...(2)$$

$$\rho(A) = 2 \qquad \text{(Number of non-zero rows)}$$

for all values of μ.

Sub-Case I. $\mu - 10 \neq 0$ *i.e.* $\mu \neq 10$

operating $R_3\left(\frac{1}{\mu - 10}\right)$ in (2) $\quad [A : B] \sim \begin{bmatrix} 1 & 1 & 1 & : & 6 \\ 0 & 1 & 2 & : & 4 \\ 0 & 0 & 0 & : & 1 \end{bmatrix}$

which is row-echelon form of [A : B].

$$\rho[A : B] = 3 \qquad \text{(Number of non-zero rows)}$$
$$\neq \rho(A) \ (= 2)$$

$\therefore$ The given system is inconsistent *i.e.* has no solution.

for $\qquad \lambda = 3$ and $\mu \neq 10$.

Sub-Case II. $\mu - 10 = 0$ *i.e.* $\mu = 10$

$\therefore$ From Eqn. (2) $\quad [A : B] = \begin{bmatrix} 1 & 1 & 1 & : & 6 \\ 0 & 1 & 2 & : & 4 \\ 0 & 0 & 0 & : & 0 \end{bmatrix}$ which is row-echelon form.

$$\rho\,[A : B] = 2 \qquad \text{(Number of non-zero rows)}$$
$$= \rho(A) = 2 = \text{Number of unknowns.}$$

$\therefore$ The given system has an infinite number of solutions for $\lambda = 3$ and $\mu = 10$.

Example 8. *Solve the equations :* $\lambda x + 2y - 2z = 1, \quad 4x + 2\lambda y - z = 2, \quad 6x + 6y + \lambda z = 3$ *considering specially the case when* $\lambda = 2$.

Sol. The matrix form of these equations is AX = B

where $\quad A = \begin{bmatrix} \lambda & 2 & -2 \\ 4 & 2\lambda & -1 \\ 6 & 6 & \lambda \end{bmatrix}, X = \begin{bmatrix} x \\ y \\ z \end{bmatrix}$ and $B = \begin{bmatrix} 1 \\ 2 \\ 3 \end{bmatrix}$

Because a_{11} = unknown λ here.

So we shall find it difficult to reduce [A : B] to row-Echelon form.

So let us find $\quad |A| = \begin{vmatrix} \lambda & 2 & -2 \\ 4 & 2\lambda & -1 \\ 6 & 6 & \lambda \end{vmatrix}$

Expanding by first row

$$|A| = \lambda(2\lambda^2 + 6) - 2\,(4\lambda + 6) - 2\,(24 - 12\lambda)$$
$$= 2\lambda^3 + 6\lambda - 8\lambda - 12 - 48 + 24\lambda$$
$$= 2\lambda^3 + 22\lambda - 60 = 2\,[\lambda^3 + 11\lambda - 30]$$
$$= 2\,[\lambda^3 - 4\lambda + 15\lambda - 30] = 2\,[\lambda(\lambda^2 - 4) + 15(\lambda - 2)]$$
$$= 2\,[\lambda(\lambda - 2)(\lambda + 2) + 15(\lambda - 2)]$$

or $\qquad |A| = 2\,(\lambda - 2)(\lambda^2 + 2\lambda + 15) \qquad$ ...(1)

Now on solving $\quad \lambda^2 + 2\lambda + 15 = 0$, we have

$$\lambda = \frac{-2 \pm \sqrt{4 - 60}}{2} = \frac{-2 \pm \sqrt{-56}}{2} = \frac{-2 \pm 2\sqrt{-14}}{2} = -1 \pm i\,\sqrt{14}$$

These values of λ are imaginary and hence rejected.

We know by Art. 35 that a system AX = B of n non-homogeneous equations in n unknowns has a **unique** solution if A is non-singular

i.e. if $|A| \neq 0$ *i.e.* if $\lambda \neq 2$ (from (1))

if $\lambda = 2$; then from (1) $|A| = 0$

Solution for $\lambda = 2$

Putting $\lambda = 2$ in the given equations, the system reduces to :

$$2x + 2y - 2z = 1, \quad 4x + 4y - z = 2, \quad 6x + 6y + 2z = 3$$

Augmented matrix is $[A : B] = \begin{bmatrix} 2 & 2 & -2 & : & 1 \\ 4 & 4 & -1 & : & 2 \\ 6 & 6 & 2 & : & 3 \end{bmatrix}$

$$\underset{R_{3,1}(-3)}{\overset{R_{2,1}(-2)}{\sim}} \begin{bmatrix} 2 & 2 & -2 & : & 1 \\ 0 & 0 & 3 & : & 0 \\ 0 & 0 & 8 & : & 0 \end{bmatrix}$$

$$\overset{R_{3,2}(-8/3)}{\sim} \begin{bmatrix} 2 & 2 & -2 & : & 1 \\ 0 & 0 & 3 & : & 0 \\ 0 & 0 & 0 & : & 0 \end{bmatrix}$$ which is Row-Echelon form.

Here $\rho(A) = \rho(A : B) = 2$ (Number of non-zero rows) < (3, the number of unknowns)

$\therefore$ An infinite number of solutions.

The reduced system of equations is $2x + 2y - 2z = 1, 3z = 0$ or $z = 0$

Taking $y = k, 2x + 2k - 0 = 1$ or $2x = 1 - 2k$

$\therefore$ $x = \frac{1}{2} - k, y = k, z = 0$

Thus the system is consistent for $\lambda = 2$ and there exists an infinite number of solutions :

$x = \frac{1}{2} - k, y = k, z = 0$, where k is an arbitrary.

EXERCISE 5

1. Discuss the consistency of the following system of equations. Find the solution set, if consistent

(*i*) $x_1 - 2x_2 + x_3 = 1$

(*ii*) $x_1 + x_2 + x_3 = 4$
$2x_1 + 5x_2 - 2x_3 = 3$

(*iii*) $x - 3y + z = 2$
$2x + y + 3z = 3$
$x + 5y + 5z = 2$

(*iv*) $x_1 + 3x_2 - x_3 = 4$
$2x_1 + x_2 + x_3 = 7$
$2x_1 - 4x_2 + 4x_3 = 6$
$3x_1 + 4x_2 = 11$

(*v*) $x + 2y - 5z + 9 = 0$
$3x - y + 2z + 5 = 0$
$2x + 3y - z - 3 = 0$
$4x - 5y + z + 3 = 0$

(*vi*) $x + 2y + 3z + 4t = 0$
$2x + 3y + 4z - 1 = 0$
$3x + 4y + t = 2$
$4x + z + 2t = 3$

(*vii*) $x + y + 2z + t = 5$
$2x + 3y - z - 2t = 2$
$4x + 5y + 3z = 7.$

2. Show that the following equations are consistent and solve the same :

$$\begin{aligned} x + 2y - 5z &= -9 \\ 3x - y + 2z &= 5 \\ 2x + 3y - z &= 3 \\ 4x - 5y + z &= -3. \end{aligned}$$

3. If $A = \begin{bmatrix} 2 & 1 & 1 \\ 1 & 1 & 1 \\ 3 & 1 & 1 \end{bmatrix}$, $B = \begin{bmatrix} b_1 \\ b_2 \\ b_3 \end{bmatrix}$, then show that the system AX = B is consistent if and only if $b_3 - 2b_1 + b_2 = 0$, where $X = \begin{bmatrix} x \\ y \\ z \end{bmatrix}$.

4. For what value of λ, does the system $\begin{bmatrix} -1 & 2 & 1 \\ 3 & -1 & 2 \\ 0 & 1 & \lambda \end{bmatrix}\begin{bmatrix} x \\ y \\ z \end{bmatrix} = B$ have (*a*) no solution (*b*) unique solution (*c*) more than one solution if :

(*i*) $B = \begin{bmatrix} 1 \\ 1 \\ 1 \end{bmatrix}$ (*ii*) $B = \begin{bmatrix} 1 \\ 1 \\ 0 \end{bmatrix}$?

5. Discuss for all values of λ, the system of equations $x + y + 4z = 6$, $x + 2y - 2z = 6$, $\lambda x + y + z = 6$ as regards the existence and nature of solutions.

6. For what value of r, the system of equations $x - 2y - z + 1 = 0$, $3x - y + 2z - 1 = 0$, $y + rz = 1$ has (*i*) no solution (*ii*) unique solution (*iii*) more than one solution.

7. For what values of a and b, the equations $x + y + 5z = 0$, $x + 2y + 3az - b = 0$, $x + 3y + az = 1$ have (*i*) no solution (*ii*) unique solution (*iii*) infinitely many solutions.

8. Show that the equations $-2x + y + z = a$, $x - 2y + z = b$, $x + y - 2z = c$ have no solution unless $a + b + c = 0$, in which case, they have infinitely many solutions. Find these solutions when $a = 1$, $b = 1$, $e = -2$.

$$\left[\textbf{Hint. } [A : B] \sim \begin{bmatrix} 1 & -2 & 1 & : & b \\ 0 & -3 & 3 & : & a + 2b \\ 0 & 0 & 0 & : & a + b + c \end{bmatrix}\right]$$

ANSWERS

1. (*i*) Consistent ; $x_1 = 1 + 2a - b$, $x_2 = a$, $x_3 = b$

(*ii*) Consistent ; $x_1 = -\dfrac{7a}{3} + \dfrac{17}{3}$, $x_2 = \dfrac{4a}{5} - \dfrac{5}{3}$, $x_3 = a$

(*iii*) Consistent ; $\left(1 - \dfrac{1}{5}, \dfrac{2}{5}\right)$ (*iv*) Consistent ; $x_1 = \dfrac{17}{5} - \dfrac{4}{5}x_3$, $x_2 = \dfrac{1}{5} + \dfrac{3}{5}x_3$

(*v*) Inconsistent (*vi*) Consistent ; $\left(\dfrac{9}{11}, \dfrac{1}{11}, -\dfrac{1}{11}, -\dfrac{1}{11}\right)$

(*vii*) Inconsistent **2.** $\dfrac{1}{2}, \dfrac{3}{2}, \dfrac{5}{2}$

4. (*i*) (*a*) $\lambda = 1$ (*b*) $\lambda \neq 1$, unique solution (*c*) never
(*ii*) (*a*) $\lambda = 1$, (*b*) $\lambda \neq 1$, (*c*) never

5. Unique solution if $\lambda \neq \frac{7}{10}$, otherwise inconsistent

6. (*i*) $r = 1$ (*ii*) $r \neq 1$ (*iii*) never

7. (*i*) $a = 1, b \neq \frac{1}{2}$ (*ii*) $a \neq 1$ and for all real b (*iii*) $a = 1, b = \frac{1}{2}$

8. $x = -1 + k, y = -1 + k, z = k$ where k is any real number.

---

37. System of Homogeneous Equations

Let AX = O be a system of m homogeneous equations in n unknowns. Since the rank of the coefficient matrix A and the augmented matrix [A : O] are same, the system is always consistent. This is also otherwise obvious, since $x_1 = x_2 = \ldots = x_n = 0$ is always a solution. However, we are interested in finding all the solutions.

Case (*i*) ρ(A) = r = n. In this case (By Art. 36) the system has X = O as the only solution. i.e. unique solution

Case (*ii*) ρ(A) = *r* < *n*. Hence By Art. 36, an infinite number of solutions, one of them zero solution and all the remaining non-zero solutions.

Hence $n—r$ variables can be selected and assigned arbitrary values and hence there are infinite number of solutions.

If $\rho(A) = r$ and $r \leq m < n$, $n—r$ variables can be selected and assigned arbitrary values.

Therefore, when the number of equations is less than the number of variables, the equations will always have an infinite number of solutions.

Example 1. *Solve the equations* $x + 2y + 3z + 4t = 0$, $8x + 5y + z + 4t = 0$, $5x + 6y + 8z + t = 0$, $8x + 3y + 7z + 2t = 0$.

Sol. Given equations are homogeneous and their matrix form is AX = O.

Here,
$$A = \begin{bmatrix} 1 & 2 & 3 & 4 \\ 8 & 5 & 1 & 4 \\ 5 & 6 & 8 & 1 \\ 8 & 3 & 7 & 2 \end{bmatrix}$$

$$\underset{R_{4,1}(-8)}{\overset{R_{2,1}(-8)}{\underset{R_{3,1}(-5)}{\sim}}} \begin{bmatrix} 1 & 2 & 3 & 4 \\ 0 & -11 & -23 & -28 \\ 0 & -4 & -7 & -19 \\ 0 & -13 & -17 & -30 \end{bmatrix}$$

$$\underset{R_{4,3}(-3)}{\overset{R_{2,3}(-3)}{\sim}} \begin{bmatrix} 1 & 2 & 3 & 4 \\ 0 & 1 & -2 & 29 \\ 0 & -4 & -7 & -19 \\ 0 & -1 & 4 & 27 \end{bmatrix}$$

$$\underset{R_{4,2}(1)}{\overset{R_{3,2}(4)}{\sim}}\begin{bmatrix}1 & 2 & 3 & 4\\0 & 1 & -2 & 29\\0 & 0 & -15 & 97\\0 & 0 & 2 & 56\end{bmatrix}$$

$$\overset{R_3\left(-\frac{1}{15}\right)}{\sim}\begin{bmatrix}1 & 2 & 3 & 4\\0 & 1 & -2 & 29\\0 & 0 & 1 & -97/15\\0 & 0 & 2 & 56\end{bmatrix}$$

$$\overset{R_{4,3}(-2)}{\sim}\begin{bmatrix}1 & 2 & 3 & 4\\0 & 1 & -2 & 29\\0 & 0 & 1 & -97/15\\0 & 0 & 0 & 1034/15\end{bmatrix}$$

$$\overset{R_4\left(-\frac{15}{1034}\right)}{\sim}\begin{bmatrix}1 & 2 & 3 & 4\\0 & 1 & -2 & 29\\0 & 0 & 1 & -97/15\\0 & 0 & 0 & 1\end{bmatrix}$$

∴ ρ(A) = 4 = number of variables.

Hence, by Art. 37 (*i*) the equations have a unique solution which is the trivial solution

$$x = 0, y = 0, z = 0, t = 0.$$

Example 2. *Solve the equations* $x + 3y - 2z = 0,\quad 2x - y + 4z = 0,\quad x - 11y + 14z = 0.$

Sol. Given equations are homogeneous and their matrix form is AX = O.

Here,
$$A = \begin{bmatrix}1 & 3 & -2\\2 & -1 & 4\\1 & -11 & 14\end{bmatrix}$$

$$\underset{R_{3,1}(-1)}{\overset{R_{2,1}(-2)}{\sim}}\begin{bmatrix}1 & 3 & -2\\0 & -7 & 8\\0 & -14 & 16\end{bmatrix}$$

$$\overset{R_{3,2}(-2)}{\sim}\begin{bmatrix}1 & 3 & -2\\0 & -7 & 8\\0 & 0 & 0\end{bmatrix}$$

Thus, ρ(A) = 2 < 3 (the number of variables)

∴ By Art. 37 (*ii*) the system has infinitely many solutions.

The system reduces to $x + 3y - 2z = 0,\ -7y + 8z = 0$

$$\Rightarrow \quad y = \frac{8}{7}z, \text{ and hence } x + 3\left(\frac{8}{7}z\right) - 2z = 0 \quad \text{or} \quad x = -\frac{24}{7}z + 2z = -\frac{10}{7}z$$

Taking $z = k$, any scalar, the general solution is given by $x = -\frac{10}{7}k, y = \frac{8}{7}k, z = k$.

Example 3. *Solve the equations $x_1 + x_2 + x_3 + x_4 = 0, x_1 + 3x_2 + 2x_3 + 4x_4 = 0, 2x_1 + x_3 - x_4 = 0$.*

Sol. Given equations are homogeneous and their matrix form is AX = O.

Here, $$A = \begin{bmatrix} 1 & 1 & 1 & 1 \\ 1 & 3 & 2 & 4 \\ 2 & 0 & 1 & -1 \end{bmatrix}$$

$$\underset{R_{3,1}(-2)}{\overset{R_{2,1}(-1)}{\sim}} \begin{bmatrix} 1 & 1 & 1 & 1 \\ 0 & 2 & 1 & 3 \\ 0 & -2 & -1 & -3 \end{bmatrix}$$

$$\underset{R_{3,2}(1)}{\sim} \begin{bmatrix} 1 & 1 & 1 & 1 \\ 0 & 2 & 1 & 3 \\ 0 & 0 & 0 & 0 \end{bmatrix}$$

$$R_2\left(\tfrac{1}{2}\right) \sim \begin{bmatrix} 1 & 1 & 1 & 1 \\ 0 & 1 & \frac{1}{2} & \frac{3}{2} \\ 0 & 0 & 0 & 0 \end{bmatrix}$$

Here, $\rho(A) = \rho_R(A) = 2$ and number of variables is 4.

∴ The system has infinitely many solutions and the general solution contains 4 – 2 = 2 arbitrary constants.

The system is equivalent to

$$x_1 + \frac{1}{2}x_3 - \frac{1}{2}x_4 = 0 \quad \text{i.e.,} \quad x_1 = -\frac{1}{2}x_3 + \frac{1}{2}x_4$$

$$x_2 + \frac{1}{2}x_3 + \frac{3}{2}x_4 = 0 \quad \text{i.e.,} \quad x_2 = -\frac{1}{2}x_3 - \frac{3}{2}x_4$$

Let $x_3 = a, x_4 = b$, then $x_2 = -\frac{1}{2}a - \frac{3}{2}b$ and $x_1 = -\frac{1}{2}a + \frac{1}{2}b$

Thus, the complete solution is

$$x_1 = -\frac{1}{2}a + \frac{1}{2}b, x_2 = -\frac{1}{2}a - \frac{3}{2}b \quad \text{and} \quad x_3 = a, x_4 = b.$$

Example 4. *For what value of λ, the system $\begin{bmatrix} 1 & 2 \\ 3 & \lambda \end{bmatrix}\begin{bmatrix} x \\ y \end{bmatrix} = \begin{bmatrix} 0 \\ 0 \end{bmatrix}$ has (i) a unique solution (ii) more than one solution ?*

Sol. Given equations are homogeneous and their matrix form is AX = O.

Here, $$A = \begin{bmatrix} 1 & 2 \\ 3 & \lambda \end{bmatrix}$$

operate $$\overset{R_{2,1}(-3)}{\sim} \begin{bmatrix} 1 & 2 \\ 0 & \lambda - 6 \end{bmatrix} \quad ...(1)$$

Case I. $\lambda - 6 \neq 0$ **i.e.** $\lambda \neq 6$

Operating $R_2\left(\frac{1}{\lambda-6}\right)$ in (1), $A \sim \begin{bmatrix} 1 & 2 \\ 0 & 1 \end{bmatrix}$ which is row-echelon form.

$$\rho(A) = 2 \quad \text{(number of non-zero rows)}$$
$$= n \text{ (Number of unknowns } x \text{ and } y)$$

$\therefore$ By Art. 37(*i*), the system has unique solution and that is trival solution ($x = 0, y = 0$)

Case II. $\lambda - 6 = 0$ **i.e.** $\lambda = 6$

Putting $\lambda = 6$ in (1), $A \sim \begin{bmatrix} 1 & 2 \\ 0 & 0 \end{bmatrix}$ which is row-echelon form.

$\rho(A) = 1 < 2$, number of unknowns

$\therefore$ By Art. 37 (*ii*), the system has an infinite number of solutions and hence more than one solution.

The reduced system of equations is $x + 2y = 0$.

Let $y = k$. Therefore $x + 2k = 0$ or $x = -2k$

$\therefore$ $x = -2k$ and $y = k$ is the general solution (k being arbitrary)

Example 5. *Find the value of k such that the system of equations* $x + ky + 3z = 0$, $4x + 3y + kz = 0$, $2x + y + 2z = 0$ *has a nontrivial solution.*

Sol. Given equations are homogeneous and their matrix form is AX = O.

Here $$A = \begin{bmatrix} 1 & k & 3 \\ 4 & 3 & k \\ 2 & 1 & 2 \end{bmatrix}$$

Let us operate $R_{1,3}$ to bring the unknown k to a lower place in the matrix.

$$\therefore \quad A \sim \begin{bmatrix} 2 & 1 & 2 \\ 4 & 3 & k \\ 1 & k & 3 \end{bmatrix} \underset{R_{3,1}\left(-\frac{1}{2}\right)}{\overset{R_{2,1}(-2)}{\sim}} \begin{bmatrix} 2 & 1 & 2 \\ 0 & 1 & k-4 \\ 0 & k-\frac{1}{2} & 2 \end{bmatrix}$$

$$\overset{R_{3,2}\left(-\left(k-\frac{1}{2}\right)\right)}{\sim} \begin{bmatrix} 2 & 1 & 2 \\ 0 & 1 & k-4 \\ 0 & 0 & 2-(k-4)\left(k-\frac{1}{2}\right) \end{bmatrix}$$

$\therefore$ for non-trivial solution, $\rho(A) < 3$ (number of unknowns) (clearly $\rho(A) \geq 2$)

$$\Rightarrow \quad 2 - (k-4)\left(k - \tfrac{1}{2}\right) = 0.$$

$$\Rightarrow \quad 2k^2 - 9k = 0 \quad \Rightarrow \quad k = 0, \frac{9}{2}.$$

EXERCISE 6

1. Solve the following systems of homogeneous equations :

(*i*) $3x - y + z = 0$
$-15x + 6y - 5z = 0$
$5x - 2y + 2z = 0$

(*ii*) $x - 2y + 3z = 0$
$2x + 5y + 6z = 0$

(*iii*) $2x - y + 3z = 0$
$3x + 2y + z = 0$
$x - 4y + 5z = 0$

(*iv*) $x + y + 3z = 0$
$x - y + z = 0$
$x - 2y = 0$
$x - 2y + z = 0$

(*v*) $x + y + z + t = 0$
$x + 3y - 2z + t = 0$
$2x - 3z + 2t = 0$
$x + y + t = 0$

(*vi*) $x + y - z + t = 0$
$x - y + 2z - t = 0$
$3x + y + t = 0.$

2. For what value of λ, does the system $\begin{bmatrix} -1 & 2 & 1 \\ 3 & -1 & 2 \\ 0 & 1 & \lambda \end{bmatrix}\begin{bmatrix} x \\ y \\ z \end{bmatrix} = O$ have (*i*) a unique solution (*ii*) more than one solution.

3. Show that the only real value of λ for which the following equations have non-zero solutions is 6
$x + 2y + 3z = \lambda x,\ 3x + y + 2z = \lambda y,\ 2x + 3y + z = \lambda z.$

$$\left[\textbf{Hint. } |A| = \begin{vmatrix} 1-\lambda & 2 & 3 \\ 3 & 1-\lambda & 2 \\ 2 & 3 & 1-\lambda \end{vmatrix} = (6-\lambda)(\lambda^2 + 3\lambda + 3) = 0 \Rightarrow \lambda = 6,\ \frac{-3 \pm i\sqrt{3}}{2}.\right]$$

4. Discuss for all values of λ the system of equations :

$$2x + 3\lambda y + (3\lambda + 4)z = 0$$
$$x + (\lambda + 4)y + (4\lambda + 2)z = 0$$
$$x + 2(\lambda + 1)y + (3\lambda + 4)z = 0.$$

ANSWERS

1. (*i*) $x = 0, y = 0, z = 0$ (*ii*) $x = -3a, y = 0, z = a$ where $a \in R$

(*iii*) $x = -y = -z = a$ (*iv*) $x = 0, y = 0, z = 0$

(*v*) $x = -a, y = 0, z = 0, t = a$ (*vi*) $x = -\frac{1}{2}b, y = \frac{3}{2}b - a, z = b, t = a$

2. (*i*) Unique solution if $\lambda \neq 1$ (*ii*) More than 1 solution if $\lambda = 1$

4. $\lambda \neq \pm 2$ null sol.

If $\lambda = 2$, then $x = 0, y = -5t, z = 3t$

If $\lambda = -2$, then $x = 4t, y = t, z = t$.

---

38. Linear Independence of Row and Column Matrices

Recall that a three dimensional vector $\vec{a} = a_1\hat{i} + a_2\hat{i} + a_3\hat{k}$ is also written as an ordered triplet (a_1, a_2, a_3) and vice versa. Generalization of this to n-dimension is an ordered n-triple $(a_1, a_2, ..., a_n)$ which is called an n-dimensional vector. Likewise, a matrix $X = [a_1, a_2 \ldots a_n]_{1 \times n}$ (called a row-matrix) can be considered an n-dimensional vector and is written as $X = (a_1, a_2, ..., a_n)$

$$\text{A matrix } X = \begin{bmatrix} a_1 \\ a_2 \\ \vdots \\ a_n \end{bmatrix}_{n \times 1} \qquad \text{(called a column-matrix)}$$

is also an n-dimensional vector and is written as $[a_1, a_2, ..., a_n]$.

When vectors are considered as row matrices or column matrices, the operations of addition of vectors will have the same properties as addition of matrices.

39. Linear Independence and Dependence of row matrices *n-row matrices $X_1, X_2, ..., X_r$ (each of type $1 \times n$) are said to be linearly dependent if there exists r scalars $\alpha_1, \alpha_2, ..., \alpha_r$ not all zero (i.e. at least one of $\alpha_1, \alpha_2 ... \alpha_r$ is non-zero) such that*

$$\alpha_1 X_1 + \alpha_2 X_2 + ... + \alpha_r X_r = O.$$

If $X_1, X_2, ..., X_r$ are not linearly dependent, we call $X_1, X_2, ..., X_r$ as linearly independent.

Note that if $X_1, X_2 ... X_r$ are linearly independent and for some scalars $\alpha_1, \alpha_2, ..., \alpha_r$,

$$\alpha_1 X_1 + \alpha_2 X_2 + ... + \alpha_r X_r = O$$

then $\alpha_1 = 0, \alpha_2 = 0, ..., \alpha_r = 0$ is the only possible set of values of $\alpha_1, \alpha_2, ... \alpha_r$.

Likewise one can define the linear independence (L.I.) and linear dependence (L.D.) of column matrices.

Example 1. *Examine the linear independence of the following sets of row-matrices :*

(i) (1, 0), (0, 1) *(ii) (1, 1, 1), (1, 2, 3), (0, 1, 2)*

(iii) (1, 1, 1), (1, 2, 3), (3, 3, 4) *(iv) (1, 2, 3), (1, 0, 0), (0, 2, 3)*

Sol. (*i*) Consider $a(1, 0) + b(0, 1) = O$ for some scalars a, b.

$\Rightarrow$ $(a, 0) + (0, b) = (0, 0) \Rightarrow a = 0, b = 0$

$\therefore$ (1, 0) and (0, 1) are linearly independent.

(*ii*) Consider $a(1, 1, 1) + b(1, 2, 3) + c(0, 1, 2) = O$

for some scalars a, b, c

$\Rightarrow$ $(a, a, a) + (b, 2b, 3b) + (0, c, 2c) = 0$

$\Rightarrow$ $(a + b, a + 2b + c, a + 3b + 2c) = (0, 0, 0)$

Equating corresponding entries

$\Rightarrow$ $a + b = 0$ ...(1)

$a + 2b + c = 0$ ...(2)

$a + 3b + 2c = 0$ ...(3)

Let us solve (1), (2), (3) simultaneously for a, b and c

From (1), $b = -a$,

Putting $b = -a$ in Eqn (2), $a - 2a + c = 0$ or $c = a$.

Putting $b = -a$ and $c = a$ in Eqn. (3), we have

$a - 3a + 2a = 0$ or $0 = 0$ which is satisfied for each a

$\therefore$ $b = -a, c = a$ is a solution for each a.

(In particular, $a = 1, b = -1, c = 1$ is a solution)

Hence, By Art. 39 given vectors (row matrices) are linearly dependent.

Second Method. Matrix form of Homogeneous Equations (1), (2) and (3) in three unknowns a, b, c is AX = O, where

$$A = \begin{bmatrix} 1 & 1 & 0 \\ 1 & 2 & 1 \\ 1 & 3 & 2 \end{bmatrix} \text{ and } X = \begin{bmatrix} a \\ b \\ c \end{bmatrix}$$

$$|A| = \begin{vmatrix} 1 & 1 & 0 \\ 1 & 2 & 1 \\ 1 & 3 & 2 \end{vmatrix} = 1(4-3) - 1(2-1) + 0(3-2) = 1 - 1 = 0$$

(Expanding by first row)

$\therefore$ Matrix A is singular

Hence $\rho(A) < n\ (= 3)$

$\therefore$ By Art. 37 (*ii*), an infinite number of sets of values of a, b, c (both zero and non-zero) exist

$\therefore$ The given row matrices are linearly dependent. (By Art. 39)

(*iii*) Consider $a\ (1, 1, 1) + b(1, 2, 3) + c(3, 3, 4) = O$ for some scalars a, b, c.

$\therefore$ $a + b + 3c = 0\ ,\ a + 2b + 3c = 0$ and $a + 3b + 4c = 0$

These are 3 homogeneous equations in 3 unknowns a, b, c. Matrix form is AX = O.

Now, $$|A| = \begin{vmatrix} 1 & 1 & 3 \\ 1 & 2 & 3 \\ 1 & 3 & 4 \end{vmatrix} = 1(8-9) - 1(4-3) + 3(3-2) = 1 \neq 0$$

$\therefore$ Matrix A is non-singular $\therefore$ $\rho(A) = n(= 3)$

$\therefore$ $a = 0, b = 0, c = 0$ is the only solution.

Hence, the given vectors are linearly independent.

(*iv*) Consider $a(1, 2, 3) + b(1, 0, 0) + c(0, 2, 3) = O$ for some scalars a, b, c.

or $(a, 2a, 3a) + (b, 0, 0) + (0, 2c, 3c) = O$

or $(a + b, 2a + 2c, 3a + 3c) = O = (0, 0, 0)$

Equating corresponding entries, we have

$a + b = 0$ or $a + b + 0c = 0$

$2a + 2c = 0$ or $2a + 0b + 2c = 0$

$3a + 3c = 0$ or $3a + 0b + 3c = 0$

Matrix form of these three homogeneous equations in three unknowns a, b, c is AX = O,

where $$A = \begin{bmatrix} 1 & 1 & 0 \\ 2 & 0 & 2 \\ 3 & 0 & 3 \end{bmatrix} \text{ and } X = \begin{bmatrix} a \\ b \\ c \end{bmatrix}$$

$$|A| = \begin{vmatrix} 1 & 1 & 0 \\ 2 & 0 & 2 \\ 3 & 0 & 3 \end{vmatrix} = 1(0-0) - 1(6-6) + 0 = 0$$

(Expanding by first row)

$\therefore$ *Matrix A is singular.*

$\therefore$ $\rho(A) < 3\ (= n)$

$\therefore$ By Art. 37 (*ii*) $\exists$ non-zero values (also) of a, b, c.

$\therefore$ The given row matrices are linearly dependent.

Example 2. *Find a if the column matrices* $\begin{bmatrix} 1 \\ -1 \\ 3 \end{bmatrix}, \begin{bmatrix} 1 \\ 2 \\ -3 \end{bmatrix}$ *and* $\begin{bmatrix} a \\ 0 \\ 1 \end{bmatrix}$ *are linearly dependent.*

Sol. Consider $\alpha \begin{bmatrix} 1 \\ -1 \\ 3 \end{bmatrix} + \beta \begin{bmatrix} 1 \\ 2 \\ -3 \end{bmatrix} + \gamma \begin{bmatrix} a \\ 0 \\ 1 \end{bmatrix} = O$ for some scalars α, β, γ.

or $\begin{bmatrix} \alpha \\ -\alpha \\ 3\alpha \end{bmatrix} + \begin{bmatrix} \beta \\ 2\beta \\ 3\beta \end{bmatrix} + \begin{bmatrix} a\gamma \\ 0\gamma \\ \gamma \end{bmatrix} = O$ or $\begin{bmatrix} \alpha + \beta + a\gamma \\ -\alpha + 2\beta + 0\gamma \\ 3\alpha - 3\beta + \gamma \end{bmatrix} = \begin{bmatrix} 0 \\ 0 \\ 0 \end{bmatrix}$

Equating corresponding entries, we have

$$\alpha + \beta + a\gamma = 0, \; -\alpha + 2\beta + 0\gamma = 0, \quad 3\alpha - 3\beta + \gamma = 0$$

Matrix form of these equations is AX = O

where $$A = \begin{bmatrix} 1 & 1 & a \\ -1 & 2 & 0 \\ 3 & -3 & 1 \end{bmatrix} \text{ and } X = \begin{bmatrix} \alpha \\ \beta \\ \gamma \end{bmatrix}$$

Because the given column matrices are linearly dependent, therefore, there exist non-zero values of α, β, γ.

$\therefore$ By Art 37 (*ii*) $\qquad \rho(A) < n\ (= 3)$.

$\therefore$ Matrix A is singular *i.e.* $\qquad |A| = 0$

i.e. $$\begin{vmatrix} 1 & 1 & a \\ -1 & 2 & 0 \\ 3 & -3 & 1 \end{vmatrix} = 0$$

Expanding by first row, $1(2 - 0) - 1(-1 - 0) + a(3 - 6) = 0$ or $3 - 3a = 0$ or $3a = 3$

$\therefore$ $\qquad a = 1$.

40. The k n-vectors $X_1, X_2, ..., X_k$ are linearly dependent if and only if the rank of the matrix $A = [X_1\ X_2 \ldots X_k]$ with given vectors as columns is less than k.

Proof. Consider the relation $\alpha_1 X_1 + \alpha_2 X_2 + \ldots + \alpha_k X_k = O$ ...(1)

for some scalars $\alpha_1, \alpha_2, ..., \alpha_k$.

If $$X_i = \begin{bmatrix} a_{1i} \\ a_{2i} \\ \vdots \\ a_{ni} \end{bmatrix}, \; i = 1, 2, ..., k, \text{ then}$$

(1) becomes $$\alpha_1\begin{bmatrix}a_{11}\\a_{21}\\\vdots\\a_{n1}\end{bmatrix}+\alpha_2\begin{bmatrix}a_{12}\\a_{22}\\\vdots\\a_{n2}\end{bmatrix}+\ldots+\alpha_k\begin{bmatrix}a_{1k}\\a_{2k}\\\vdots\\a_{nk}\end{bmatrix}=O$$

$$\Rightarrow \quad \begin{bmatrix}a_{11}\alpha_1+a_{12}\alpha_2+\ldots+a_{1k}\alpha_k\\a_{21}\alpha_1+a_{22}\alpha_2+\ldots+a_{2k}\alpha_k\\\ldots\ldots\ldots\ldots\ldots\ldots\ldots\ldots\\a_{n1}\alpha_1+a_{n2}\alpha_2+\ldots+a_{nk}\alpha_k\end{bmatrix}=O$$

$$\Rightarrow \quad \begin{bmatrix}a_{11}&a_{12}&\ldots&a_{1k}\\a_{21}&a_{22}&\ldots&a_{2k}\\\ldots&\ldots&\ldots&\ldots\\a_{n1}&a_{n2}&\ldots&a_{nk}\end{bmatrix}\begin{bmatrix}\alpha_1\\\alpha_2\\\vdots\\\alpha_k\end{bmatrix}=O \Rightarrow AX=O \qquad \ldots(2)$$

(*i*) **Let $X_1, X_2, \ldots, X_k$ be linearly dependent**. Then, from (1) scalars $\alpha_1, \alpha_2, \ldots, \alpha_k$ are not all zero.

$\therefore$ the homogeneous linear equations given by (2) have a non-null solution. $\therefore$ $\rho(A) < k$.

(*ii*) **Conversely, let $\rho(A) < k$**

Then, the homogeneous equations given by (2) have a non-null solution.

$\Rightarrow$ $X \neq O \Rightarrow$ $\alpha_1, \alpha_2, \ldots, \alpha_k$ are not all zero.

$\therefore$ from (1), vectors $X_1, X_2, \ldots, X_k$ are linearly dependent.

Cor. The k n-vectors $X_1, X_2, \ldots, X_k$ are linearly independent if the rank of the matrix $A = [X_1, X_2, \ldots, X_k]$ is equal to k.

(Proceed as in the above theorem and the result follows from (2))

41. Theorem.

A square matrix A is singular if and only if its columns (rows) are linearly dependent.

Proof. Let A be an *n*-square matrix having $X_1, X_2, \ldots, X_n$ as its columns.

i.e., $$A = [X_1\ X_2\ \ldots\ X_n]$$

Consider the relation

$$\alpha_1X_1 + \alpha_2X_2 + \ldots + \alpha_nX_n = O. \qquad \ldots(1)$$

for some scalars $\alpha_1, \alpha_2, \ldots, \alpha_n$

Proceeding as in the theorem 40, we have

$$\begin{bmatrix}a_{11}&a_{12}&\ldots&a_{1n}\\a_{21}&a_{22}&\ldots&a_{2n}\\\vdots&\vdots&\vdots&\vdots\\a_{n1}&a_{n2}&\ldots&a_{nn}\end{bmatrix}\begin{bmatrix}\alpha_1\\\alpha_2\\\vdots\\\alpha_n\end{bmatrix}=O$$

or $$AX = O \qquad \ldots(2)$$

(*i*) *Let the vectors $X_1, X_2, \ldots, X_n$ be linearly dependent.*

$\therefore$ from (1) scalars $\alpha_1, \alpha_2, \ldots, \alpha_n$ are not all zero. Therefore the homogeneous equations given by (2) have a non-null solution.

$\therefore \quad \rho(A) < n$

$\Rightarrow \quad |A| = 0 \Rightarrow$ A is singular.

(ii) Conversely, let A be singular.

$\Rightarrow \quad |A| = 0$

$\Rightarrow$ From (2), the equations have a non-null solution.

$\Rightarrow$ $X \neq O$.

$\therefore$ The scalars $\alpha_1, \alpha_2, ..., \alpha_n$ are not all zero.

Hence, from (1), $X_1, X_2, ..., X_n$ are linearly dependent.

EXERCISE 7

1. Examine for linear independence or dependence of the following sets of vectors :

   (i) {(1, 2, 3), (2, – 2, 0)} *(ii)* {(1, 2, 3), (3, – 2, 1), (1, – 6, – 5)}

   (iii) {(1, 3, 2), (5, – 2, 1), (– 7, 13, 14)} *(iv)* {(1, 1, 1), (1, 2, 3), (2, 3, 8)}

   (v) {(2, 3, 1), (– 1, 4, – 2), (1, 18, – 4)}

   (vi) {(3, 0, – 3), (– 1, 1, 2), (4, 2, – 2), (2, 1, 1)}.

2. Show that three row vectors as well as the three column vectors of the matrix $\begin{bmatrix} 2 & 3 & 1 \\ 7 & -6 & 17 \\ 5 & 2 & 7 \end{bmatrix}$ are linearly dependent.

ANSWERS

1. *(i)* L.I. *(ii)* L.D. *(iii)* L.D. *(iv)* L.I. *(v)* L.D *(vi)* L.D.

5

Characteristic Roots and Characteristic Vectors

1. Definitions

Let $A = [a_{ij}]$ be an n-square matrix and λ some scalar.

(*i*) **Characteristic Matrix.** *The matrix*

$$A - \lambda I = \begin{bmatrix} a_{11} - \lambda & a_{12} & \ldots\ldots & a_{1n} \\ a_{21} & a_{22} - \lambda & \ldots\ldots & a_{2n} \\ \ldots & \ldots & \ldots\ldots & \ldots\ldots \\ a_{n1} & a_{n2} & \ldots\ldots & a_{nn} - \lambda \end{bmatrix}$$

is called the characteristic matrix of A.

(*ii*) **Characteristic Polynomial or Function.** *The determinant*

$$|A - \lambda I| = \begin{vmatrix} a_{11} - \lambda & a_{12} & \ldots\ldots & a_{1n} \\ a_{21} & a_{22} - \lambda & \ldots\ldots & a_{2n} \\ \ldots & \ldots & \ldots\ldots & \ldots \\ a_{n1} & a_{n2} & \ldots\ldots & a_{nn} - \lambda \end{vmatrix}$$

when expanded is a polynomial $\phi(\lambda)$ of degree n in λ and is called the characteristic polynomial (or function) of the matrix A.

(*iii*) **Characteristic Equation.** *The equation $|A - \lambda I| = 0$ (i.e., $\phi(\lambda) = 0$) is called the characteristic equation of A.*

(*iv*) **Characteristic Roots.** *The roots of the characteristic equation $|A - \lambda I| = 0$ are called the characteristic roots of matrix A. Characteristic roots are also called* **characteristic values, eigen values, latent roots or proper values.**

Since $|A - \lambda I| = 0$ is of degree n, every n-square matrix A has n characteristic roots, which of course, may not all be different and may not be all real.

(*v*) **Spectrum.** *The set of characteristic roots of a matrix A is called the spectrum of the matrix A.*

(*vi*) **Characteristic Vectors.** *If λ is a characteristic root of an n-square matrix A, then a non-zero solution $X = (x_1, x_2 \ldots\ldots, x_n)$ of the matrix equation $AX = \lambda X$ is called a* ***characteristic vector*** *of matrix A.*

A characteristic vector of the matrix A is also called latent vector, eigen vector, invarient vector or **pole** *of the matrix A.*

Remark : In fact $X = \begin{bmatrix} x_1 \\ x_2 \\ \vdots \\ x_n \end{bmatrix}$ is a column matrix. To save space we write it as $X = [x_1\ x_2 \ldots x_n]$

2. Theorem

λ is a characteristic root of a matrix A if and only if there exists a non-zero vector X such that AX = λX.

Proof. Let λ be a characteristic root of the matrix A.

$\Rightarrow \quad | A - \lambda I | = 0 \quad \Rightarrow$ The matrix $(A - \lambda I)$ is singular

$\Rightarrow$ the equation $(A - \lambda I)X = O$ has a non-zero solution X *i.e.*, there exists a non-zero vector X such that $(A - \lambda I) X = 0$ *i.e.*, $AX - \lambda I X = 0$ *i.e.*, $AX - \lambda X = 0$ *i.e.*, $AX = \lambda X$.

Conversely, let there exist a non-zero vector X such that $AX = \lambda X$.

$\Rightarrow$ there exists a non-zero solution X of the equation

$$(A - \lambda I)X = O. \text{ Therefore } \rho(A - \lambda I) < n$$

$\Rightarrow$ $A - \lambda I$ is a singular matrix.

$\Rightarrow$ the determinant of the coefficient matrix $A - \lambda I$ is zero

i.e.,
$$| A - \lambda I | = 0$$

$\Rightarrow$ λ is a characteristic root of the matrix A.

3. Theorem

A characteristic vector X of a matrix A cannot correspond to more than one characteristic root of A.

Proof. Let, if possible, a characteristic vector X of a matrix A correspond to two different characteristic roots λ_1 and λ_2.

Then,
$$AX = \lambda_1 X \quad \text{and} \quad AX = \lambda_2 X$$

$$\Rightarrow \quad \lambda_1 X = \lambda_2 X \quad \Rightarrow \quad (\lambda_1 - \lambda_2)X = O$$

$$\Rightarrow \quad \lambda_1 - \lambda_2 = 0 \qquad (\because \quad X \neq O)$$

$$\Rightarrow \quad \lambda_1 = \lambda_2.$$

However, the converse of the above theorem is false. A characteristic vector corresponding to a characteristic root is undetermined to the extent of a scalar multiplier.

For, if X is a characteristic vector of a matrix A corresponding to a characteristic root λ, then
$$AX = \lambda X.$$

$$\Rightarrow \quad A(kX) = \lambda(kX), \text{ for a scalar } k.$$

$\Rightarrow$ corresponding to a characteristic root λ, corresponds a characteristic vector kX, $\forall$ $k \neq 0$.

Example 1. *Find the characteristic roots and the spectrum of the matrix*

$$A = \begin{bmatrix} 1 & 2 & 3 \\ 0 & -4 & 2 \\ 0 & 0 & 7 \end{bmatrix}.$$

Sol.
$$| A - \lambda I | = \begin{vmatrix} 1-\lambda & 2 & 3 \\ 0 & -4-\lambda & 2 \\ 0 & 0 & 7-\lambda \end{vmatrix} = (1 - \lambda)(-4 - \lambda)(7 - \lambda)$$

$\therefore$ the characteristic equation is

$$| A - \lambda I | = (1 - \lambda)(-4 - \lambda)(7 - \lambda) = 0$$

Its roots are
$$\lambda = 1, -4, 7.$$

Thus, characteristic roots are 1, – 4, 7 and the spectrum of the matrix A is {1, – 4, 7}.

Example 2. *Find the characteristic roots and the associated characteristic vectors of the matrix*

$$A = \begin{bmatrix} 8 & -6 & 2 \\ -6 & 7 & -4 \\ 2 & -4 & 3 \end{bmatrix}.$$

Sol.
$$|A - \lambda I| = \begin{vmatrix} 8-\lambda & -6 & 2 \\ -6 & 7-\lambda & -4 \\ 2 & -4 & 3-\lambda \end{vmatrix}$$

$$= (8-\lambda)\{(3-\lambda)(7-\lambda) - 16\} + 6\{-6(3-\lambda) + 8\} + 2\{24 - 2(7-\lambda)\}$$

$$= -\lambda^3 + 18\lambda^2 - 45\lambda \qquad \text{(on simplification)}$$

∴ Characteristic equation of the matrix A is

$$|A - \lambda I| = 0 \quad \text{or} \quad -\lambda^3 + 18\lambda^2 - 45\lambda = 0$$

Its roots are $\lambda = 0, 3, 15$.

∴ Characteristic roots of the matrix A are 0, 3, 15.

To Find Characteristic Vectors

A characteristic vector of A corresponding to a characteristic root λ is a non-zero solution of (A – λI)X = O.

or
$$\begin{bmatrix} 8-\lambda & -6 & 2 \\ -6 & 7-\lambda & -4 \\ 2 & -4 & 3-\lambda \end{bmatrix}\begin{bmatrix} x_1 \\ x_2 \\ x_3 \end{bmatrix} = O, \text{ where } X = (x_1, x_2, x_3). \qquad ...(1)$$

(*i*) characteristic vector corresponding to $\lambda = 0$ is given by

$$\begin{bmatrix} 8-0 & -6 & 2 \\ -6 & 7-0 & -4 \\ 2 & -4 & 3-0 \end{bmatrix}\begin{bmatrix} x_1 \\ x_2 \\ x_3 \end{bmatrix} = O.$$

$$\Rightarrow \quad \underset{R_{2,3}(3)}{\overset{R_{1,3}(-4)}{\sim}} \begin{bmatrix} 0 & 10 & -10 \\ 0 & -5 & 5 \\ 2 & -4 & 3 \end{bmatrix}\begin{bmatrix} x_1 \\ x_2 \\ x_3 \end{bmatrix} = O$$

$$\Rightarrow \quad \underset{R_2\left(-\frac{1}{5}\right)}{\overset{R_{1,3}}{\sim}} \begin{bmatrix} 2 & -4 & 3 \\ 0 & 1 & -1 \\ 0 & 10 & -10 \end{bmatrix}\begin{bmatrix} x_1 \\ x_2 \\ x_3 \end{bmatrix} = O$$

$$\Rightarrow \quad \overset{R_{3,2}(-10)}{\sim} \begin{bmatrix} 2 & -4 & 3 \\ 0 & 1 & -1 \\ 0 & 0 & 0 \end{bmatrix}\begin{bmatrix} x_1 \\ x_2 \\ x_3 \end{bmatrix} = O$$

$$\Rightarrow \quad 2x_1 - 4x_2 + 3x_3 = 0,\ x_2 - x_3 = 0 \ \Rightarrow\ x_2 = x_3,\ x_1 = \frac{1}{2}x_3.$$

Taking $x_3 = 2k$, we find
$$X = \begin{bmatrix} k \\ 2k \\ 2k \end{bmatrix} = k\begin{bmatrix} 1 \\ 2 \\ 2 \end{bmatrix}$$

For different values of k, we have different characteristic vectors of A corresponding to the characteristic root 0.

(*ii*) When $\lambda = 3$, substituting in (1), the corresponding characteristic vector is given by

$$\begin{bmatrix} 8-3 & -6 & 2 \\ -6 & 7-3 & -4 \\ 2 & -4 & 3-3 \end{bmatrix}\begin{bmatrix} x_1 \\ x_2 \\ x_3 \end{bmatrix} = O \quad \text{or} \quad \begin{bmatrix} 5 & -6 & 2 \\ -6 & 4 & -4 \\ 2 & -4 & 0 \end{bmatrix}\begin{bmatrix} x_1 \\ x_2 \\ x_3 \end{bmatrix} = O$$

$$\Rightarrow \quad \underset{R_{2,3}(3)}{\overset{R_{1,3}(-2)}{\sim}} \begin{bmatrix} 1 & 2 & 2 \\ 0 & -8 & -4 \\ 2 & -4 & 0 \end{bmatrix}\begin{bmatrix} x_1 \\ x_2 \\ x_3 \end{bmatrix} = O$$

$$\Rightarrow \quad \overset{R_{3,1}(-2)}{\sim} \begin{bmatrix} 1 & 2 & 2 \\ 0 & -8 & -4 \\ 0 & -8 & -4 \end{bmatrix}\begin{bmatrix} x_1 \\ x_2 \\ x_3 \end{bmatrix} = O$$

$$\Rightarrow \quad \overset{R_{3,2}(-1)}{\sim} \begin{bmatrix} 1 & 2 & 2 \\ 0 & -8 & -4 \\ 0 & 0 & 0 \end{bmatrix}\begin{bmatrix} x_1 \\ x_2 \\ x_3 \end{bmatrix} = O$$

$$\Rightarrow \quad x_1 + 2x_2 + 2x_3 = 0,\ -8x_2 - 4x_3 = 0$$

$$\Rightarrow \quad x_1 = -2x_2 - 2x_3,\ x_2 = \frac{x_3}{-2} = k \text{ (say)}$$

$$\Rightarrow \quad x_1 = -2k + 4k = 2k,\ x_2 = k,\ x_3 = -2k$$

$$\Rightarrow \quad x_1 = 2k,\ x_2 = k,\ x_3 = -2k$$

$\therefore \quad X = k\begin{bmatrix} 2 \\ 1 \\ -2 \end{bmatrix}$ is a characteristic vector for each $k \neq 0$, corresponding to $\lambda = 3$.

(*iii*) When $\lambda = 15$, substituting in (1), the corresponding characteristic vector is given by

$$\begin{bmatrix} -7 & -6 & 2 \\ -6 & -8 & -4 \\ 2 & -4 & -12 \end{bmatrix}\begin{bmatrix} x_1 \\ x_2 \\ x_3 \end{bmatrix} = O$$

Applying successively $R_{1,3}(4)$, $R_{2,3}(3)$, $R_{3,1}(-2)$ and $R_{3,2}(2)$, we have

$$\sim \begin{bmatrix} 1 & -22 & -46 \\ 0 & -20 & -40 \\ 0 & 0 & 0 \end{bmatrix}\begin{bmatrix} x_1 \\ x_2 \\ x_3 \end{bmatrix} = O$$

$$\Rightarrow \quad x_1 - 22x_2 - 46x_3 = 0,\ -20x_2 - 40x_3 = 0$$

$$\Rightarrow \quad x_1 - 22x_2 - 46x_3 = 0$$

$$x_2 = -2x_3 \quad \text{or} \quad \frac{x_2}{-2} = \frac{x_3}{1} = k$$

$$\Rightarrow \quad x_2 = -2k,\ x_3 = k,\ x_1 = 22x_2 + 46x_3 = 2k$$

$\therefore \quad X = k\begin{bmatrix} 2 \\ -2 \\ 1 \end{bmatrix}$ is a characteristic vector for each $k \neq 0$, corresponding to $\lambda = 15$.

Example 3. *Find the characteristic roots and the corresponding characteristic vectors for the matrix* $A = \begin{bmatrix} 6 & -2 & 2 \\ -2 & 3 & -1 \\ 2 & -1 & 3 \end{bmatrix}$.

Sol.
$$|A - \lambda I| = \begin{vmatrix} 6-\lambda & -2 & 2 \\ -2 & 3-\lambda & -1 \\ 2 & -1 & 3-\lambda \end{vmatrix}$$

$$\overset{C_{3,2}(1)}{\sim} \begin{vmatrix} 6-\lambda & -2 & 0 \\ -2 & 3-\lambda & 2-\lambda \\ 2 & -1 & 2-\lambda \end{vmatrix}$$

$$\overset{R_{2,3}(-1)}{\sim} \begin{vmatrix} 6-\lambda & -2 & 0 \\ -4 & 4-\lambda & 0 \\ 2 & -1 & 2-\lambda \end{vmatrix}$$

$$= (2-\lambda)[(6-\lambda)(4-\lambda) - 8]$$
$$= (2-\lambda)(\lambda^2 - 10\lambda + 16) = (2-\lambda)(\lambda-2)(\lambda-8)$$

$\therefore$ The characteristic equation of the matrix A is

$$(2-\lambda)(\lambda-2)(\lambda-8) = 0 \Rightarrow \lambda = 2, 2, 8.$$

$\therefore$ The characteristic roots of A are 2, 2, 8. Characteristic vectors are non-zero solutions of $(A - \lambda I)X = O$

or
$$\begin{bmatrix} 6-\lambda & -2 & 2 \\ -2 & 3-\lambda & -1 \\ 2 & -1 & 3-\lambda \end{bmatrix} \begin{bmatrix} x_1 \\ x_2 \\ x_3 \end{bmatrix} = O \qquad ...(1)$$

corresponding to λ.

(*i*) For $\lambda = 2$, the corresponding characteristic vector is given by (Putting $\lambda = 2$ in (1))

$$\begin{bmatrix} 4 & -2 & 2 \\ -2 & 1 & -1 \\ 2 & -1 & 1 \end{bmatrix} \begin{bmatrix} x_1 \\ x_2 \\ x_3 \end{bmatrix} = O$$

$$\underset{R_{2,3}(1)}{\overset{R_{1,3}(-2)}{\sim}} \begin{bmatrix} 0 & 0 & 0 \\ 0 & 0 & 0 \\ 2 & -1 & 1 \end{bmatrix} \begin{bmatrix} x_1 \\ x_2 \\ x_3 \end{bmatrix} = O$$

$$\Rightarrow \quad 2x_1 - x_2 + x_3 = 0$$

Taking $x_2 = k_1$ and $x_3 = k_2$, we get $x_1 = \frac{1}{2}(k_1 - k_2)$

Thus, characteristic vector corresponding to $\lambda = 2$ is given by

$$X = \begin{bmatrix} \frac{1}{2}(k_1 - k_2) \\ k_1 \\ k_2 \end{bmatrix} = k_1 \begin{bmatrix} \frac{1}{2} \\ 1 \\ 0 \end{bmatrix} + k_2 \begin{bmatrix} -\frac{1}{2} \\ 0 \\ 1 \end{bmatrix} = (\tfrac{1}{2}k_1 - \tfrac{1}{2}k_2, k_1, k_2).$$

(*ii*) For $\lambda = 8$, the corresponding characteristic vector is given by (Putting $\lambda = 8$ in (1))

$$\begin{bmatrix} -2 & -2 & 2 \\ -2 & -5 & -1 \\ 2 & -1 & -5 \end{bmatrix} \begin{bmatrix} x_1 \\ x_2 \\ x_3 \end{bmatrix} = O$$

$$\begin{matrix} R_{2,1}(-1) \\ \sim \\ R_{3,1}(1) \end{matrix} \begin{bmatrix} -2 & -2 & 2 \\ 0 & -3 & -3 \\ 0 & -3 & -3 \end{bmatrix} \begin{bmatrix} x_1 \\ x_2 \\ x_3 \end{bmatrix} = O$$

$$\begin{matrix} R_{3,2}(-1) \\ \sim \end{matrix} \begin{bmatrix} -2 & -2 & 2 \\ 0 & -3 & -3 \\ 0 & 0 & 0 \end{bmatrix} \begin{bmatrix} x_1 \\ x_2 \\ x_3 \end{bmatrix} = O$$

$\Rightarrow$ $$-2x_1 - 2x_2 + 2x_3 = 0,\ -3x_2 - 3x_3 = 0$$

$\Rightarrow$ $$\frac{x_2}{-1} = \frac{x_3}{1} = k \text{ (say)},\ 2x_1 = -2x_2 + 2x_3 = 2k + 2k = 4k$$

$\Rightarrow$ $$x_1 = 2k,\ x_2 = -k,\ x_3 = k$$

Thus, $X = k \begin{bmatrix} 2 \\ -1 \\ 1 \end{bmatrix}$, for each $k \neq 0$, is a characteristic vector corresponding to $\lambda = 8$.

Example 4. *Find the characteristic roots and the corresponding characteristic vectors for the matrix* $A = \begin{bmatrix} 1 & 0 & 0 \\ 3 & 1 & 0 \\ -7 & 5 & 1 \end{bmatrix}$.

Sol. $$|A - \lambda I| = \begin{vmatrix} 1-\lambda & 0 & 0 \\ 3 & 1-\lambda & 0 \\ -7 & 5 & 1-\lambda \end{vmatrix} = (1-\lambda)^3$$

$\therefore$ the characteristic equation of A is $(1 - \lambda)^3 = 0$ and characteristic roots are $\lambda = 1, 1, 1$.

The characteristic vector corresponding to $\lambda = 1$ is given by the non-zero solution of $(A - \lambda I)X = O$, where $\lambda = 1$.

or $$\begin{bmatrix} 0 & 0 & 0 \\ 3 & 0 & 0 \\ -7 & 5 & 0 \end{bmatrix} \begin{bmatrix} x_1 \\ x_2 \\ x_3 \end{bmatrix} = O$$

$$\begin{matrix} R_{3,2}\left(\frac{7}{3}\right) \\ \sim \end{matrix} \begin{bmatrix} 0 & 0 & 0 \\ 3 & 0 & 0 \\ 0 & 5 & 0 \end{bmatrix} \begin{bmatrix} x_1 \\ x_2 \\ x_3 \end{bmatrix} = O$$

$$\begin{matrix} R_2\left(\frac{1}{3}\right), R_3\left(\frac{1}{5}\right) \\ \sim \end{matrix} \begin{bmatrix} 0 & 0 & 0 \\ 1 & 0 & 0 \\ 0 & 1 & 0 \end{bmatrix} \begin{bmatrix} x_1 \\ x_2 \\ x_3 \end{bmatrix} = O$$

$\therefore$ Non-matrix form of these equations is $x_1 = 0$, $x_2 = 0$.

Let $$x_3 = k. \quad \therefore \quad X = \begin{bmatrix} 0 \\ 0 \\ k \end{bmatrix}$$

or $X = k \begin{bmatrix} 0 \\ 0 \\ 1 \end{bmatrix}$ for $k \neq 0$ is a characteristic vector corresponding to the characteristic root 1.

Example 5. *Show that the characteristic roots of a triangular matrix are the diagonal elements.*

Sol. We prove the care for an upper triangular matrix. Similarly, one can prove for lower triangular matrix.

Let $A = \begin{bmatrix} a_{11} & a_{12} & \dots & a_{1n} \\ 0 & a_{22} & \dots & a_{2n} \\ \dots & \dots & \dots & \dots \\ 0 & 0 & \dots & a_{nn} \end{bmatrix}$ be an upper triangular matrix of order n.

The characteristic equation of A is $|A - \lambda I| = 0$

$$\Rightarrow \begin{vmatrix} a_{11}-\lambda & a_{12} & \dots & a_{1n} \\ 0 & a_{22}-\lambda & \dots & a_{2n} \\ \dots & \dots & \dots & \dots \\ 0 & 0 & \dots & a_{nn}-\lambda \end{vmatrix} = 0$$

$$\Rightarrow \quad (a_{11}-\lambda)(a_{22}-\lambda) \dots\dots (a_{nn}-\lambda) = 0$$

$\Rightarrow \lambda = a_{11}, a_{22}, \dots\dots, a_{nn}$ and these are the diagonal elements of A.

Hence, the result.

EXERCISE 1

1. Find the characteristic roots and the spectrum of the following matrices :

(*i*) $\begin{bmatrix} 1 & 1 & -2 \\ -1 & 2 & 1 \\ 0 & 1 & -1 \end{bmatrix}$ (*ii*) $\begin{bmatrix} 0 & 1 & 2 \\ 1 & 0 & -1 \\ 2 & -1 & 0 \end{bmatrix}$

(*iii*) $\begin{bmatrix} 2 & 3 & 11 \\ 0 & 3 & 17 \\ 0 & 0 & -2 \end{bmatrix}$ (*iv*) $\begin{bmatrix} a & h & g \\ 0 & b & f \\ 0 & 0 & c \end{bmatrix}$ (*v*) $\begin{bmatrix} 1 & 0 & 0 \\ 0 & 2 & 0 \\ 0 & 0 & 3 \end{bmatrix}$.

[**Hint.** It may be noted that matrices in (*iii*) and (*iv*) are upper triangular.]

2. Find the characteristic roots and the associated characteristic vectors :

(*i*) $\begin{bmatrix} -2 & -1 \\ 5 & 4 \end{bmatrix}$ (*ii*) $\begin{bmatrix} 3 & 1 & 4 \\ 0 & 2 & 6 \\ 0 & 0 & 5 \end{bmatrix}$

(*iii*) $\begin{bmatrix} 1 & 0 & -1 \\ 1 & 2 & 1 \\ 2 & 2 & 3 \end{bmatrix}$ (*iv*) $\begin{bmatrix} 2 & 1 & 1 \\ 1 & 2 & 1 \\ 0 & 0 & 1 \end{bmatrix}$

(*v*) $\begin{bmatrix} -3 & -9 & -12 \\ 1 & 3 & 4 \\ 0 & 0 & 1 \end{bmatrix}$ (*vi*) $\begin{bmatrix} 0 & 1 & 0 \\ 0 & 0 & 1 \\ 1 & -3 & 3 \end{bmatrix}$

(*vii*) $\begin{bmatrix} 3 & 1 & 1 \\ 2 & 4 & 2 \\ 1 & 1 & 3 \end{bmatrix}$.

3. Show that the characteristic roots of a diagonal matrix are given by its diagonal elements.

4. Show that the unit latent vectors of the symmetric matrix $\begin{bmatrix} 2 & 0 & 0 \\ 0 & -3 & 1 \\ 0 & 1 & -3 \end{bmatrix}$ are form (1, 0, 0),

$\left(0, \frac{1}{\sqrt{2}}, \frac{1}{\sqrt{2}}\right)$ and $\left(0, \frac{1}{\sqrt{2}}, \frac{-1}{\sqrt{2}}\right)$.

[**Hint.** To find a unit vector along (a, b, c), devide each entry by $\sqrt{a^2 + b^2 + c^2}$.]

ANSWERS

1. (*i*) {– 1, 1, 2} (*ii*) $\{2, -1 \pm \sqrt{3}\}$ (*iii*) {2, 3, – 2} (*iv*) $\{a, b, c\}$ (*v*) {1, 2, 3}

2. (*i*) – 1, (1, – 1) ; 3, (1, – 5) (*ii*) 2, (1, – 1, 0) ; 3, (1, 0, 0) ; 5, (3, 2, 1)

(*iii*) 1, (1, – 1, 0) ; 2, (2, – 1, – 2) ; 3, (1, – 1, – 2)

(*iv*) 1, k_1(1, 0, – 1) + k_2(0, 1, – 1) ; 3, (1, 1, 0) (*v*) 0, k(3, – 1, 0) ; 1, (12, – 4, – 1)

(*vi*) 1, (1, 1, 1) (*vii*) 2, k_1 (– 1, 1, 0) + k_2(– 1, 0, 1) ; 6, k(1, 2, 1).

---

4. Theorem

The characteristic roots of a square matrix A and its transpose A′ are same.

Proof. Let

$$A = \begin{bmatrix} a_{11} & a_{12} & \dots & a_{1n} \\ a_{21} & a_{22} & \dots & a_{2n} \\ \vdots & & & \\ a_{n1} & a_{n2} & \dots & a_{nn} \end{bmatrix}$$

be an n-square matrix.

$$\therefore \quad A' = \begin{bmatrix} a_{11} & a_{21} & \dots & a_{n1} \\ a_{12} & a_{22} & \dots & a_{n2} \\ \dots & \dots & \dots & \dots \\ a_{1n} & a_{2n} & \dots & a_{nn} \end{bmatrix}$$

The characteristic equation of A is given by

$$|A - \lambda I| = \begin{vmatrix} a_{11} - \lambda & a_{12} & \dots & a_{1n} \\ a_{21} & a_{22} - \lambda & \dots & a_{2n} \\ \dots & \dots & \dots & \dots \\ a_{n1} & a_{n2} & \dots & a_{nn} - \lambda \end{vmatrix} = 0 \quad \dots(1)$$

Also, the characteristic equation of A′ is given by

$$|A' - \lambda I| = \begin{vmatrix} a_{11} - \lambda & a_{21} & \dots & a_{n1} \\ a_{12} & a_{22} - \lambda & \dots & a_{n2} \\ \dots & \dots & \dots & \dots \\ a_{1n} & a_{2n} & \dots & a_{nn} - \lambda \end{vmatrix} = 0 \quad \dots(2)$$

Since the value of a determinant remains unchanged when its rows and columns are interchanged, therefore the characteristic equations (1) and (2) are same.

Hence the characteristic roots of A and A′ are same.

Remark. Though the characteristic roots of the matrices A and A′ are same, their characteristic vectors will be different unless A is symmetric.

5. Theorem

The matrices A and $P^{-1}AP$ have same characteristic roots, P being an invertible matrix of the same order as A.

Proof. $| P^{-1} AP - \lambda I | = | P^{-1} AP - \lambda P^{-1}P |$

$$= | P^{-1}(A - \lambda I)P | = | P^{-1} | . | A - \lambda I | . | P |$$

$$= | A - \lambda I | . | P^{-1}P | = | A - \lambda I | . | I | = | A - \lambda I |$$

$\therefore$ $| P^{-1}AP - \lambda I | = 0$ and $| A - \lambda I | = 0$ are same. *i.e.*, the characteristic equations of A and $P^{-1}AP$ are same. Hence, A and $P^{-1}AP$ have same characteristic roots.

6. Theorem

The characteristic roots of a Hermitian matrix are all real.

Proof. Let A be a Hermitian matrix.

$\therefore$ $$A^\theta = A \quad \text{...(1)}$$

Let λ be a characteristic root of A and X the corresponding characteristic vector.

$\therefore$ $$AX = \lambda X \quad \text{...(2)}$$

Taking conjugate transposes on both sides of eqn. (2), we have

$$(AX)^\theta = (\lambda X)^\theta \quad \text{or} \quad X^\theta A^\theta = \bar{\lambda} X^\theta$$

or $$X^\theta A = \bar{\lambda} X^\theta \qquad [\because \text{ By (1), } A^\theta = A]$$

Post-multiplying both sides by X, we have $X^\theta AX = \bar{\lambda} X^\theta X$

Putting $AX = \lambda X$ from (2), we get

$$X^\theta \lambda X = \bar{\lambda} X^\theta X$$

$\Rightarrow$ $$\lambda X^\theta X = \bar{\lambda} X^\theta X \quad \Rightarrow \quad (\lambda - \bar{\lambda}) X^\theta X = O$$

But $$X^\theta X \neq O \qquad (\because \text{ X is a non-zero vector})$$

$\therefore$ $$\lambda - \bar{\lambda} = 0 \quad i.e., \quad \lambda = \bar{\lambda}.$$ Therefore λ is real.

Cor. *The characteristic roots of a real symmetric matrix are all real.*

Proof. Let A be a real symmetric matrix.

$\therefore$ $\bar{A} = A$ and $A' = A$ $\qquad$ ($\because$ A is real and symmetric)

$\therefore$ $A^\theta = (\bar{A})' = A' = A$

$\therefore$ A is a Hermitian matrix.

$\therefore$ By Art. 6, the characteristic roots of real symmetric matrix A are all real.

7. Theorem

The characteristic roots of a skew-Hermitian matrix are either zero or purely imaginary numbers.

Proof. Let A be a skew-Hermitian matrix. Then, one can easily see that iA is Hermitian. If λ is a characteristic root of A and X the corresponding characteristic vector, then

$$AX = \lambda X$$

$\Rightarrow$ $$i(AX) = i(\lambda X) \quad \Rightarrow \quad (iA)X = (i\lambda)X$$

$\Rightarrow$ $i\lambda$ is a characteristic root of iA

$\Rightarrow$ $i\lambda$ is real $\qquad$ ($\because$ iA is a Hermitian matrix)

$\Rightarrow$ λ is either 0 or a purely imaginary number.

Cor. *The characteristic roots of a real skew-symmetric matrix are either zero or purely imaginary numbers.*

Proof. Let A be a real skew-symmetric matrix.

$\therefore$ $\overline{A} = A$ and $A' = -A$

$\therefore$ $A^{\theta} = (\overline{A})' = A' = -A$

$\therefore$ A is a skew-Hermitian Matrix

$\therefore$ By Art. 7, the characteristic roots of real symmetric matrix A are either zero or purely imaginary.

8. Theorem

If A is a n square matrix over a field F, then $0 \in F$ is a characteristic root of A iff A is singular.

Proof. Let $0 \in F$ be a characteristic root of matrix A.

$\therefore$ **$\lambda = 0$ satisfies the characteristic equation $|A - \lambda I| = 0$**

$\therefore$ $|A| = 0$

$\therefore$ Matrix A is singular.

Conversely. Let A be a singular matrix.

$$\therefore \quad |A| = 0 \text{ or } |A - 0I| = 0$$

Comparing with $|A - \lambda I| = 0$, $\lambda = 0$ is a characteristic root of the matrix A.

Cor. Therefore 0 can never be a characteristic root of a non-singular matrix A.

9. Theorem

If A is a non-singular matrix, then prove that the characteristic roots of A^{-1} are the reciprocals of the characteristic roots of A.

Proof. If λ is a characteristic root of A and X, the corresponding characteristic vector, then

$$AX = \lambda X.$$

Premultiplying by A^{-1}, we get

$$X = A^{-1}(\lambda X)$$

$\Rightarrow$ $\dfrac{1}{\lambda} X = A^{-1}X$ *i.e.,* $A^{-1}X = \dfrac{1}{\lambda} X$ $\quad$ [$\because$ A is non-singular, we have $\lambda \neq 0$]

$\Rightarrow$ $\dfrac{1}{\lambda}$ is a characteristic root of A^{-1} and X the corresponding characteristic vector.

Also, if λ is a characteristic root of A^{-1} then $\dfrac{1}{\lambda}$ is a characteristic root of

$$(A^{-1})^{-1} \text{ i.e, } A \text{ and } \lambda = \frac{1}{1/\lambda}.$$

$\therefore$ The characteristic roots of A^{-1} are the recisrocals of the characteristic roots of A.

10. Theorem

If λ is a characteristic root of a non-singular matrix A, then prove that $\dfrac{|A|}{\lambda}$ is a characteristic root of adj. A.

Proof. λ is a characteristic root of non-singular matrix A.

$\therefore$ $\lambda \neq 0$ (By cor. Art. 8) and there exists a non-zero vector X such that $AX = \lambda X$.

Pre-multiplying both sides by Adj. A, we get

$$(\text{adj. A})(\text{AX}) = (\text{adj. A})(\lambda \text{X})$$

$$\Rightarrow \quad ((\text{adj. A})\text{A})\,\text{X} = \lambda(\text{adj. A})\,\text{X}$$

$$\Rightarrow \quad (|\,\text{A}\,|\,\text{I})\,\text{X} = \lambda(\text{adj. A})\,\text{X} \qquad [\because \ (\text{adj. A})\,\text{A} = |\,\text{A}\,|\,\text{I})$$

Dividing both sides by $\lambda \neq 0$,

$$\Rightarrow \quad (\text{adj. A})\,\text{X} = \left(\frac{|\text{A}|}{\lambda}\right)\text{X}.$$

$\Rightarrow$ Comparing with BX = λX, we conclude that $\frac{|\text{A}|}{\lambda}$ is a characteristic root of the matrix adj. A.

11. Theorem

Any two characteristic vectors corresponding to two different characteristic roots of a Hermitian matrix are orthogonal.

(Note. That two vectors X_1, X_2 are said to be orthogonal if $X_1^\theta X_2 = O$.)

Proof. Let X_1, X_2 be two characteristic verctors corresponding to different characteristic roots λ_1, λ_2 of a Hermitian matrix A (*i.e.*, $A^\theta = A$)

$$\therefore \quad AX_1 = \lambda_1 X_1 \qquad ...(1)$$

and

$$AX_2 = \lambda_2 X_2 \qquad ...(2)$$

Since characteristic roots of a Hermitian matrix are real (By Art. 6), therefore, λ_1, λ_2 are real and thus $\bar{\lambda}_1 = \lambda_1$ and $\bar{\lambda}_2 = \lambda_2$.

Pre-multiplying (1) and (2) by X_2^θ and X_1^θ respectively,

$$\therefore \quad X_2^\theta AX_1 = X_2^\theta \lambda_1 X_1 \qquad ...(3)$$

and

$$X_1^\theta AX_2 = X_1^\theta \lambda_2 X_2 \qquad ...(4)$$

Taking conjugate Transposes on both sides of Eqn (3)

$$(X_2^\theta AX_1)^\theta = (\lambda_1 X_2^\theta X_1)^\theta$$

$$\Rightarrow \quad X_1^\theta A^\theta X_2 = \bar{\lambda}_1 X_1^\theta X_2 \qquad \text{(Reversal Law)}$$

$$\Rightarrow \quad X_1^\theta AX_2 = \lambda_1 X_1^\theta X_2 \qquad (\because \ A^\theta = A \text{ and } \lambda_1 \text{ is real})$$

$$\Rightarrow \quad X_1^\theta \lambda_2 X_2 = \lambda_1 X_1^\theta X_2 \qquad \text{(By (4))}$$

$$\Rightarrow \quad (\lambda_2 - \lambda_1)\,X_1^\theta X_2 = O \Rightarrow X_1^\theta X_2 = O \qquad (\because \ \lambda_1 \neq \lambda_2)$$

$\Rightarrow$ X_1 and X_2 are orthogonal.

12. Theorem

If $\lambda_1, \lambda_2, ..., \lambda_n$ are different characteristic roots of a matrix A and if $X_1, X_2, ..., X_n$ are non-zero characteristic vectors associated with there roots, then $X_1, X_2, ..., X_n$ are linearly independent.

Proof. Let if possible, $X_1, X_2, ..., X_n$ be not linearly independent. Therefore there exist scalars $a_1, a_2, ..., a_n$ not all zero such that

$$a_1 X_1 + a_2 X_2 + ... + a_n X_n = O$$

Pre-multiplying by A, we have

$$a_1 AX_1 + a_2 AX_2 + ... + a_n AX_n = O \qquad ...(1)$$

Since $$AX_1 = \lambda_1 X_1$$
$$AX_2 = \lambda_2 X_2 \qquad ...(2)$$
$$\dots\dots\dots\dots$$
$$AX_n = \lambda_n X_n$$

$\therefore$ Equation (1) becomes

$$a_1 \lambda_1 X_1 + a_2 \lambda_2 X_2 + \dots\dots + a_n \lambda_n X_n = O \qquad ...(3)$$

Again pre-multiplying (3) by A and using relations (2), we get

$$a_1 \lambda_1^2 X_1 + a_2 \lambda_2^2 X_2 + \dots\dots + a_n \lambda_n^2 X_n = O \qquad ...(4)$$

Repeating the process, we get

$$a_1 \lambda_1^3 X_1 + a_2 \lambda_2^3 X_2 + \dots\dots + a_n \lambda_n^3 X_n = O$$
$$\dots\dots\dots\dots\dots\dots\dots\dots\dots\dots$$
$$\dots\dots\dots\dots\dots\dots\dots\dots\dots\dots$$
$$a_1 \lambda_1^{n-1} X_1 + a_2 \lambda_2^{n-1} X_2 + \dots\dots + a_n \lambda_n^{n-1} X_n = O$$

System of above equations can be written in matrix form as

$$\begin{bmatrix} 1 & 1 & \dots & 1 \\ \lambda_1 & \lambda_2 & \dots & \lambda_n \\ \lambda_1^2 & \lambda_2^2 & \dots & \lambda_n^2 \\ \dots & \dots & \dots & \dots \\ \lambda_1^{n-1} & \lambda_2^{n-1} & \dots & \lambda_n^{n-1} \end{bmatrix} \begin{bmatrix} a_1 X_1 \\ a_2 X_2 \\ \dots \\ \dots \\ a_n X_n \end{bmatrix} = O \qquad ...(5)$$

As λ's are different, therefore determinant of pre-factor of L.H.S. of (5) $\neq 0$

$$\therefore \qquad \begin{bmatrix} a_1 X_1 \\ a_2 X_2 \\ \dots \\ \dots \\ a_n X_n \end{bmatrix} = O$$

which shows that $a_1 = a_2 = \dots\dots = a_n = 0$ $\qquad [\because \ X_i = O]$

which is a contradiction and therefore our supposition is wrong.

Hence $X_1, X_2, \dots\dots X_n$ are linearly independent.

EXERCISE 2

1. Prove that characteristic roots of $\overline{A}$ and of A^θ are the conjugates of the characteristic roots of A.

[**Hint.** We have $|A^\theta - \bar{\lambda} I| = |A^\theta - \bar{\lambda} I^\theta| = |(A - \lambda I)^\theta|$

$$= |\overline{(A - \lambda I)'}| = |\overline{A - \lambda I}|\,\overline{|A - \lambda I|}.$$

$\therefore$ $\bar{\lambda}$ is a characteristic root of A^θ

iff $\qquad |A^\theta - \bar{\lambda} I| = 0$ *i.e.* iff $\overline{|A - \lambda I|} = 0$

i.e. $\qquad$ iff $|A - \lambda I| = 0$

i.e. $\qquad$ iff λ is a characteristic root of matrix A.]

2. If $\lambda_1, \lambda_2, \ldots\ldots, \lambda_n$ are n characteristic roots of n-square matrix A and if k is a scalar, then prove

(*i*) $k\lambda_1, k\lambda_2, \ldots\ldots, k\lambda_n$ are characteristic roots of kA.

(*ii*) $\lambda_1 - k, \lambda_2 - k, \ldots\ldots, \lambda_n - k$ are the characteristic roots of $A - kI$.

3. If A and B are n-square matrices and A is non-singular, then show that $A^{-1}B$ and BA^{-1} have the same characteristic roots.

[**Hint.** We know that characteristic equation of the matrix $A^{-1}B$ is $|A^{-1}B - \lambda I| = 0$

or $\quad |A^{-1}BI - \lambda I| = 0$ or $|A^{-1}BI - \lambda II| = 0$

or $\quad |A^{-1}B A^{-1}A - \lambda A^{-1} AI| = 0$

or $\quad |A^{-1}(B A^{-1} - \lambda I)| |A| = 0$

or $\quad |A^{-1}| |BA^{-1} - \lambda I| |A| = 0$

or $\quad |BA^{-1} - \lambda I| |A^{-1}| |A| = 0$

or $\quad |BA^{-1} - \lambda I| |A^{-1}A| = 0$

or $\quad |BA^{-1} - \lambda I| |I| = 0$ or $|BA^{-1} - \lambda I| = 0$

which is the characteristic equation of the matrix BA^{-1}.

$\therefore$ Matrices $A^{-1}B$ and BA^{-1} have the same characteristic equation and hence same characteristic roots.]

---

13. Matric Polynomial

Recall that an expression of the form $a_nx^n + a_{n-1}x^{n-1} + \ldots\ldots + a_0$, $a_n \neq 0$ is called a polynomial in x (or a scalar polynomial in x) of degree n with coefficients $a_0, a_1, \ldots\ldots, a_n$ belonging to some number system.

Likewise, *an expression of the form* $F(x) \equiv A_nx^n + A_{n-1}x^{n-1} + \ldots\ldots + A_1x + A_0$, *where* $A_n, A_{n-1}, \ldots\ldots, A_0$ *all square matrices of same order, is called a Matric Polynomial of degree n if* $A_n \neq O$. The coefficient A_n of highest power of x, is called the leading coefficient. If each of matric coefficients $A_n, A_{n-1}, \ldots\ldots, A_0$, is of order m, we say that the matric polynomial is m-rowed (or of order m). If the matric coefficients are all over a field F, we say that the matric polynomial is over F. The symbol 'x' is called *indeterminate* and is assumed to be commutative with every matrix coefficient. The matrices themselves are matric polynomials with zero degree.

Two matric polynomials are equal iff the coefficients of like powers of x are same.

Operations of addition and multiplication on matric polynomials are to be done parallel to those on ordinary polynomials. One can easily see that the set of all n-rowed matric polynomials over any field is a ring w.r.t. the two operations of polynomial addition and multiplication.

14. Scalar Matric Polynomial

A matric polynomial is said to be a scalar matric polynomial if its coefficients are all scalar matrices.

Thus an m-rowed scalar matric polynomial is of the form

$$\begin{bmatrix} a_n & & O \\ & a_n & \\ O & & a_n \end{bmatrix} x^n + \begin{bmatrix} a_{n-1} & & O \\ & a_{n-1} & \\ O & & a_{n-1} \end{bmatrix} x^{n-1} + \ldots$$

$$+ \begin{bmatrix} a_1 & & O \\ & a_1 & \\ O & & a_1 \end{bmatrix} x + \begin{bmatrix} a_0 & & O \\ & a_0 & \\ O & & a_0 \end{bmatrix}$$

$$= a_n \mathrm{I} x^n + a_{n-1} \mathrm{I} x^{n-1} + \ldots\ldots + a_1 \mathrm{I} x + a_0 \mathrm{I}$$
$$= (a_n x^n + a_{n-1} x^{n-1} + \ldots\ldots + a_0)\mathrm{I}$$

where I is an m-rowed unit matrix and $a_n, a_{n-1} \ldots\ldots, a_0$ are all numbers.

EXISTENCE OF MINIMAL EQUATION OF A MATRIX

15. Theorem

If $m_1(x)$ and $m_2(x)$ are two scalar polynomials of lowest degree such that $m_1(A) = O$, $m_2(A) = O$, then each of $m_1(x)$ and $m_2(x)$, is a scalar multiple of the other.

Proof. Since $m_1(x)$ and $m_2(x)$ are polynomials of lowest degree, their degrees are equal. By division algorithm, there exists a number q, and a scalar polynomial $r(x)$ such that

$$m_2(x) = qm_1(x) + r(x) \qquad \ldots(1)$$

where either $r(x) = 0$ or the degree of $r(x)$ is less than the degree of $m_1(x)$.

If $r(x) \neq 0$, then from (1),

$$m_2(\mathrm{A}) = qm_1(\mathrm{A}) + r(\mathrm{A})$$

$$\Rightarrow \quad \mathrm{O} = q \,.\, \mathrm{O} + r(\mathrm{A}) \quad \Rightarrow \quad r(\mathrm{A}) = \mathrm{O},$$

so that A satisfies an equation of degree lower than that of $m_1(x)$.

Thus, we arrive at a contradiction.

$\therefore$ $r(x) = 0$

$\therefore$ From (1), $m_2(x) = qm_1(x)$.

As a result of the above theorem, we define the minimal polynomial of A as below :

16. Minimal Polynomial and Minimal Equation

If $m(x)$ is a scalar polynomial of lowest degree with leading coefficient unity such that $m(x) = 0$ is satisfied by the matrix A, i.e., $m(A) = O$, then the polynomial $m(x)$ is called the minimal polynomial of the matrix A. The equation $m(x) = 0$ is called the equation of A.

Note. The degree of the minimal equation of an n-square matrix is always $\leq n$.

17. Derogatory and Non-derogatory Matrices

An n-square matrix is said to be derogatory or non-derogatory according as the degree of its minimal equation is less than or equal to n.

18. Theorem

If $f(x)$ is a scalar polynomial such that $f(A) = O$, and if $m(x) = 0$ is the minimal equation of A, then there exists a polynomial $q(x)$ such that

$$f(x) = m(x)q(x)$$

Proof. By division algorithm, there exist two scalar polynomials $q(x)$ and $r(x)$ such that

$$f(x) = m(x)q(x) + r(x),$$

where $r(x) = 0$ or its degree is less than the degree of $m(x)$.

Let, if possible, $\quad r(x) \neq 0.$

$\therefore \quad f(A) = m(A)q(A) + r(A)$

$\Rightarrow \quad O = O + r(A) \quad \Rightarrow \quad r(A) = O.$

$\Rightarrow$ A satisfies an equation of degree smaller than that of the minimal equation.

Thus, we arrive at a contradiction.

$\therefore \quad r(x) = 0$

Hence, $\quad f(x) = m(x)q(x).$

Cor. Minimal Polynomial is always unique.

Let, if possible, $m_1(x)$ and $m_2(x)$ be two minimal polynomials of a square matrix A.

Since $m_1(x)$ is an annihilating polynomial of A and $m_2(x)$ is the minimal polynomial.

$\therefore \quad m_2(x)$ divides $m_1(x)$.

Similarly, $m_1(x)$ divides $m_2(x)$.

$\therefore \quad m_1(x) = km_2(x)$, for some scalar k. But, being minimal polynomials, $m_1(x)$ and $m_2(x)$ have both leading coefficients unity.

$\therefore \quad k = 1 \quad \therefore \quad m_1(x) = m_2(x).$

The following theorem gives the *relationship between the roots of a characteristic equation and the minimal equation of a square matrix.*

19. Theorem

The distinct roots of the characteristic equation $\phi(\lambda) = 0$ of a matrix A are also the distinct roots of the minimal equation $m(\lambda) = 0$ of A and conversely.

Proof. By the above theorem, $\phi(\lambda) = m(\lambda) \, . \, q(\lambda)$ for some scalar polynomial $q(\lambda)$.

This shows that every root of $m(\lambda) = 0$ is a root of $\phi(\lambda) = 0$.

Conversely, let λ_1 be a root of the characteristic equation $\phi(\lambda) = 0$.

Divide $m(\lambda)$ by $\lambda - \lambda_1$.

By division algorithm, we have $m(\lambda) = (\lambda - \lambda_1) \, . \, q(\lambda) + r$, where r is a constant.

Substituting A for λ, we have $m(A) = (A - \lambda_1 I) \, q(A) + rI$.

But $m(A) = O$.

$\therefore \quad (A - \lambda_1 I) \, q(A) + rI = O$

If $r \neq 0$, then $(A - \lambda_1 I)\left(\frac{-1}{r} q(A)\right) = I$.

$\Rightarrow$ $-\frac{1}{r}\, q(A)$ is the multiplicative inverse of $A - \lambda_1 I$.

But $A - \lambda_1 I$ is singular ($\because$ $|A - \lambda_1 I| = 0$ as λ_1 is a characteristic root of A).

$\therefore$ We arrive at a contradiction.

Hence, $r = 0$.

$$\therefore \quad m(\lambda) = (\lambda - \lambda_1)\, q(\lambda)$$

$\Rightarrow$ λ_1 is a root of $m(\lambda) = 0$ and the theorem is proved.

Counting each repeated root only once, the aggregate of the roots of $\phi(\lambda) = 0$ is the same as that of $m(\lambda) = 0$. The roots of the two equations can only differ in respect of their multiplicities.

If the characteristic roots of a n-square matrix are all different, then its minimal equation is also of nth degree and in fact, apart from the constant factor $(-1)^n$, it coincides with the characteristic equation. i.e., If the characteristic roots of A are different and, $\phi(\lambda)$ is the characteristic polynomial and $m(\lambda)$ is the minimal polynomial, then

$$\phi(\lambda) = (-1)^n\, m(\lambda).$$

One might be tempted to conclude at this point that the minimal polynomial of a matrix is just the product of different factors of the characteristic equation. But, this is not the case as is clear from the following example :

Let
$$A = \begin{bmatrix} 1 & 1 & 1 & \cdots & 1 \\ 0 & 1 & 1 & \cdots & 1 \\ 0 & 0 & 1 & \cdots & 1 \\ \cdots & \cdots & \cdots & \cdots & \cdots \\ 0 & 0 & 0 & \cdots & 1 \end{bmatrix}_{n \times n}$$

Characteristic equation of A is $|A - \lambda I| = 0$

$\Rightarrow$ $(-1)^n (\lambda - 1)^n = 0$.

Then, $A - I \neq O$ so that $\lambda - 1$ is not the minimal polynomial. However, one can verify that $(A - I)^n = O$ while no lesser power of $A - I$ is O.

Hence, minimal polynomial is $(\lambda - 1)^n$.

20. Theorem

A square matrix A is scalar iff its minimal polynomial is of degree one.

Proof. Let A be a scalar matrix.

$\therefore$ There exists a real number α such that $A = \alpha I$ or $A - \alpha I = 0$

$\therefore$ The matrix A satisfies the equation $x - \alpha = 0$ which is a polynomial equation of degree one (and hence is of lowest degree) and hence is the minimial equation of matrix A.

Converse. Let minimal equation of matrix A be of degree one say $x - \alpha = 0$.

$\therefore$ Matrix A satisfies it *i.e.* $A - \alpha I = 0$ or $A = \alpha I$

$\therefore$ A is a scalar matrix.

Example 1. *Find the characteristic equation and minimal equation of the matrix*

$$A = \begin{bmatrix} 7 & 4 & -1 \\ 4 & 7 & -1 \\ -4 & -4 & 4 \end{bmatrix}$$

Is the matrix derogatory ?

Sol.
$$| A - \lambda I | = \begin{vmatrix} 7-\lambda & 4 & -1 \\ 4 & 7-\lambda & -1 \\ -4 & -4 & 4-\lambda \end{vmatrix}$$

$$= -\lambda^3 + 18\lambda^2 - 81\lambda + 108 \quad \text{(on simplification)}$$

∴ Characteristic equation of A is

$$-\lambda^3 + 18\lambda^2 - 81\lambda + 108 = 0$$

or
$$\lambda^3 - 18\lambda^2 + 81\lambda - 108 = 0.$$

To find minimal equation, let us examine whether a second degree equation $\lambda^2 + a\lambda + b = 0$ (say) is satisfied by A.

i.e.,
$$A^2 + aA + bI = O.$$

Now
$$A^2 = \begin{bmatrix} 7 & 4 & -1 \\ 4 & 7 & -1 \\ -4 & -4 & 4 \end{bmatrix}\begin{bmatrix} 7 & 4 & -1 \\ 4 & 7 & -1 \\ -4 & -4 & 4 \end{bmatrix}$$

$$= \begin{bmatrix} 69 & 60 & -15 \\ 60 & 69 & -15 \\ -60 & -60 & 24 \end{bmatrix} \quad \text{(on actual multiplication)}$$

∴ $A^2 + aA + bI = O$

$$\Rightarrow \begin{bmatrix} 69 & 60 & -15 \\ 60 & 69 & -15 \\ -60 & -60 & 24 \end{bmatrix} + a\begin{bmatrix} 7 & 4 & -1 \\ 4 & 7 & -1 \\ -4 & -4 & 4 \end{bmatrix} + b\begin{bmatrix} 1 & 0 & 0 \\ 0 & 1 & 0 \\ 0 & 0 & 1 \end{bmatrix} = \begin{bmatrix} 0 & 0 & 0 \\ 0 & 0 & 0 \\ 0 & 0 & 0 \end{bmatrix}$$

$$\Rightarrow \quad 69 + 7a + b = 0$$
$$60 + 4a + 0b = 0$$

Solving, we have $a = -15, b = 36.$

One may verify that all other elements of $A^2 + aA + bI = O$ are satisfied by

$$a = -15, b = 36.$$

Thus, $A^2 - 15A + 36I = O.$

Since minimal equation of A cannot be of first degree,

$$\therefore \quad \lambda^2 - 15\lambda + 36 = 0$$

is the minimal equation of A.

Since, 3-square matrix A satisfies a second degree equation, therefore A is derogatory.

Remark. One can verify that in the above theorem,

$$-x^3 + 18x^2 - 81x + 108 = (x^2 - 15x + 36)(-x + 3).$$

Example 2. *Find the characteristic equation and minimal equation of the matrix*

$$A = \begin{bmatrix} 1 & 2 & 3 \\ 2 & 3 & 1 \\ 3 & 1 & 2 \end{bmatrix}.$$

Sol. Characteristic equation of A is given by

$$| A - \lambda I | = 0$$

$$\Rightarrow \quad \begin{vmatrix} 1-\lambda & 2 & 3 \\ 2 & 3-\lambda & 1 \\ 3 & 1 & 2-\lambda \end{vmatrix} = 0$$

$$\Rightarrow \quad \lambda^3 - 6\lambda^2 - 3\lambda + 18 = 0 \qquad \text{(Verify)}$$

$$\Rightarrow \quad \lambda^3 - 3\lambda - 6\lambda^2 + 18 = 0$$

$$\Rightarrow \quad \lambda(\lambda^2 - 3) - 6(\lambda^2 - 3) = 0$$

$$\Rightarrow \quad (\lambda - 6)(\lambda^2 - 3) = 0 \ \Rightarrow \ \lambda = 6, \pm\sqrt{3}.$$

Since the characteristic roots of A are all different.

$\therefore$ Minimal equation of A is $\lambda^3 - 6\lambda^2 - 3\lambda + 18 = 0$.

Example 3. *Find the characteristic polynomial and the minimal polynomial of*

$$A = \begin{bmatrix} 6 & -2 & 2 \\ -2 & 3 & -1 \\ 2 & -1 & 3 \end{bmatrix}.$$

Sol $\quad | A - \lambda I | = \begin{vmatrix} 6-\lambda & -2 & 2 \\ -2 & 3-\lambda & -1 \\ 2 & -1 & 3-\lambda \end{vmatrix}$

$$\overset{C_{3,2}(1)}{\sim} \begin{vmatrix} 6-\lambda & -2 & 0 \\ -2 & 3-\lambda & 2-\lambda \\ 2 & -1 & 2-\lambda \end{vmatrix}$$

$$= (2-\lambda) \begin{vmatrix} 6-\lambda & -2 & 0 \\ -2 & 3-\lambda & 1 \\ 2 & -1 & 1 \end{vmatrix} \quad \text{Taking } (2-\lambda) \text{ common from } C_3$$

$$\overset{R_{2,3}(-1)}{\sim} (2-\lambda) \begin{vmatrix} 6-\lambda & -2 & 0 \\ -4 & 4-\lambda & 0 \\ 2 & -1 & 1 \end{vmatrix}$$

$$= (2-\lambda)[(6-\lambda)(4-\lambda) - 8]$$

$$= (2-\lambda)(\lambda^2 - 10\lambda + 16) = (2-\lambda)(\lambda-2)(\lambda-8)$$

$$= -\lambda^3 + 12\lambda^2 - 36\lambda + 32 \qquad \text{(on multiplication)}$$

$\therefore$ Characteristic polynomial is $-\lambda^3 + 12\lambda^2 - 36\lambda + 32$.

Characteristic roots are 2, 2, 8.

Since distinct characteristic roots of A are also roots of the minimal equation of A.

$\therefore$ 2, 8 are roots of the minimal equation of A.

$\therefore$ The minimal polynomial of A is either $(\lambda - 2)(\lambda - 8)$ *i.e.*, $\lambda^2 - 10\lambda + 16$ or $\lambda^3 - 12\lambda^2 + 36\lambda - 32$ (with leading coefficient 1)

Now $A^2 - 10A + 16I$

$$= \begin{bmatrix} 6 & -2 & 2 \\ -2 & 3 & -1 \\ 2 & -1 & 3 \end{bmatrix}^2 - 10\begin{bmatrix} 6 & -2 & 2 \\ -2 & 3 & -1 \\ 2 & -1 & 3 \end{bmatrix} + 16\begin{bmatrix} 1 & 0 & 0 \\ 0 & 1 & 0 \\ 0 & 0 & 1 \end{bmatrix}$$

$$= \begin{bmatrix} 44 & -20 & 20 \\ -20 & 14 & -10 \\ 20 & -10 & 14 \end{bmatrix} - \begin{bmatrix} 60 & -20 & 20 \\ -20 & 30 & -10 \\ 20 & -10 & 30 \end{bmatrix} + \begin{bmatrix} 16 & 0 & 0 \\ 0 & 16 & 0 \\ 0 & 0 & 16 \end{bmatrix}$$

$$= \begin{bmatrix} 0 & 0 & 0 \\ 0 & 0 & 0 \\ 0 & 0 & 0 \end{bmatrix} = O$$

$\therefore$ Minimal polynomial of A is $\lambda^2 - 10\lambda + 16$.

EXERCISE 3

1. Find the characteristic polynomial and the minimal polynomial of the following matrices :

(*i*) $\begin{bmatrix} 1 & 2 & 3 \\ 0 & -4 & 2 \\ 0 & 0 & 7 \end{bmatrix}$ (*ii*) $\begin{bmatrix} 1 & 0 & 0 \\ 3 & 1 & 0 \\ -7 & 5 & 1 \end{bmatrix}$

(*iii*) $\begin{bmatrix} 1 & 1 & 1 \\ 0 & 1 & 1 \\ 0 & 0 & 1 \end{bmatrix}$.

2. Find the minimal and the characteristic equations of the following matrices :

(*i*) $\begin{bmatrix} 8 & -6 & 2 \\ -6 & 7 & -4 \\ 2 & -4 & 3 \end{bmatrix}$ (*ii*) $\begin{bmatrix} 2 & 3 & 4 \\ 0 & 2 & -1 \\ 0 & 0 & 1 \end{bmatrix}$

(*iii*) $\begin{bmatrix} 5 & -6 & -6 \\ -1 & 4 & 2 \\ 3 & -6 & -4 \end{bmatrix}$.

ANSWERS

1. (i) $-\lambda^3 + 4\lambda^2 + 25\lambda - 28$, $\lambda^3 - 4\lambda^2 - 25\lambda + 28$

(ii) $(1-\lambda)^3$, $(\lambda - 1)^3$ (iii) $(1-\lambda)^3$, $(\lambda-1)^3$

2.

Minimal equation :	Characteristic equation
(i) $x^3 - 18x^2 + 45x = 0$	$x^3 - 18x^2 + 45x = 0$
(ii) $(x-1)(2-x)^2 = 0$	$(2-x)^2(1-x) = 0$
(iii) $\lambda^2 - 3\lambda + 2 = 0$	$\lambda^3 - 5\lambda^2 + 12 = 0.$

21. MATRIC POLYNOMIAL OF DEGREE TWO

Before we deal with Cayley-Hamilton theorem, we consider a third order matrix whose elements are polynomials of degree at the most two (say).

$$\begin{bmatrix} 2\lambda^2+3\lambda+1 & \lambda^2+1 & 3\lambda+2 \\ 4\lambda+9 & \lambda^2+\lambda+2 & \lambda^2+3 \\ 9\lambda^2+1 & 9\lambda^2+\lambda+3 & 11\lambda^2+5 \end{bmatrix} = \begin{bmatrix} 2\lambda^2+3\lambda+1 & \lambda^2+0\lambda+1 & 0\lambda^2+3\lambda+2 \\ 0\lambda^2+4\lambda+9 & \lambda^2+\lambda+2 & \lambda^2+0\lambda+3 \\ 9\lambda^2+0\lambda+1 & 9\lambda^2+\lambda+3 & 11\lambda^2+0\lambda+5 \end{bmatrix}$$

$$= \begin{bmatrix} 2 & 1 & 0 \\ 0 & 1 & 1 \\ 9 & 9 & 11 \end{bmatrix}\lambda^2 + \begin{bmatrix} 3 & 0 & 3 \\ 4 & 1 & 0 \\ 0 & 1 & 0 \end{bmatrix}\lambda + \begin{bmatrix} 1 & 1 & 2 \\ 9 & 2 & 3 \\ 1 & 3 & 5 \end{bmatrix}$$

which is a matric polynomial in λ of degree two.

Thus, any matrix whose elements are polynomials in λ of degree at the most n (here $n = 2$), can always be expressed as a matric polynomial of degree at most n (here $n = 2$) in λ.

22. Theorem *(Cayley-Hamilton Theorem)*

Every square matrix satisfies its characteristic equation

i.e., if $$|A - \lambda I| = a_0\lambda^n + a_1\lambda^{n-1} + \ldots\ldots + a_n = 0 \quad \ldots(1)$$

is the characteristic equation of an n-square matrix A, then

$$a_0A^n + a_1A^{n-1} + \ldots\ldots + a_nI = O.$$

Proof. Since the elements of $A - \lambda I$ are at the most of first degree in λ, the co-factors of elements of $|A - \lambda I|$ are polynomials in λ of degree at the most $n - 1$. Hence, the elements of adj.$(A - \lambda I)$ are polynomials in λ of degree $\le n - 1$. Therefore, adj. $(A - \lambda I)$ can be written as a matric polynomial of degree $n - 1$. (See Art. 21)

Let adj. $(A - \lambda I)$ be written as

$$\text{adj. }(A - \lambda I) = B_0\lambda^{n-1} + B_1\lambda^{n-2} + \ldots\ldots + B_{n-1} \quad \ldots(2)$$

where $B_0, B_1, \ldots\ldots, B_{n-1}$ are n-square matrices whose elements are functions of elements of A.

Since for a square matrix A,

$$A(\text{adj. }A) = (\text{adj. }A)A = |A| \,.\, I$$

therefore for the matrix $A - \lambda I$, we have

$$(A - \lambda I)\text{ adj. }(A - \lambda I) = |A - \lambda I| \,.\, I \quad \ldots(3)$$

Using (1) and (2) in (3), we have

$$(A - \lambda I)(B_0\lambda^{n-1} + B_1\lambda^{n-2} + B_2\lambda^{n-3} + \ldots\ldots + B_{n-1})$$
$$= (a_0\lambda^n + a_1\lambda^{n-1} + a_2\lambda^{n-2} \ldots\ldots + a_n)I$$

Equating the co-efficients of like powers of λ on both sides, we have

$$-B_0 = a_0 I,$$
$$AB_0 - B_1 = a_1 I,$$
$$AB_1 - B_2 = a_2 I$$
$$\ldots\ldots\ldots\ldots\ldots\ldots$$
$$AB_{n-1} = a_n I.$$

Pre-multiplying the above equations by A^n, A^{n-1}, A^{n-2}, ... I, respectively and adding, we have

$$O = a_0A^n + a_1A^{n-1} + a_2A^{n-2} \ldots\ldots + a_n I,$$

for the terms on the left cancel in pairs. Hence theorem is proved.

Remark. If A is a non-singular matrix, then multiplying the equation $a_0A^n + a_1A^{n-1} + \ldots\ldots + a_nI = O$ by A^{-1}, we have

$$a_0A^{n-1} + a_1A^{n-2} + \ldots\ldots + a_nA^{-1} = O.$$

$$\Rightarrow \qquad A^{-1} = \frac{-1}{a_n}(a_0A^{n-1} + a_1A^{n-2} + \ldots\ldots + a_{n-1}I)$$

which can be evaluated after substituting the values of A^{n-1}, A^{n-2}, ... etc. Thus, the theorem can be used to compute the inverse of a matrix.

Example 1. *Verify Cayley-Hamilton theorem for the matrix*

$$A = \begin{bmatrix} 1 & 0 & 2 \\ 0 & 2 & 1 \\ 2 & 0 & 3 \end{bmatrix}. \text{ Hence find } A^{-1}.$$

Sol. $$|A - \lambda I| = \begin{vmatrix} 1-\lambda & 0 & 2 \\ 0 & 2-\lambda & 1 \\ 2 & 0 & 3-\lambda \end{vmatrix}$$

$$= (2-\lambda)\begin{vmatrix} 1-\lambda & 2 \\ 2 & 3-\lambda \end{vmatrix}$$

$$= -\lambda^3 + 6\lambda^2 - 7\lambda - 2 \qquad \text{(on simplification)}$$

$\therefore$ The characteristic equation of A is

$$|A - \lambda I| = -\lambda^3 + 6\lambda^2 - 7\lambda - 2 = 0$$

or $$\lambda^3 - 6\lambda^2 + 7\lambda + 2 = 0 \qquad \ldots(1)$$

By Cayley-Hamilton theorem, the matrix A satisfies (1)

i.e., $$A^3 - 6A^2 + 7A + 2I = O \qquad \ldots(2)$$

Verification

$$A^2 = \begin{bmatrix} 1 & 0 & 2 \\ 0 & 2 & 1 \\ 2 & 0 & 3 \end{bmatrix}\begin{bmatrix} 1 & 0 & 2 \\ 0 & 2 & 1 \\ 2 & 0 & 3 \end{bmatrix} = \begin{bmatrix} 5 & 0 & 8 \\ 2 & 4 & 5 \\ 8 & 0 & 13 \end{bmatrix}$$

$$A^3 = A^2 \,.\, A = \begin{bmatrix} 5 & 0 & 8 \\ 2 & 4 & 5 \\ 8 & 0 & 13 \end{bmatrix}\begin{bmatrix} 1 & 0 & 2 \\ 0 & 2 & 1 \\ 2 & 0 & 3 \end{bmatrix} = \begin{bmatrix} 21 & 0 & 34 \\ 12 & 8 & 23 \\ 34 & 0 & 55 \end{bmatrix}$$

$\therefore \quad A^3 - 6A^2 + 7A + 2I$

$$= \begin{bmatrix} 21 & 0 & 34 \\ 12 & 8 & 23 \\ 34 & 0 & 55 \end{bmatrix} - 6\begin{bmatrix} 5 & 0 & 8 \\ 2 & 4 & 5 \\ 8 & 0 & 13 \end{bmatrix} + 7\begin{bmatrix} 1 & 0 & 2 \\ 0 & 2 & 1 \\ 2 & 0 & 3 \end{bmatrix} + 2\begin{bmatrix} 1 & 0 & 0 \\ 0 & 1 & 0 \\ 0 & 0 & 1 \end{bmatrix}$$

$$= \begin{bmatrix} 21 & 0 & 34 \\ 12 & 8 & 23 \\ 34 & 0 & 55 \end{bmatrix} + \begin{bmatrix} -30 & 0 & -48 \\ -12 & -24 & -30 \\ -48 & 0 & -78 \end{bmatrix} + \begin{bmatrix} 7 & 0 & 14 \\ 0 & 14 & 7 \\ 14 & 0 & 21 \end{bmatrix} + \begin{bmatrix} 2 & 0 & 0 \\ 0 & 2 & 0 \\ 0 & 0 & 2 \end{bmatrix}$$

$$= \begin{bmatrix} 21-30+7+2 & 0+0+0+0 & 34-48+14+0 \\ 12-12+0+0 & 8-24+14+2 & 23-30+7+0 \\ 34-48+14+0 & 0+0+0+0 & 55-78+21+2 \end{bmatrix}$$

$$= \begin{bmatrix} 0 & 0 & 0 \\ 0 & 0 & 0 \\ 0 & 0 & 0 \end{bmatrix} = O.$$

Hence, Cayley-Hamilton theorem is verified.

To find A^{-1}

Pre-multiplying (2) by A^{-1}, we have

$$A^2 - 6A + 7I + 2A^{-1} = O$$

$$\Rightarrow \qquad A^{-1} = \tfrac{1}{2}\,(-A^2 + 6A - 7I)$$

$$= \tfrac{1}{2}\left\{-\begin{bmatrix} 5 & 0 & 8 \\ 2 & 4 & 5 \\ 8 & 0 & 13 \end{bmatrix} + 6\begin{bmatrix} 1 & 0 & 2 \\ 0 & 2 & 1 \\ 2 & 0 & 3 \end{bmatrix} - 7\begin{bmatrix} 1 & 0 & 0 \\ 0 & 1 & 0 \\ 0 & 0 & 1 \end{bmatrix}\right\}$$

$$= \begin{bmatrix} -3 & 0 & 2 \\ -1 & \frac{1}{2} & \frac{1}{2} \\ 2 & 0 & -1 \end{bmatrix}.$$

Example 2. *Prove that* $A = \begin{bmatrix} 2 & 6 & 1 \\ 0 & 1 & -6 \\ 3 & 4 & 2 \end{bmatrix}$ *satisifies its characteristic equation. Also, find its inverse, if exists.*

Sol. $|A - \lambda I| = 0 \Rightarrow \begin{bmatrix} 2-\lambda & 6 & 1 \\ 0 & 1-\lambda & -6 \\ 3 & 4 & 2-\lambda \end{bmatrix} = 0$

$\Rightarrow \quad \lambda^3 - 5\lambda^2 + 29\lambda + 59 = 0$ (on simplification)

which is the characteristic equation of A.

Now $A = \begin{bmatrix} 2 & 6 & 1 \\ 0 & 1 & -6 \\ 3 & 4 & 2 \end{bmatrix}, \quad A^2 = \begin{bmatrix} 7 & 22 & -32 \\ -18 & -23 & -18 \\ 12 & 30 & -17 \end{bmatrix}$

and $A^3 = A \cdot A^2 = \begin{bmatrix} -82 & -64 & -189 \\ -90 & -203 & 84 \\ -27 & 34 & -202 \end{bmatrix}$ (on calculation)

Now, $A^3 - 5A^2 + 29A + 59I$

$$= \begin{bmatrix} -82 & -64 & -189 \\ -90 & -203 & 84 \\ -27 & 34 & -202 \end{bmatrix} - 5\begin{bmatrix} 7 & 22 & -32 \\ -18 & -23 & -18 \\ 12 & 30 & -17 \end{bmatrix} + 29\begin{bmatrix} 2 & 6 & 1 \\ 0 & 1 & -6 \\ 3 & 4 & 2 \end{bmatrix} + 59\begin{bmatrix} 1 & 0 & 0 \\ 0 & 1 & 0 \\ 0 & 0 & 1 \end{bmatrix}$$

$$= \begin{bmatrix} -82-35+58+59 & -64-110+174+0 & -189+160+29+0 \\ -90+90+0+0 & -203+115+29+59 & 84+90-174+0 \\ -27-60+87+0 & 34-150+116+0 & -202+85+58+59 \end{bmatrix}$$

$$= \begin{bmatrix} 0 & 0 & 0 \\ 0 & 0 & 0 \\ 0 & 0 & 0 \end{bmatrix} = O$$

i.e., $A^3 - 5A^2 + 29A + 59I = O$

Hence A satisfies its characteristic equation.

To compute the inverse :

Multiplying (2) by A^{-1}, we get

$$A^2 - 5A + 29I + 59A^{-1} = O$$

$$\Rightarrow \quad A^{-1} = \frac{1}{59}[-29I + 5A - A^2]$$

$$\therefore \quad A^{-1} = \frac{1}{59}\left\{-29\begin{bmatrix} 1 & 0 & 0 \\ 0 & 1 & 0 \\ 0 & 0 & 1 \end{bmatrix} + 5\begin{bmatrix} 2 & 6 & 1 \\ 0 & 1 & -6 \\ 3 & 4 & 2 \end{bmatrix} - \begin{bmatrix} 7 & 22 & -32 \\ -18 & -23 & -18 \\ 12 & 30 & -17 \end{bmatrix}\right\}$$

$$= \frac{1}{59}\begin{bmatrix} -29+10-7 & 0+30-22 & 0+5+32 \\ 0+0+18 & -29+5+23 & 0-30+18 \\ 0+15-12 & 0+20-30 & -29+10+17 \end{bmatrix}$$

$$= \frac{1}{59}\begin{bmatrix} -26 & 8 & 37 \\ 18 & -1 & -12 \\ 3 & -10 & -2 \end{bmatrix}$$

Remark. One can verify that $AA^{-1} = I$.

EXERCISE 4

1. Verify Cayley-Hamilton theorem for the matrix A. Hence find A^{-1} when

(*i*) $A = \begin{bmatrix} 0 & 0 & 1 \\ 3 & 1 & 0 \\ -2 & 1 & 4 \end{bmatrix}$ (*ii*) $= \begin{bmatrix} 2 & -1 & 1 \\ -1 & 2 & -1 \\ 1 & -1 & 2 \end{bmatrix}$.

2. Verify Cayley-Hamilton theorem for the matrix $A = \begin{bmatrix} 1 & 2 & 3 \\ 2 & 3 & 1 \\ 3 & 1 & 2 \end{bmatrix}$.

3. (*a*) Use Cayley-Hamilton theorem to show that if

$$A = \begin{bmatrix} 2 & -1 & 2 \\ 5 & -3 & 3 \\ -1 & 0 & -2 \end{bmatrix},\ A^3 + 3A^2 + 3A + I = O.$$

Hence evaluate A^{-1}.

(*b*) Verify Cayley-Hamilton theorem for the matrix $A = \begin{bmatrix} 2 & -1 & 2 \\ 5 & -3 & 3 \\ -1 & 0 & -2 \end{bmatrix}$.

Hence find out A^{-1}.

(*c*) Verify Cayley Hamilton theorem for the matrix A and find its inverse if exists

$$A = \begin{bmatrix} 2 & 1 & 2 \\ 5 & 3 & 3 \\ -1 & 0 & -2 \end{bmatrix}.$$

4. If $A = \begin{bmatrix} 3 & 1 \\ -1 & 2 \end{bmatrix}$, use Cayley-Hamilton theorem to express $2A^5 - 3A^4 + A^2 - 4I$ as a linear polynomial in A.

[**Hint.** By Cayley-Hamilton theorem $A^2 - 5A + 7I = O$.]

5. Prove that $A = \begin{bmatrix} 2 & 6 & 1 \\ 0 & 1 & -6 \\ 3 & 4 & -2 \end{bmatrix}$ satisfies its characteristic equation. Also find its inverse, if exists.

6. Find the characteristic equation of $A = \begin{bmatrix} 1 & 1 & 3 \\ 5 & 2 & 6 \\ -2 & -1 & -3 \end{bmatrix}$ and show that A satisfies this equation.

7. Find the characteristic equation of the matrix $A = \begin{bmatrix} 2 & \sqrt{2} \\ \sqrt{2} & 1 \end{bmatrix}$ and verify that A satisfies it. Also find the characteristic roots and the corresponding characteristic vectors of A.

ANSWERS

1. (*i*) $\begin{bmatrix} 4/5 & 1/5 & -1/5 \\ -12/5 & 2/5 & 3/5 \\ 1 & 0 & 0 \end{bmatrix}$ (*ii*) $\frac{1}{4}\begin{bmatrix} 3 & 1 & -1 \\ 1 & 3 & 1 \\ -1 & 1 & 3 \end{bmatrix}$

3. (*a*) $\begin{bmatrix} -6 & 2 & -3 \\ -7 & 2 & -4 \\ 3 & -1 & 1 \end{bmatrix}$ (*b*) $\begin{bmatrix} -6 & 2 & -3 \\ -7 & 2 & -4 \\ 3 & -1 & 1 \end{bmatrix}$ (*c*) $\begin{bmatrix} -6 & 2 & -3 \\ 7 & -2 & 4 \\ 3 & -1 & 1 \end{bmatrix}$

4. $138A - 403I_2$ **5.** $A^{-1} = \frac{-1}{67}\begin{bmatrix} 22 & 16 & -37 \\ -18 & -7 & 12 \\ -3 & 10 & 2 \end{bmatrix}$ **6.** $\lambda^3 = 0$

7. $\lambda^2 - 3\lambda = 0$; For $\lambda = 0$; $\begin{bmatrix} -1 \\ \sqrt{2} \\ 1 \end{bmatrix}$, For $\lambda = 3$; $\begin{bmatrix} \sqrt{2} \\ 1 \end{bmatrix}$.

SECTION II

6

Relations between the Roots and Co-efficients of an Equation

1. INTRODUCTION

We are already familiar with linear and quadratic equations. Here we propose to study equations of higher degrees. With this goal in view, we shall first define a polynomial.

2. POLYNOMIAL

Definition. Any expression of the form $a_0x^n + a_1x^{n-1} + a_2x^{n-2} + \ldots\ldots + a_{n-1}x + a_n$, $n \geq 0$ is an integer, $a_0, a_1, a_2, \ldots\ldots, a_n$ *belong* to some number system F, is called **a polynomial** in the variable x over the number system F. We shall denote a polynomial by $f(x)$ or $g(x)$ etc.

(*i*) $a_0, a_1, a_2, \ldots\ldots, a_n$ are called the **co-efficients** of the polynomial $f(x)$.

(*ii*) The terms of a polynomial can be arranged in any other way also, the two common ways of writing $f(x)$ are

$$f(x) = a_0x^n + a_1x^{n-1} + a_2x^{n-2} + \ldots\ldots + a_{n-1}x + a_n$$

and

$$f(x) = p_0 + p_1x + p_2x^2 + \ldots\ldots + p_nx^n.$$

The first is in *descending order of the powers of x,* while the second is in *ascending order of the powers of x.*

(*iii*) If all the co-efficients of a polynomial $f(x)$ are real, then the polynomial $f(x)$ is said to be a *real polynomial*.

3. ZERO POLYNOMIAL

A polynomial $f(x)$, all of whose co-efficients are zero is called a **Zero Polynomial.** All other polynomials are known as **Non-zero Polynomials.**

4. DEGREE OF A NON-ZERO POLYNOMIAL

The highest *index* of the variable x occurring in the terms of a polynomial $f(x)$ is called the **degree of f(x).** Thus if $f(x) = a_0x^n + a_1x^{n-1} + \ldots\ldots + a_n$ with $(a_0 \neq 0)$.

Then the integer $n (\geq 0)$ is **defined** as the **degree** of the polynomial $f(x)$ and we write **deg. f(x) = n.**

For example,

(*i*) The polynomial $f(x) = 5 + 7x + 8x^3$ is of degree **three.**

(*ii*) The constant $c = cx^0$ is a polynomial of degree **zero.**

(*iii*) The polynomial $f(x) = 8x^4 + 5$ is of degree **four**.

Remarks. A polynomial $f(x)$ is said to be :

Linear, Quadratic, Cubic, Bi-quadratic according as it is of the degree one, two, three, four respectively.

5. EQUALITY OF TWO POLYNOMIALS

Two polynomials

$$f(x) = a_0 + a_1x + a_2x^2 + \ldots\ldots + a_nx^n$$

and $$g(x) = b_0 + b_1x + b_2x^2 + \ldots\ldots + b_mx^m$$

with coefficients over the same number system F are said to be **equal polynomials** and written as $f(x) = g(x)$ if the coefficients of same powers of x in the two polynomial are equal.

6. COMPLETE AND INCOMPLETE POLYNOMIAL

A polynomial of nth degree is said to be a **complete polynomial** if it contains all powers of the variable from 0 to n, otherwise, it is said to be an **incomplete polynomial.**

$2x^4 - x - 7$ is an incomplete polynomial, and it can be made complete by supplying zero coefficients to the missing terms. The above incomplete polynomial can be made complete by writing it as $2x^4 + 0x^3 + 0x^2 - x - 7$.

7. ZERO OF A POLYNOMIAL

A number 'α' is called a **zero** of the polynomial $f(x)$ if and only if $f(\alpha) = 0$. For example, 2 and 3 are zeros of the polynomial $f(x) = x^2 - 5x + 6$, because $f(2) = 0$ and $f(3) = 0$.

8. THEOREM (DIVISION ALGORITHM WITHOUT PROOF)

If $f(x)$ and $g(x)$ be any two non-zero polynomials, then there exist *unique* polynomials $q(x)$ and $r(x)$ such that $f(x) = q(x) \cdot g(x) + r(x)$, ...(1)
where either $r(x)$ is a zero polynomial or degree of $r(x) <$ degree of $g(x)$.

The polynomial $q(x)$ is called the **quotient** and $r(x)$ *the* **remainder,** when $f(x)$ is divided by $g(x)$

and degree of $q(x)$ = degree $f(x)$ – degree $g(x)$.

Particular case. When $g(x) = x - h$ *i.e.*, when degree of $g(x) = 1$ then, by division algorithm degree of remainder $r(x) <$ degree of $g(x)$ (= 1).

∴ Either $r(x) = 0$ or degree of $r(x) = 0$ *i.e.*, $r(x)$ is a constant polynomial, so we can take $r(x) = \mathrm{R}$.

∴ In this case, we have

$$\mathbf{f(x) = (x - h)\ q(x) + R,}$$

where degree $q(x)$ = degree $f(x) - 1$ and R is a constant.

9. POLYNOMIAL EQUATION

If $f(x)$ is a polynomial then $f(x) = 0$ is called a **polynomial equation** and degree of the polynomial equation $f(x) = 0$ is the same as that of the polynomial $f(x)$.

10. ROOT OF AN EQUATION

A number 'α' is called a **root** of the polynomial equation $f(x) = 0$, if and only if $f(\alpha) = 0$, *i.e.*, if α satisfies the polynomial equation $f(x) = 0$ and conversely.

11. REMAINDER THEOREM

If a polynomial f(x) is divided by x – h, then the remainder is equal to f(h).

Proof. By division algorithm for the polynomials $f(x)$ and $(x - h)$, we have

$$f(x) = q(x)(x - h) + \mathrm{R}, \qquad \ldots(1)$$

where $q(x)$ is the quotient and R, the remainder.

Since (1) holds for all values of x.

$\therefore$ Put $x = h$ in (1),

$\therefore$ $f(h) = q(h)(h - h) + R$ *i.e.,* $\mathbf{f(h) = R}$

Hence the remainder when f(x) is divided by (x – h) is f(h).

Example 1. *Without actual division, find the remainder when the polynomial* $2x^4 - x^3 - 6x^2 + 4x - 8$ *is divided by* $x + 2$.

Sol. Here $f(x) = 2x^4 - x^3 - 6x^2 + 4x - 8$

Comparing $x + 2$ with $x - h$, we get $h = -2$.

$\therefore$ Remainder $R = f(h) = f(-2)$ $[\because\ h = -2]$

$= 2(-2)^4 - (-2)^3 - 6(-2)^2 + 4(-2) - 8$

$= 32 + 8 - 24 - 8 - 8 = 0.$

Thus remainder, here is zero.

Note. In this case, we say that $f(x)$ is **exactly divisible by (x + 2).**

12. FACTOR THEOREM

If h is a root of the equation f(x) = 0, then (x – h) is a factor of f(x) and conversely

Proof. Using division algorithm for $f(x)$ and $(x - h)$, we have

$$f(x) = q(x)(x - h) + R \quad ...(1)$$

where $q(x)$ is the quotient and R, the remainder.

Putting $x = h$ in (1), we have

$$f(h) = q(h)(h - h) + R$$

i.e., $f(h) = R$

But 'h' is a root of the equation $f(x) = 0$ $\therefore$ $f(h) = 0$.

Hence $R = f(h) = 0$

$\therefore$ (1) becomes $f(x) = q(x)(x - h)$

which shows that $(x - h)$ is a *factor* of $f(x)$.

Conversely *we shall prove that if (x – h) is a factor of f(x) then h is a root of the equation f(x) = 0.*

$\because$ $(x - h)$ is a factor of $f(x)$.

$\therefore$ $(x - h)$ divides $f(x)$ exactly (without remainder).

$\therefore$ **Dividing $f(x)$ by $(x - h)$, we get**

$f(x) = q(x)(x - h)$, where $q(x)$ is quotient.

Putting $x = h$

$\therefore$ $f(h) = q(h)(h - h) = 0$

i.e., $f(h) = 0$

which shows that 'h' is a root of the equation f(x) = 0.

13. HORNERS METHOD OF SYNTHETIC DIVISION

It is a rule of the detached co-efficients with the help of which we can find quickly the *quotient* and the *remainder* when a polynomial $f(x)$ is divided by a polynomial of the form $x - h$ or $ax + b$ $(a \neq 0)$.

The rule is explained as under :

(*i*) *In the first line write down in descending order the* **co-efficients of given equation** *(co-efficients of missing terms are to be written as zeros).*

(*ii*) Put the divisor $(x - h) = 0$ and find the value of x *i.e.*, $x = h$, which is called the **multiplier** and may be written at the left hand corner.

The first coefficient is same and which is multiplied by the multiplier and so on.

(*iii*) The further process is explained by an example given below :

Example 2. *Find by use of synthetic division, the quotient and the remainder when the polynomial* $3x^4 - 5x^3 + 10x^2 + 11x - 61$ *is divided by* $x - 3$.

Sol. Put the divisor = 0

i.e., $x - 3 = 0 \quad \therefore \quad x = 3$

i.e., 3 is the *multiplier*. Writing down the co-efficients of given equation and using the rule, we get

3)	3	– 5	10	11	– 61
		9	12	66	231
	3	4	22	77	| 170 = Remainder

Hence the *remainder* = 170 and the *quotient* is $3x^3 + 4x^2 + 22x + 77$.

14. TO FIND THE QUOTIENT AND THE REMAINDER WHEN f(x) IS DIVIDED BY ax – b

Sol. The divisor can be written as $a\left(x - \dfrac{b}{a}\right)$

Divide $f(x)$ by $\left(x - \dfrac{b}{a}\right)$, let $q(x)$ be the quotient and R the remainder.

$$\therefore \qquad f(x) = q(x) \cdot \left(x - \frac{b}{a}\right) + R = q(x) \cdot \frac{(ax - b)}{a} + R = \frac{1}{a} \cdot q(x)\,(ax - b) + R$$

This shows that when $f(x)$ is divided by $(ax - b)$ instead of $\left(x - \dfrac{b}{a}\right)$, the **remainder R remains unchanged,** whereas the **quotient q(x) is to be divided by "a" the co-efficient of x in the divisor ax – b.**

Example 3. *Find the quotient and the remainder, when* $2x^4 - 5x^2 + 11x - 7$ *is divided by* $(2x - 3)$.

Sol. Write the divisor $2x - 3$ as $2(x - \frac{3}{2})$

We first divide by $x - \frac{3}{2}$.

Put $x - \frac{3}{2} = 0 \quad \therefore \quad x = \frac{3}{2}$ is a multiplier.

Writing down the co-efficients of the given polynomial, **supplying zero co-efficients for missing terms** *i.e.*,

$(\frac{3}{2})$	2	0	– 5	11	– 7
		3	$\frac{9}{2}$	$-\frac{3}{4}$	$\frac{123}{8}$
	2	3	$-\frac{1}{2}$	$\frac{41}{4}$	$\frac{67}{8}$

This gives the remainder and the quotient when $f(x)$ is divided by $(x - \frac{3}{2})$.

Therefore, the remainder and quotient when $f(x)$ is divided by $(2x - 3)$ are as follows :

$$\text{remainder} = \frac{67}{8} \qquad \text{[Remainder remains same]}$$

$$\text{quotient} = \tfrac{1}{2} \text{ (quotient obtained above)}$$

$$= \frac{1}{2}\left(2x^3 + 3x^2 - \frac{1}{2}x + \frac{41}{4}\right) = x^3 + \frac{3}{2}x^2 - \frac{1}{4}x + \frac{41}{8}.$$

Example 4. *Given that – 8 is a root of the equation $2x^3 + 31x^2 + 112x - 64 = 0$. Solve the equation completely.*

Sol. Given equation is $2x^3 + 31x^2 + 112x - 64 = 0$ ...(1)

$\because$ – 8 is a root of the equation

$\therefore$ $(x + 8)$ is a factor of the L.H.S. of (1)

$\therefore$ dividing L.H.S. of (1) by $(x + 8)$

– 8)	2	31	112	– 64
		– 16	– 120	64
	2	15	– 8	zero

$\therefore$ quotient is $2x^2 + 15x - 8$. The other two roots can be found by putting the quotient = 0

i.e., $$2x^2 + 15x - 8 = 0$$

This being a quadratic in x, can be solved easily

$$\therefore \quad x = \frac{-15 \pm \sqrt{225 + 64}}{4} = \frac{-15 \pm 17}{4}$$

i.e., $x = -8, \tfrac{1}{2}$. Hence roots are $-8, -8, \tfrac{1}{2}$.

Remark. In a cubic whenever one root is known, the other two roots can be found by the above method.

Example 5. *Use method of Synthetic Division to express $f(x) = x^4 + 2x^3 + 6x^2 - 8x + 4$ as a polynomial in powers of $(x - 3)$.*

Sol. Given polynomial is $f(x) = x^4 + 2x^3 + 6x^2 - 8x + 4$.

When expressed as polynomials in powers of $(x - 3)$, let

$$f(x) = A_0(x - 3)^4 + A_1(x - 3)^3 + A_2(x - 3)^2 + A_3(x - 3) + A_4$$

To find co-efficients A_0, A_1, A_2, A_3, A_4 divide $f(x)$ successively by $x - 3$, using synthetic division, the remainders obtained will be the values of A_4, A_3, A_2, A_1, A_0 respectively. Thus,

3)	1	2	6	– 8	4
		3	15	63	165
	1	5	21	55	169 = A_4
		3	24	135	
	1	8	45	190 = A_3	
		3	33		
	1	11	78 = A_2		
		3			
	1	14 = A_1			
	1 = A_0				

Hence required polynomial is

$$f(x) = (x-3)^4 + 14(x-3)^3 + 78(x-3)^2 + 190(x-3) + 169.$$

Example 6. *If α is a real root of the equation $x^3 + px^2 + qx + r = 0$ where p, q and r are real, show that the other two roots are real if $p^2 - 4q - 2p\alpha - 3\alpha^2 \ge 0$.*

Sol. Since α is a root of $x^3 + px^2 + qx + r = 0$

$$\therefore \qquad \alpha^3 + p\alpha^2 + q\alpha + r = 0$$

Dividing $x^3 + px^2 + qx + r$ by $x - \alpha$, using synthetic division, we have

α	1	p	q	r
		α	$p\alpha + \alpha^2$	$q\alpha + p\alpha^2 + \alpha^3$
	1	$p + \alpha$	$q + p\alpha + \alpha^2$	$r + q\alpha + p\alpha^2 + \alpha^3 = 0$

The depressed equation is $x^2 + (p + \alpha)x + (q + p\alpha + \alpha^2) = 0$

The roots of this equation are real if $(p + \alpha)^2 - 4(q + p\alpha + \alpha^2) \ge 0$, $\qquad (b^2 - 4ac \ge 0)$

i.e., if $\qquad (p^2 + \alpha^2 + 2p\alpha) - 4(q + p\alpha + \alpha^2) \ge 0$

i.e., if $\qquad p^2 - 4q - 2p\alpha - 3\alpha^2 \ge 0.$

EXERCISE 1

1. Without actual division, find the remainder when
   (*i*) $x^4 - 3x^3 - 4x^2 - 6x + 15$ is divided by $(x - 1)$
   (*ii*) $x^5 + x^2 - 10x + 113$ is divided by $(x + 2)$.
2. Find by use of synthetic division, the quotient and the remainder, when
   (*i*) $2x^3 - 3x^2 + 4x + 5$ is divided by $x - 3$
   (*ii*) $5x^5 + 4x^4 - 3x^3 + 8x^2 + 2$ is divided by $x + 1$
   (*iii*) $3x^4 + 2x^2 + x + 1$ is divided by $2x + 1$.
3. Use synthetic division to compute $f(2)$, where $f(x) = x^4 - 12x^2 + 40x - 71$.
4. Use synthetic division to show that $(x - 1)^2$ is a factor of $x^6 - x^5 - x + 1$. Find the quotient.
5. Show without actual division that $x^4 - 2x^3 - x + 2$ is exactly divisible by $x^2 - 3x + 2$.
6. Find the values of a and b so that $(x - 3)$ and $(x - 1)$ may exactly divide the polynomial
   $$2x^4 - 7x^2 + ax + b.$$
7. If $ax^3 + bx + c$ has a factor of the form $x^2 + kx + 1$, show that $a^2 - c^2 = ab$.
8. Given that -4 is a root of the equation $2x^3 + 6x^2 + 7x + 60 = 0$
   Find the other roots.
9. Express $f(x) = x^3 - 6x^2 + 8x + 7$ as a polynomial in powers of $x - 2$.

ANSWERS

1. (*i*) 3 $\qquad$ (*ii*) 105
2. (*i*) 44 ; $2x^2 + 3x + 13$ $\qquad$ (*ii*) 12 ; $5x^4 - x^3 - 2x^2 + 10x - 10$
   (*iii*) $1\frac{3}{16}$; $\frac{3}{2}x^3 - \frac{3}{4}x^2 + \frac{11}{8}x - \frac{3}{16}$
3. -23 $\qquad$ **4.** $x^4 + x^3 + x^2 + x + 1$ $\qquad$ **6.** $a = -52$, $b = 57$
8. $\dfrac{1 \pm \sqrt{-29}}{2}$ $\qquad$ **9.** $(x-2)^3 - 4(x-2) + 7$.

15. FUNDAMENTAL THEOREM OF ALGEBRA (*Statement*)

Every polynomial equation $f(x) = 0$ of degree $n \geq 1$, has at least one root, real or imaginary.

(The proof of the above theorem is beyond the scope of this book and therefore, we assume the above theorem without any proof.)

16. THEOREM

Every equation f(x) = 0 of degree n has n roots, and no more.

Proof. Let the given equation be

$$f(x) = a_0x^n + a_1x^{n-1} + a_2x^{n-2} + ... + a_{n-1}x + a_n = 0 \qquad (a_0 \neq 0)$$

$\therefore$ By fundamental theorem of algebra, $f(x) = 0$ has a root real or imaginary, let it be α_1

$\therefore$ By the factor theorem, $(x - \alpha_1)$ is a factor of $f(x)$

i.e., $(x - \alpha_1)$ divides $f(x)$ exactly.

$$\therefore \quad f(x) = (x - \alpha_1)f_1(x) \qquad ...(1)$$

where the quotient $f_1(x)$ is polynomial of degree $(n - 1)$.

Again, $f_1(x) = 0$ is an equation of degree $(n - 1)$

$\therefore$ It has a root say α_2

$\therefore$ $(x - \alpha_2)$ is a factor of $f_1(x)$ ...(2)

$$\therefore \quad f_1(x) = (x - \alpha_2)f_2(x)$$

where $f_2(x)$ is a polynomial of degree $(n - 2)$.

Substituting for $f_1(x)$ from (2) in (1), we get $f(x) = (x - \alpha_1)(x - \alpha_2)(f_2(x))$

where $f_2(x)$ is a polynomial of degree $(n - 2)$.

Proceeding like this, after n steps, we shall get

$$f(x) = (x - \alpha_1)(x - \alpha_2) ..... (x - \alpha_n)f_n(x)$$

where $f_n(x)$ is a polynomial of degree $(n - n)$ *i.e.*, of zero degree *i.e.*, $f_n(x)$ is a **constant** polynomial say $f_n(x) = k$.

Thus, $f(x) = k\ (x - \alpha_1)(x - \alpha_2) ..... (x - \alpha_n)$ ...(3)

i.e., $a_0x^n + a_1x^{n-1} + ... + a_{n-1}x + a_n = k(x - \alpha_1)(x - \alpha_2) ......(x - \alpha_n)$.

Since two equal polynomials have the coefficients of like terms equal.

$\therefore$ Equating the co-efficients x^n, we get $a_0 = k$.

Thus, (3) can be written as $f(x) = a_0(x - \alpha_1)(x - \alpha_2) ...... (x - \alpha_n)$ ...(4)

Now $f(x)$ becomes zero, when any one of the value $\alpha_1, \alpha_2, \alpha_3, ..., \alpha_n$ is substituted for x *i.e.*, $f(a_i) = 0$ (for $i = 1, 2, ..., n$)

Hence $\alpha_1, \alpha_2, \alpha_3, ..., \alpha_n$ all are the roots of the equation $f(x) = 0$

$\therefore$ **f(x) = 0 has n roots.**

If possible, let $f(x) = 0$ has more than n roots.

Let α be the $(n + 1)$th root

$$\therefore \quad f(x) = a_0(x - \alpha)(x - \alpha_1)(x - \alpha_2) ...... (x - \alpha_n)$$

Now L.H.S. *i.e.*, $f(x)$ is of degree n

and R.H.S. is of degree $(n + 1)$.

This is not possible and thus our supposition is wrong.

Hence nth degree equation has only n roots and no more. This proves the theorem.

Remark. In the above theorem any root which is repeated 'k' times is to be counted k-times.

Cor. 1. *The converse of the above theorem.* From the above theorem, the equation whose roots are $\alpha_1, \alpha_2, \alpha_3, \ldots\ldots, \alpha_n$ is $a_0(x-\alpha_1)(x-\alpha_2)(x-\alpha_3)\ldots\ldots(x-\alpha_n)=0$ or dividing by $a_0 \neq 0$, we get $(x-\alpha_1)(x-\alpha_2)(x-\alpha_3)\ldots\ldots(x-\alpha_n)=0$ the equation whose roots are $\alpha_1, \alpha_2, \ldots\ldots, \alpha_n$.

Thus, to get an equation whose roots are given, equate to zero, the product of factors corresponding to the given roots.

Cor. 2. *If an nth degree equation f(x) = 0 is satisfied by more than n distinct values of x, then f(x) = 0 is an identity, i.e., all the coefficients of f(x) must vanish.*

Proof. Let $\alpha_1, \alpha_2, \ldots\ldots, \alpha_n$ be n distinct roots of the equation

$$f(x) = a_0x^n + a_1 x^{n-1} + \ldots\ldots + a_n = 0$$

$$\therefore \quad f(x) = a_0(x-\alpha_1)(x-\alpha_2)\ldots\ldots(x-\alpha_n)$$

Let α be another root different from

$$\alpha_1, \alpha_2, \ldots\ldots, \alpha_n \text{ of } f(x) = 0$$

$$\therefore \quad f(\alpha) = 0$$

$$\Rightarrow \quad a_0(\alpha-\alpha_1)(\alpha-\alpha_2)\ldots\ldots(\alpha-\alpha_n) = 0$$

But, $\quad \alpha \neq \alpha_i,\ i = 1, 2, \ldots\ldots, n$

$\therefore \quad a_0 = 0$ and the equation $f(x) = 0$

reduces to $a_1x^{n-1} + a_2x^{n-2} + \ldots\ldots + a_n = 0$, which is an equation of $(n-1)$th degree having n roots.

Applying the same argument, we find $a_1 = 0$. Continuing like this, we have

$$a_0 = a_1 = \ldots\ldots a_n = 0$$

Hence, $f(x) = 0$ is an identity.

17. THEOREM

In an equation with real co-efficients non-real complex (imaginary) roots occur in conjugate pairs.

Proof. The theorem in fact, states that if $\alpha + i\beta(\beta \neq 0)$ is a root of $f(x) = 0$ then $\alpha - i\beta$ is also a root of $f(x) = 0$.

Let the equation be $\quad f(x) = a_0x^n + a_1 x^{n-1} + \ldots\ldots + a_n = 0$

where $a_0, a_1, a_2 \ldots\ldots a_n$ are all real numbers.

Divide the polynomial $f(x)$ by the product $(x-\alpha-i\beta)(x-\alpha+i\beta)$.

Let Q(x) be the quotient and the remainder, if any, will be at the most of 1st degree in x ($\because$ divisor is of second degree).

Let the remainder be Rx + R′.

$\therefore$ By division algorithm, we get

$$f(x) = Q(x)(x-\alpha-i\beta)(x-\alpha+i\beta) + Rx + R' \quad \ldots(1)$$

Put $\quad x = \alpha + i\beta$

$$\therefore \quad f(\alpha+i\beta) = Q(\alpha+i\beta)(\alpha+i\beta-\alpha-i\beta)(\alpha+i\beta-\alpha+i\beta) + R(\alpha+i\beta) + R'$$

i.e., $\quad f(\alpha+i\beta) = R(\alpha+i\beta) + R'$

or $\quad f(\alpha+i\beta) = (R\alpha + R') + iR\beta$

Since $(\alpha + i\beta)$ is a root of $f(x) = 0$ (given)

$$\therefore \quad f(\alpha+i\beta) = 0$$

$\therefore$ We get $\quad 0 = (R\alpha + R') + iR\beta$

$\therefore \quad R\alpha + R' = 0 \quad \ldots(i)$ [$\because$ If a complex number $a + ib = 0$, then $a = 0$ and $b = 0$]

$\quad R\beta = 0 \quad \ldots(ii)$

Since $\beta \neq 0$ $\therefore$ $R = 0$ and

$\therefore$ From (*i*), $R' = 0$

i.e., R and R′ are both zero.

$\therefore$ (1) becomes

$$f(x) = Q(x)(x - \alpha - i\beta)(x - \alpha + i\beta)$$

which shows that $(x - \alpha - i\beta)$ and $(x - \alpha + i\beta)$ both are factors of $f(x)$ *i.e.*,

$x = \alpha + i\beta$ and $x = \alpha - i\beta$ both are the roots of $f(x) = 0$ which proves the theorem.

Note 1. Since whenever $\alpha + i\beta$ $(\beta \neq 0)$ is a root of $f(x) = 0$ then $\alpha - i\beta$ is also a root of $f(x) = 0$. Hence the total number of non-real complex roots in any equation with real co-efficients must always be even. *Thus every odd degree equation with real co-efficients must have at least one real root.*

Note 2. If a root $\alpha + i\beta$ $(\beta \neq 0)$ of $f(x) = 0$ is repeated k-times, then $(\alpha - i\beta)$ is also repeated k-times where $f(x)$ is a polynomial having all its co-efficients real numbers.

18. THEOREM

If an equation f(x) = 0, whose co-efficients are all rational quantities, has a root the irrational number $\alpha + \sqrt{\beta}$ (where α, β rational, $\beta > 0$, and not a square of a rational number), it must also have a root the conjugate irrational number $\alpha - \sqrt{\beta}$.

Proof. Let $f(x) = 0$ be the given polynomial, whose co-efficients are all rational numbers. Let $\alpha + \sqrt{\beta}$ (where $\beta > 0$ and is not a square of rational number) be a root of the equation $f(x) = 0$.

$$\therefore \qquad f(\alpha + \sqrt{\beta}) = 0 \qquad \text{...(1)}$$

We shall prove that $(\alpha - \sqrt{\beta})$ is also a root.

Now divide $f(x)$ by $(x - \alpha - \sqrt{\beta})(x - \alpha + \sqrt{\beta})$.

Suppose $Q(x)$ is the quotient and if there is any remainder, it can be at the most linear.

[$\because$ the divisor is of 2nd degree]

Let the remainder be $Rx + R'$.

$$\therefore \qquad f(x) = Q(x)\,(x - \alpha - \sqrt{\beta})(x - \alpha + \sqrt{\beta}) + Rx + R' \qquad \text{...(2)}$$

Put $x = \alpha + \sqrt{\beta}$

$$\therefore \qquad f(\alpha + \sqrt{\beta}) = 0 + R(\alpha + \sqrt{\beta}) + R' = (R\alpha + R') + R\sqrt{\beta}.$$

But $f(\alpha + \sqrt{\beta}) = 0$ from (1)

$$\therefore \qquad 0 = (R\alpha + R') + R\sqrt{\beta}$$

$R\alpha + R' = 0$ ...(*i*)

$R \cdot \sqrt{\beta} = 0$ ...(*ii*)

[$\because$ if an irrational number of the type $m + n\sqrt{d} = 0$ then $m = 0$ and $n\sqrt{d} = 0$]

Since $\beta \neq 0$ *i.e.*, $\sqrt{\beta} \neq 0$.

$\therefore$ $R = 0$ and, therefore from (1), $R' = 0$

Thus result (2) becomes

$$f(x) = Q(x)(x - \alpha - \sqrt{\beta})(x - \alpha + \sqrt{\beta})$$

which shows that $(x - \alpha - \sqrt{\beta})$ and $(x - \alpha + \sqrt{\beta})$ both are *factors* of $f(x)$ and in particular $x - \alpha + \sqrt{\beta}$ is a factor of $f(x)$

$\therefore$ $\alpha - \sqrt{\beta}$ is a root of $f(x) = 0$, which proves the theorem.

Notes. 1. If an equation $f(x) = 0$ with rational coefficients, has an irrational root $\alpha + \sqrt{\beta}$ repeated r times, then it has also $(\alpha - \sqrt{\beta})$ as a root repeated r times.

2. If the irrational root of the equation $f(x) = 0$ is of the type $\sqrt{\alpha} + \sqrt{\beta}$, where α, β, are both not perfect squares, then $\sqrt{\alpha} - \sqrt{\beta}$, $-\sqrt{\alpha} + \sqrt{\beta}$, $-\sqrt{\alpha} - \sqrt{\beta}$, are also roots of $f(x) = 0$.

3. If one root of $f(x) = 0$ is of the type $\sqrt{\alpha} + i\beta$, then $\sqrt{\alpha} - i\beta$, $-\sqrt{\alpha} + i\beta$, $-\sqrt{\alpha} - i\beta$ are also roots of $f(x) = 0$.

Example 1. *Find an equation of the lowest degree with real co-efficients which has $1 + 2i$ and $3 - i$ as two of its roots.*

Sol. Given roots are $1 + 2i$, $3 - i$. Since these are non-real complex roots of an equation with real co-efficients, therefore, they occur in conjugate pairs.

$\therefore$ $1 - 2i$ and $3 + i$ are also the roots of the equation.

$\therefore$ Roots of the equation are $(1 + 2i), (1 - 2i), (3 - i), (3 + i)$.

$\therefore$ The corresponding equation is

$$(x - 1 - 2i)(x - 1 + 2i)(x - 3 + i)(x - 3 - i) = 0$$

or $$[(x - 1)^2 - (2i)^2][(x - 3)^2 - (i)^2] = 0$$

or $$(x^2 - 2x + 1 + 4)(x^2 - 6x + 9 + 1) = 0$$

or $$(x^2 - 2x + 5)(x^2 - 6x + 10) = 0$$

i.e., $$x^4 - 6x^3 + 10x^2 - 2x^3 + 12x^2 - 20x + 5x^2 - 30x + 50 = 0$$

or $$x^4 - 8x^3 + 27x^2 - 50x + 50 = 0$$

is the required equation.

Example 2. *Form the equation with rational co-efficients one of whose roots is $\sqrt{2} + \sqrt{3}$.*

Sol. First Method

One of the given roots is $\sqrt{2} + \sqrt{3}$.

Since the quadratic surd roots occur in pairs.

$\therefore$ Roots are $\sqrt{2} + \sqrt{3}$; $\sqrt{2} - \sqrt{3}$; $-\sqrt{2} + \sqrt{3}$; $-\sqrt{2} - \sqrt{3}$

$\therefore$ The corresponding factors are $x - \sqrt{2} - \sqrt{3}$; $x - \sqrt{2} + \sqrt{3}$; $x + \sqrt{2} - \sqrt{3}$; $x + \sqrt{2} + \sqrt{3}$.

$\therefore$ The required equation is $(x - \sqrt{2} - \sqrt{3})(x - \sqrt{2} + \sqrt{3})(x + \sqrt{2} - \sqrt{3})(x + \sqrt{2} + \sqrt{3}) = 0$

Multiplying the factors in pairs, we have $[(x - \sqrt{2})^2 - 3][(x + \sqrt{2})^2 - 3] = 0$

or $$(x^2 - 2\sqrt{2}\,x - 1)(x^2 + 2\sqrt{2}\,x - 1) = 0$$

or $$(x^2 - 1)^2 - (2\sqrt{2}\,x)^2 = 0 \quad \text{or} \quad x^4 - 2x^2 + 1 - 8x^2 = 0$$

or $$x^4 - 10x^2 + 1 = 0.$$

Second Method

Let $$x = \sqrt{2} + \sqrt{3} \quad \therefore \quad x - \sqrt{2} = \sqrt{3}$$

$$\therefore \quad (x - \sqrt{2})^2 = 3 \quad \text{or} \quad x^2 - 2\sqrt{2}x + 2 - 3 = 0 \quad \text{or} \quad x^2 - 1 = 2\sqrt{2}x$$

Square again to remove radical sign

$$\therefore \quad (x^2 - 1)^2 = 8x^2 \text{ or } x^4 - 2x^2 + 1 - 8x^2 = 0 \quad \text{or} \quad x^4 - 10x^2 + 1 = 0$$

is the required equation.

Example 3. *Solve the equation $x^4 + x^3 - 25x^2 + 41x + 66 = 0$, given that one root is equal to $3 + i\sqrt{2}$.*

Sol. Given equation is $x^4 + x^3 - 25x^2 + 41x + 66 = 0$ ...(1)

One root is $3 + i\sqrt{2}$

$\therefore$ The other root is $3 - i\sqrt{2}$

Equation corresponding to these two roots is $(x - 3 - i\sqrt{2})(x - 3 + i\sqrt{2})$

or $(x-3)^2 - (i\sqrt{2})^2$ or $x^2 - 6x + 9 + 2$ or $x^2 - 6x + 11$

Divide the L.H.S. of (1) by $x^2 - 6x + 11$

i.e.,

$$\begin{array}{r|l} x^2 - 6x + 11 & x^4 + x^3 - 25x^2 + 41x + 66 \;(x^2 + 7x + 6 \\ & x^4 - 6x^3 + 11x^2 \\ & - \quad + \quad - \\ \hline & 7x^3 - 36x^2 + 41x \\ & 7x^3 - 42x^2 + 77x \\ & - \quad + \quad - \\ \hline & 6x^2 - 36x + 66 \\ & 6x^2 - 36x + 66 \\ \hline & \times \end{array}$$

The quotient is $x^2 + 7x + 6$.

Putting the quotient = 0, we get

$$x^2 + 7x + 6 = 0 \quad \text{or} \quad (x + 6)(x + 1) = 0$$

which gives $x = -1$ and $x = -6$.

Hence the roots are $-1, -6, 3 + i\sqrt{2}\,;\, 3 - i\sqrt{2}$.

Example 4. *If 1, $\alpha, \beta, \gamma, \delta$ are the roots of the equation $x^5 - 1 = 0$, then prove that*

$$(1 - \alpha)(1 - \beta)(1 - \gamma)(1 - \delta) = 5$$

Sol. $\because$ $1, \alpha, \beta, \gamma, \delta$ are the roots of $x^5 - 1 = 0$

$\therefore$ $x^5 - 1 = (x - 1)(x - \alpha)(x - \beta)(x - \gamma)(x - \delta)$

dividing both sides by $(x - 1)$, we get

$$\frac{x^5 - 1}{x - 1} = (x - \alpha)(x - \beta)(x - \gamma)(x - \delta)$$

i.e., $$x^4 + x^3 + x^2 + x + 1 = (x - \alpha)(x - \beta)(x - \gamma)(x - \delta)$$

This being true for all values of x.

Put $x = 1$

$\therefore$ $$1 + 1 + 1 + 1 + 1 = (1 - \alpha)(1 - \beta)(1 - \gamma)(1 - \delta)$$

Hence $(1 - \alpha)(1 - \beta)(1 - \gamma)(1 - \delta) = 5$.

Example 5. *Show that the equation $\frac{A^2}{x-a} + \frac{B^2}{x-b} + \frac{C^2}{x-c} = k$ where A, B, C, a, b, c are all real and different cannot have any imaginary root.*

Sol. Given equation is $\dfrac{A^2}{x-a}+\dfrac{B^2}{x-b}+\dfrac{C^2}{x-c}=k$ ...(1)

If possible let it have an imaginary root $(\alpha + i\beta)$.

Now (1) is an equation with real co-efficients

$\therefore$ There must be a root $\alpha - i\beta$ of (1).

Putting $x = \alpha + i\beta$ and $x = \alpha - i\beta$ in (1), we have

$$\frac{A^2}{(\alpha+i\beta)-a}+\frac{B^2}{(\alpha+i\beta)-b}+\frac{C^2}{(\alpha+i\beta)-c}=k$$

and $$\frac{A^2}{(\alpha-i\beta)-a}+\frac{B^2}{(\alpha-i\beta)-b}+\frac{C^2}{(\alpha-i\beta)-c}=k \quad ...(2)$$

Re-writing these equations, we get

$$\frac{A^2}{(\alpha-a)+i\beta}+\frac{B^2}{(\alpha-b)+i\beta}+\frac{C^2}{(\alpha-c)+i\beta}=k \quad ...(3)$$

and $$\frac{A^2}{(\alpha-a)-i\beta}+\frac{B^2}{(\alpha-b)-i\beta}+\frac{C^2}{(\alpha-c)-i\beta}=k \quad ...(4)$$

Subtract (4) from (3),

$$\therefore \quad A^2\left[\frac{1}{(\alpha-a)+i\beta}-\frac{1}{(\alpha-a)-i\beta}\right]+B^2\left[\frac{1}{(\alpha-b)+i\beta}-\frac{1}{(\alpha-b)-i\beta}\right]$$
$$+C^2\left[\frac{1}{(\alpha-c)+i\beta}-\frac{1}{(\alpha-c)-i\beta}\right]=0$$

$$A^2\left[\frac{\alpha-a-i\beta-\alpha+a-i\beta}{(\alpha-a)^2+\beta^2}\right]+B^2\left[\frac{\alpha-b-i\beta-\alpha+b-i\beta}{(\alpha-b)^2+\beta^2}\right]$$
$$+C^2\left[\frac{\alpha-c-i\beta-\alpha+c-i\beta}{(\alpha-c)^2+\beta^2}\right]=0$$

i.e., $$-2i\beta\left[\frac{A^2}{(\alpha-a)^2+\beta^2}+\frac{B^2}{(\alpha-b)^2+\beta^2}+\frac{C^2}{(\alpha-c)^2+\beta^2}\right]=0.$$

Since A, B, C, α, β, a, b, c are all real and different.

$\therefore$ The expression inside the bracket being the sum of squares cannot be zero.

$\therefore$ $i\beta = 0$ or $\beta = 0$

$\therefore$ The root $\alpha + i\beta$ no longer remains imaginary.

Hence the given equation cannot have any imaginary root.

EXERCISE 2

1. Find the equation of the lowest degree with rational co-efficients which has 1, $2+\sqrt{3}$, $\sqrt{2}-1$ as of roots.

2. (*a*) Form an equation of the lowest degree with rational co-efficients if one of its roots is $\sqrt{2}+\sqrt{5}$.

   (*b*) Construct a polynomial equation over rationals of degree 4 whose roots are $\sqrt{3}$ and $1+2i$.

3. If the roots of the equation $x^n - 1 = 0$ be 1, α, β, γ, ...; show that
$$(1 - \alpha)(1 - \beta)(1 - \gamma) \ldots\ldots = n.$$

4. Solve the following equations completely :

(i) $x^3 + 6x + 20 = 0$, one root being $1 + 3i$.

(ii) $x^4 + 2x^3 - 16x^2 - 22x + 7 = 0$, given that one root is $2 + \sqrt{3}$.

(iii) $x^4 + 4x^3 + 6x^2 + 4x + 5 = 0$ if one root is $\sqrt{-1}$.

(iv) $x^6 - x^5 - 8x^4 + 2x^3 + 21x^2 - 9x - 54 = 0$, given that one root is $\sqrt{2} + i$.

5. If b and c are real and $2 - \sqrt{-3}$ is root of the equation $x^3 + x^2 + bx + c = 0$, what are its other roots and what is the value of c ?

6. Form an equation whose roots are $m + n$, $m\omega + n\omega^2$, $m\omega^2 + n\omega$, it being given that 'ω' is one of the complex cube roots of unity.

7. If α, β, γ are roots of the equation $(a - x)(b - x)(c - x) + 1 = 0$, then prove that a, b, c are roots of the equation $(x - \alpha)(x - \beta)(x - \gamma) + 1 = 0$.

8. Show that the equation $\dfrac{A^2}{x - a} + \dfrac{B^2}{x - b} + \dfrac{C^2}{x - c} + \ldots\ldots + \dfrac{L^2}{x - l} = x - m$, where A, B, C, ... a, b, c are all real and different can't have any imaginary root.

ANSWERS

1. $x^5 - 3x^4 - 6x^3 + 14x^2 - 7x + 1 = 0$

2. (a) $x^4 - 14x^2 + 9 = 0$ (b) $x^4 - 2x^3 + 2x^2 + 6x - 15 = 0$

4. (i) $1 + 3i, 1 - 3i, -2$ (ii) $2 \pm \sqrt{3}, -3 \pm \sqrt{2}$

(iii) $\pm i, -2 \pm i$ (iv) $\sqrt{2} \pm i, -\sqrt{2} \pm i, 3, -2$

5. $2 \pm \sqrt{-3}, -5, c = 35$ 6. $x^3 - 3mnx - (m^3 + n^3) = 0$.

19. ESTABLISH THE RELATIONS BETWEEN THE ROOTS AND COEFFICIENTS OF A GENERAL POLYNOMIAL EQUATION OF THE nth DEGREE

Let the given nth degree equation be
$$f(x) \equiv a_0x^n + a_1x^{n-1} + a_2x^{n-2} + \ldots\ldots + a_{n-1}x + a_n = 0 \qquad (a_0 \neq 0)$$
Let $\alpha_1, \alpha_2, \alpha_3, \ldots\ldots, \alpha_n$ be its n roots.
$$\therefore \quad f(x) = a_0(x - \alpha_1)(x - \alpha_2) \ldots (x - \alpha_n).$$
Equating the two expressions for the polynomial $f(x)$, we have
$$a_0x^n + a_1x^{n-1} + a_2x^{n-2} + \ldots\ldots + a_{n-1}x + a_n = a_0(x - \alpha_1)(x - \alpha_2) \ldots\ldots (x - \alpha_n).$$
Dividing by a_0 and multiplying the factors on the R.H.S., we have
$$x^n + \frac{a_1}{a_0}x^{n-1} + \frac{a_2}{a_0}x^{n-2} + \ldots + \frac{a_{n-1}}{a_0}x + \frac{a_n}{a_0}$$
$$= x^n - \Sigma\alpha_1x^{n-1} + \Sigma\alpha_1\alpha_2x^{n-2} - \Sigma\alpha_1\alpha_2\alpha_3x^{n-3} + \ldots\ldots + (-1)^n\alpha_1\alpha_2\ldots\ldots\alpha_n \quad \ldots(1)$$
where $\Sigma\alpha_1$ = the sum of all the roots taken one by one

and $\Sigma\alpha_1\alpha_2$ = the sum of the products of the roots taken *two* at a time

$\Sigma\alpha_1\alpha_2\alpha_3$ = the sum of the products of the roots taken *three* at a time etc.

Equating the co-efficients of like powers of x on both sides of (1), we get

$$\left.\begin{aligned}
\Sigma\alpha_1 &= \frac{-a_1}{a_0} = (-1)^1 . \frac{a_1}{a_0} \\
\Sigma\alpha_1\alpha_2 &= \frac{a_2}{a_0} = (-1)^2 . \frac{a_2}{a_0} \\
\Sigma\alpha_1\alpha_2\alpha_3 &= \frac{-a_3}{a_0} = (-1)^3 . \frac{a_3}{a_0} \\
&\dots\dots\dots\dots \\
\Sigma\alpha_1\alpha_2 \dots\dots \alpha_r &= (-1)^r . \frac{a_r}{a_0} \\
\alpha_1\alpha_2 \dots\dots \alpha_n &= (-1)^n . \frac{a_n}{a_0}
\end{aligned}\right\} \quad \dots(2)$$

Thus, results (2) give the relations between the roots and the co-efficients of the equation $f(x) = 0$.

Note. We shall denote $\Sigma\alpha_1, \Sigma\alpha_1\alpha_2, \Sigma\alpha_1\alpha_2 \dots\dots \alpha_r$ by $\sigma_1, \sigma_2, \dots\dots \sigma_r$ respectively.

Remember. In the complete equation $f(x) = 0$

$$\begin{aligned}
\sigma_r &= \Sigma\alpha_1\alpha_2 \dots \alpha_r \\
&= \text{(Sum of the products of roots taken } r \text{ at a time)} \\
&= (-1)^r . \frac{a_r}{a_0} \\
&= (-1)^r . \frac{\textbf{co-efficient of (r + 1)th term}}{\textbf{co-efficient of 1st term}}
\end{aligned}$$

Particular Cases :

(*i*) If α, β are the roots of the equation

$$ax^2 + bx + c = 0, \quad \text{then } \sigma_1 \equiv \alpha + \beta = -\frac{b}{a} \quad \text{and} \quad \sigma_2 \equiv \alpha\beta = \frac{c}{a}$$

(*ii*) If α, β, γ are the roots of the equation

$$ax^3 + bx^2 + cx + d = 0, \quad \text{then } \sigma_1 \equiv \alpha + \beta + \gamma = -\frac{b}{a}$$

$$\sigma_2 \equiv \alpha\beta + \alpha\gamma + \beta\gamma = \frac{c}{a}, \ \sigma_3 = \alpha\beta\gamma = -\frac{d}{a}$$

Remark. It is sometimes convenient to write as $\sigma_2 = \alpha(\beta + \gamma) + \beta\gamma$.

(*iii*) If $\alpha, \beta, \gamma, \delta$ are the roots of the equation

$$ax^4 + bx^3 + cx^2 + dx + e = 0,$$

then

$$\sigma_1 \equiv \alpha + \beta + \gamma + \delta = -\frac{b}{a}$$

$$\sigma_2 \equiv \alpha\beta + \alpha\gamma + \alpha\delta + \beta\gamma + \beta\delta + \gamma\delta = \frac{c}{a}$$

$$\sigma_3 \equiv \alpha\beta\gamma + \alpha\beta\delta + \alpha\gamma\delta + \beta\gamma\delta = -\frac{d}{a}$$

$$\sigma_4 = \alpha\beta\gamma\delta = \frac{e}{a}.$$

Remarks. It is useful to write as

$$\sigma_2 = (\alpha + \beta)(\gamma + \delta) + \alpha\beta + \gamma\delta$$
$$\sigma_3 = \alpha\beta(\gamma + \delta) + \gamma\delta(\alpha + \beta).$$

Note 1. When sum of roots is known, it is always convenient to take :

(*i*) **three roots in A.P. as**

$$\alpha - \beta, \alpha, \alpha + \beta.$$

(*ii*) **four roots in A.P. as**

$$\alpha - 3\beta, \alpha - \beta, \alpha + \beta, \alpha + 3\beta.$$

When product of roots is known, it is convenient to take :

(*iii*) **three roots in G.P. as**

$$\frac{\alpha}{\beta}, \alpha, \alpha\beta.$$

(*iv*) **four roots in G.P. as**

$$\frac{\alpha}{\beta^3}, \frac{\alpha}{\beta}, \alpha\beta, \alpha\beta^3.$$

Note 2. (*i*) When four roots are in A.P., then the sum of 1st and 4th root = sum of 2nd and 3rd.

(*ii*) When four roots are in G.P., then the product of 1st and 4th = product of 2nd and 3rd.

Example 1. *Solve the equation $x^3 - 5x^2 - 16x + 80 = 0$ it being given that the sum of its two roots is zero.*

Sol. Given equation is $x^3 - 5x^2 - 16x + 80 = 0$ ...(1)

Let its three roots be α, β, γ such that $\beta + \gamma = 0$.

Now $\sigma_1 = \alpha + \beta + \gamma = 5$ $\left[-\dfrac{\text{co-effi. of } x^2}{\text{co-effi. of } x^3}\right]$

Since $\beta + \gamma = 0 \therefore \alpha = 5$

i.e., one root of the equation (1) is 5.

$\therefore$ Dividing L.H.S. of (1) by $(x - 5)$, we have

5)	1	– 5	– 16	80
		5	0	– 80
	1	0	– 16	| zero

$\therefore$ The quotient is $x^2 - 16$.

Putting the quotient to zero, *i.e.*, $x^2 - 16 = 0$

$\therefore$ $x^2 = 16$ *i.e.*, $x = \pm 4$

Hence the three roots are – 4, 4, 5.

Example 2. *Solve the equation $x^3 - 9x^2 + 23x - 15 = 0$, the roots being in A.P.*

Sol. Given equation is $x^3 - 9x^2 + 23x - 15 = 0$ ...(1)

Let its roots in A.P. be $(\alpha - \beta)$, α, $(\alpha + \beta)$

$\therefore$ $\sigma_1 = \alpha - \beta + \alpha + \alpha + \beta = 9$

i.e., $3\alpha = 9$ or $\alpha = 3$.

But 'α' is a root of the equation (1)

i.e., one root of (1) is 3.

$\therefore$ Dividing L.H.S. of (1) by $(x - 3)$, we get

3)	1	− 9	23	− 15
		3	− 18	15
	1	− 6	5	| zero

The quotient is $x^2 - 6x + 5$.

Put the quotient $= 0$ *i.e.*, $x^2 - 6x + 5 = 0$ or $(x - 5)(x - 1) = 0$

i.e., $x = 1$ and $x = 5$

Hence the roots of (1) are 1, 3, 5.

Example 3. *Solve the equation* $2x^3 - 9x^2 + 12x - 4 = 0$, *given that two of its roots are equal.*

Sol. Given equation is $2x^3 - 9x^2 + 12x - 4 = 0$

Let its roots be α, α, β

$\therefore$ $\sigma_1 = \alpha + \alpha + \beta = 9/2$ *i.e.*, $2\alpha + \beta = \frac{9}{2}$ ...(1)

$\sigma_2 = \alpha.\alpha + \alpha.\beta + \alpha.\beta = \frac{12}{2}$ *i.e.*, $\alpha^2 + 2\alpha\beta = 6$ ...(2)

$\sigma_3 = \alpha.\alpha.\beta = \frac{4}{2}$ *i.e.*, $\alpha^2\beta = 2$ ...(3)

From (1), $\beta = \frac{9}{2} - 2\alpha$

Putting in (2), we have

$\alpha^2 + 2\alpha(\frac{9}{2} - 2\alpha) = 6$ or $\alpha^2 + 9\alpha - 4\alpha^2 = 6$

i.e., $3\alpha^2 - 9\alpha + 6 = 0$ or $\alpha^2 - 3\alpha + 2 = 0$ or $(\alpha - 2)(\alpha - 1) = 0$

$\therefore$ Either $\alpha = 1$ or $\alpha = 2$

Put $\alpha = 1$ in the given equation

$\therefore$ $2 - 9 + 12 - 4 \neq 0$.

$\therefore$ $\alpha = 1$ is not a root of the given equation and is, therefore rejected.

$\therefore$ $\alpha = 2$

$\therefore$ From (3), $\beta = \frac{1}{2}$

Hence the roots are 2, 2, $\frac{1}{2}$.

Example 4. *Solve the cubic* $x^3 - 7x^2 + 14x - 8 = 0$; *it being given that its roots are in G.P.*

Sol. Given equation is $x^3 - 7x^2 + 14x - 8 = 0$ ...(1)

Let its three roots be $\frac{\alpha}{\beta}$, α, $\alpha\beta$

$\therefore$ $\sigma_3 = \frac{\alpha}{\beta} . \alpha . \alpha\beta = 8$

or $\alpha^3 = 8$ *i.e.*, $\alpha = 2$.

But 'α' is one of the roots of (1), $\therefore$ one root of (1) is = 2.

Divide L.H.S. of (1) by the factor $(x - 2)$.

2)	1	− 7	14	− 8
		2	− 10	8
	1	− 5	4	| zero

$\therefore$ The quotient is $x^2 - 5x + 4$

Put the quotient = 0 *i.e.*, $x^2 - 5x + 4 = 0$ or $(x-1)(x-4) = 0, x = 1, 4$

Hence the roots of (1) are 1, 2, 4.

Example 5. *Solve the equation $4x^4 + 8x^3 + 13x^2 + 2x + 3 = 0$. It being given that the sum of two of its roots is zero.*

Or

Solve the equation $4x^4 + 8x^3 + 13x^2 + 2x + 3 = 0$, given that one of its roots is purely imaginary.

Sol. Given equation is $4x^4 + 8x^3 + 13x^2 + 2x + 3 = 0$ ...(1)

Let its roots be $\alpha, \beta, \gamma, \delta$ with $\alpha + \beta = 0$

Now $\alpha + \beta + \gamma + \delta = (\text{sum of roots}) = -\frac{8}{4} = -2$

But $\alpha + \beta = 0 \quad \therefore \quad \gamma + \delta = -2$

Let $\alpha\beta = k$ and $\gamma\delta = k'$

Thus α and β are the roots of $x^2 - (0)x + k = 0$

and γ and δ are the roots of $x^2 - (-2)x + k' = 0$

[x^2 – (sum of roots) x + product of roots = 0]

Thus bi-quadratic (1) resolves into two quadratic factors.

$$\therefore \quad 4x^4 + 8x^3 + 13x^2 + 2x + 3 = 4(x^2 + k)(x^2 + 2x + k')$$

[Note that coefficients of highest degree terms must be same on both sides]

$\therefore$ Equating the coefficients of like powers of x, *i.e.*, of x^2, x and constant terms, we have

$$13 = 4(k' + k) \qquad \therefore \quad k + k' = 13/4 \qquad ...(2)$$

$$2 = 8k \qquad \therefore \quad k = \tfrac{1}{4} \qquad ...(3)$$

$$3 = 4kk' \qquad \therefore \quad kk' = \tfrac{3}{4} \qquad ...(4)$$

$$k = \tfrac{1}{4} \qquad \therefore \quad \text{From (2), } k' = 3$$

These two values of k and k' satisfy (4) also.

Hence (1) breaks into two quadratics which are

$x^2 + \frac{1}{4} = 0$ and $x^2 + 2x + 3 = 0$

$\therefore \quad x^2 = -\frac{1}{4}$

i.e., $x = \pm\sqrt{-\frac{1}{4}} = \pm\frac{i}{2}$

$$x = \frac{-2 \pm \sqrt{4 - 12}}{2} = \frac{-2 \pm 2\sqrt{2}i}{2} = -1 \pm i\sqrt{2}$$

Hence the roots of the given equation are $\frac{i}{2}, -\frac{i}{2}, -1 + i\sqrt{2}, -1 - i\sqrt{2}$.

Second Method

Equation is $4x^4 + 8x^3 + 13x^2 + 2x + 3 = 0$ ...(*i*)

Let its roots be $\alpha, \beta, \gamma, \delta$ with $\alpha + \beta = 0$

$$\sigma_1 = \alpha + \beta + \gamma + \delta = -\frac{8}{4} = -2$$

But $\alpha + \beta = 0 \quad \therefore \quad \gamma + \delta = -2$ ...(*ii*)

$$\sigma_2 = (\alpha + \beta)(\gamma + \delta) + \alpha\beta + \gamma\delta = \frac{13}{4}$$

$$\therefore \quad \alpha\beta + \gamma\delta = \frac{13}{4} \quad \text{...(iii)}$$

$$\sigma_3 = \alpha\beta(\gamma + \delta) + \gamma\delta(\alpha + \beta) = -\frac{2}{4} = -\frac{1}{2}$$

$$\therefore \quad \alpha\beta\,(-2) = -\tfrac{1}{2} \quad \text{or} \quad \alpha\beta = \tfrac{1}{4} \quad \text{...(iv)}$$

From (*iii*), $\gamma\delta = \frac{13}{4} - \frac{1}{4} = 3.$

Thus $\alpha + \beta = 0,\ \alpha\beta = \frac{1}{4}$

and $\gamma + \delta = -2,\ \gamma\delta = 3$

$\therefore$ α, β are the roots of $x^2 - x(\text{S}) + \text{P} = 0$, *i.e.*, $x^2 + \frac{1}{4} = 0$

$\Rightarrow \quad x^2 = -\frac{1}{4} \quad \therefore \quad x = \pm i\,\frac{1}{2}$

γ, δ are the roots of $x^2 - x(-2) + 3 = 0$

or $x^2 + 2x + 3 = 0$

$$\therefore \quad x = \frac{-2 \pm \sqrt{4 - 12}}{2} = \frac{-2 \pm 2\sqrt{2}i}{2} = -1 \pm i\sqrt{2}.$$

Hence roots are $\pm \frac{i}{2}, -1 \pm i\sqrt{2}$.

Example 6. *Solve the equation $x^4 - 8x^3 + 14x^2 + 8x - 15 = 0$ the roots being in A.P.*

Sol. Given equation is $x^4 - 8x^3 + 14x^2 + 8x - 15 = 0$ ...(*i*)

The roots are in A.P.

$\therefore$ Sum of 1st and 4th root = sum of 2nd and 3rd root

$\therefore$ If the roots be $\alpha, \beta, \gamma, \delta$, we have $\alpha + \delta = \beta + \gamma$

But $\alpha + \beta + \gamma + \delta = 8 \quad \therefore \quad \alpha + \delta = \beta + \gamma = 4$

i.e., $\alpha + \delta = 4$ and $\beta + \gamma = 4$

Let $\alpha\delta = k$ and $\beta\gamma = k'$

$\therefore$ α and δ are the roots of $x^2 - 4x + k = 0$ and β and γ are the roots of $x^2 - 4x + k' = 0$

Thus given equation resolves into the two quadratic equations.

$$\therefore \quad x^4 - 8x^3 + 14x^2 + 8x - 15 = (x^2 - 4x + k)(x^2 - 4x + k')$$

Equating the co-efficients of like powers of x, *i.e.*, of x^2, x and constant terms, we get

$$14 = k + k' + 16 \quad \therefore \quad k + k' = -2$$

$$8 = -4\,(k + k') \quad \therefore \quad k + k' = -2$$

$$-15 = kk' \quad \therefore \quad kk' = -15$$

$\therefore$ Solving for k and k', we get $k = -5$ and $k' = 3$.

Thus the two quadratic equations are

$x^2 - 4x - 5 = 0$ and $x^2 - 4x + 3 = 0$

$\therefore$ $(x - 5)(x + 1) = 0$ and $(x - 3)(x - 1) = 0$

i.e., $x = -1, 5$ and $x = 1$ and 3.

Hence the roots are $-1, 1, 3$ and 5.

Example 7. *Solve the equation $3x^4 - 40x^3 + 130x^2 - 120x + 27 = 0$ given that the product of two of its roots is equal to the product of the other two roots.*

Sol. Given equation is $3x^4 - 40x^3 + 130x^2 - 120x + 27 = 0$ ...(1)

Let its roots be α, β, γ and δ

and let. $\alpha\beta = \gamma\delta$

Product of all the roots $= \alpha\beta\gamma\delta = \frac{27}{3} = 9$

$\therefore \quad \alpha\beta = \gamma\delta = \pm 3$

Let us take $\alpha\beta = 3 \quad \therefore \quad \gamma\delta = 3$

let $\alpha + \beta = k$ and $\gamma + \delta = k'$

$\therefore$ α, β are the roots of $x^2 - kx + 3 = 0$ | γ, δ are the roots of $x^2 - k'x + 3 = 0$

$\therefore \quad 3x^4 - 40x^3 + 130x^2 - 120x + 27 = 3(x^2 - kx + 3)(x^2 - k'x + 3).$

[Note the step of multiplying the R.H.S. by 3, to make co-efficients of leading terms equal.]

Equating the co-efficients of like powers of x

$$\therefore \quad -40 = -3(k' + k) \quad \therefore \quad k + k' = \frac{40}{3} \qquad ...(i)$$

$$130 = 3(3 + 3 + kk') \quad \therefore \quad kk' = \frac{130}{3} - 6 = \frac{112}{3} \qquad ...(ii)$$

$$-120 = 3(-3k - 3k') \quad \therefore \quad k + k' = \frac{40}{3} \qquad ...(iii)$$

Since equations (*i*) and (*iii*) are same, therefore, our choice $\alpha\beta = 3$ is correct.

Thus $k + k' = \frac{40}{3}$ and $kk' = \frac{112}{3}$

$\therefore$ k, k' are the roots of $t^2 - \frac{40}{3}t + \frac{112}{3} = 0$

$$\therefore \quad 3t^2 - 40t + 112 = 0$$

$$\therefore \quad t = \frac{40 \pm \sqrt{1600 - 1344}}{6} = \frac{40 \pm 16}{6} = \frac{28}{3} \text{ and } 4.$$

Thus let $k = 4 \quad \therefore \quad k' = \frac{28}{3}$

$\therefore$ The two quadratic equations are

$x^2 - 4x + 3 = 0$ and $x^2 - \frac{28}{3}x + 3 = 0$

$\therefore \quad (x - 3)(x - 1) = 0$ and $3x^2 - 28x + 9 = 0$

$\therefore \quad x = 1, 3$ and $3x^2 - 27x - x + 9 = 0$

i.e., $3x(x - 9) - 1(x - 9) = 0$

i.e., $(x - 9)(3x - 1) = 0$

i.e., $x = 9$ and $x = \frac{1}{3}$

Hence the roots are $\frac{1}{3}$, 1, 3 and 9.

Remark 1. In the solution of problems as the above one, on equating the co-efficients of like powers, we usually get three equations in k and k'. If the value of k and k' obtained from any two satisfy the third equation, then our taking $\alpha\beta = \gamma\delta = 3$ is justified. In case these equations in k and k' are not all satisfied, we shall take $\alpha\beta = \gamma\delta = -3$.

Remark 2. In a bi-quadratic when the roots are given to be in G.P., we can follow the above method as we know that if four numbers are in G.P., the product of 1st and 4th is equal to the product of second and third.

EXERCISE 3

1. Solve the following equations :
   (*i*) $f(x) = 2x^3 - 7x^2 - 4x + 14 = 0$, it being given that $\sqrt{2}$ is a root of it.
   (*ii*) $x^3 - 5x^2 - 2x + 24 = 0$, the product of two roots being 12.
2. Solve the equations :
   (*i*) $32x^3 - 48x^2 + 22x - 3 = 0$, the roots being in A.P.
   (*ii*) $x^3 + 3x^2 - 6x - 8 = 0$, the roots being in G.P.
   (*iii*) $4x^3 - 4x^2 - 15x + 18 = 0$, two of its roots being equal.
3. Solve the equations :
   (*i*) $x^3 - 7x^2 + 36 = 0$, one root being double the other.
   (*ii*) $x^3 - 13x^2 + 15x + 189 = 0$, one root exceeds the other by 2.
   (*iii*) $x^3 - 9x^2 + 14x + 24 = 0$, two of its roots being in the ratio 2 : 3.
4. (*i*) Solve the equation $12x^3 - 8x^2 + kx + 18 = 0$ and find the value of k, given that one root is negative of another.
   (*ii*) Find the value of k in order that the roots of the equation $2x^3 + 6x^2 + 5x + k = 0$ are in A.P.
5. Solve the equation $x^4 + x^3 + 7x^2 + 4x + 12 = 0$, given that one of its roots is purely imaginary.
6. (*i*) Solve the equation $x^4 - 8x^3 + 23x^2 + 28x + 12 = 0$, the sum of two roots being equal to the sum of the other two.
   (*ii*) Solve the equation $x^4 - 8x^3 + 23x^2 - 28x + 12 = 0$, given that the difference of two of the roots is equal to the difference of the other two.
   [**Hint.** If the roots be α, β, γ and δ, it is given that $\alpha - \beta = \gamma - \delta$ which is same as $\alpha + \delta = \beta + \gamma$.]
7. (*i*) The sum of two roots of the equation $x^4 - 8x^3 + 19x^2 + 4\lambda x + 2 = 0$ is equal to the sum of the other two roots. Find λ and solve the equation.
   (*ii*) Solve the equation $x^4 + 2x^3 - 21x^2 - 22x + 40 = 0$, the roots being in A.P.
8. The roots of the equation $x^4 - 6x^3 + 18x^2 - 30x + 25 = 0$, are of the form $\alpha + i\beta$, $\beta + i\alpha$, where α, β are real. Find all the roots.
9. The equation $x^4 + 4x^3 - 2x^2 - 12x + 9 = 0$ has two pairs of equal roots ; find them.
10. (*a*) Solve the equation $x^4 + x^3 - 16x^2 - 4x + 48 = 0$ having given that the product of two roots is 6.
    (*b*) Solve $3x^4 + 17x^3 - 5x^2 + 8x + 12 = 0$ given that the product of two roots is unity.
11. Solve the equation $x^4 - 5x^3 + 10x^2 - 10x + 4 = 0$, given that product of its two roots is equal to the product of the other two.
12. Solve the equation $x^4 - 6x^3 + 11x^2 - 18x + 9 = 0$, given that the ratio of two roots is equal to the ratio of the other two.
    $\left[\textbf{Hint. } \frac{\alpha}{\beta} = \frac{\gamma}{\delta} \text{ i.e., } \alpha\delta = \beta\gamma.\right]$
13. Solve the equation $x^4 + 5x^3 - 30x^2 - 40x + 64 = 0$, whose roots are in G.P.
14. Solve the equation $27x^4 - 195x^3 + 494x^2 - 520x + 192 = 0$, the roots being in G.P.
    $\left[\textbf{Hint. } \text{To solve } 27t^2 - 195t + 350 = 0, \text{ factors are } 27t^2 - 105t - 90t + 350 = 0 \therefore t = \frac{10}{3}, \frac{35}{9}.\right]$

ANSWERS

1. (*i*) $\pm\sqrt{2}, \frac{7}{2}$ (*ii*) – 2, 3, 4

2. (*i*) $\frac{1}{4}, \frac{1}{2}, \frac{3}{4}$ (*ii*) – 1, 2, – 4 (*iii*) $\frac{3}{2}, \frac{3}{2}, -2$

3. (*i*) – 2, 3, 6 (*ii*) – 3, 7, 9 (*iii*) 4, 6, – 1

4. (*i*) $k = -27$, roots are $\pm\frac{3}{2}, \frac{2}{3}$ (*ii*) $k = 1$

5. $\pm 2i, \frac{-1\pm\sqrt{-11}}{2}$

6. (*i*) $\frac{1}{2}(3\pm\sqrt{5}), \frac{1}{2}(5\pm\sqrt{5})$ (*ii*) 1, 2, 2, 3

7. (*i*) $\lambda = -3, 2\pm\sqrt{3}, 2\pm\sqrt{2}$ (*ii*) – 5, – 2, 1, 4

8. $2\pm i, 1\pm 2i$

9. – 3, – 3, 1, 1

10. (*a*) 3, 2, – 2, – 4

10. (*b*) $-6, -\frac{2}{3}, \frac{1\pm\sqrt{-3}}{2}$,

11. $1, 2, 1\pm i$

12. $\frac{5\pm\sqrt{13}}{2}, \frac{1\pm\sqrt{11}\,i}{2}$

13. 1, – 2, 4, – 8.

14. $\frac{8}{9}, \frac{4}{3}$, 2, 3.

---

20. TO FIND CONDITIONS THAT THE ROOTS OF AN EQUATION MAY SATISFY A GIVEN RELATION AMONG THE ROOTS

Example 1. *Find the condition that the roots of the equation $ax^3 + bx^2 + cx + d = 0$, $(a \neq 0)$ may be in A.P.*

Sol. Given equation is $ax^3 + bx^2 + cx + d = 0$ ...(1)

Let its roots in A.P. be $\alpha - \beta, \alpha, \alpha + \beta$

$\therefore$ Sum of roots $= (\alpha - \beta) + \alpha + (\alpha + \beta) = -\frac{b}{a}$

or $3\alpha = -\frac{b}{a} \quad \therefore \quad \alpha = -\frac{b}{3a}$

But α is a root of (1) $\therefore$ It must satisfy equation (1).

$\therefore$ Putting $\alpha = -\frac{b}{3a}$ in (1), we have

$$a\left(-\frac{b}{3a}\right)^3 + b\left(-\frac{b}{3a}\right)^2 + c\left(-\frac{b}{3a}\right) + d = 0$$

or
$$-\frac{b^3}{27a^2} + \frac{b^3}{9a^2} - \frac{bc}{3a} + d = 0$$

or
$$-b^3 + 3b^3 - 9abc + 27a^2d = 0$$

or
$$2b^3 - 9abc + 27a^2d = 0$$

which is the required condition.

Example 2. *Find the condition that the roots of the equation $x^3 - px^2 + qx - r = 0$ be in geometrical progression.*

Hence solve the equation $27x^3 + 42x^2 - 28x - 8 = 0$.

Sol. Given equation is $x^3 - px^2 + qx - r = 0$ ...(1)

Let the three roots in G.P. be $\frac{\alpha}{\beta}, \alpha, \alpha\beta$.

$\therefore \quad \sigma_3 = \text{product of roots} = \dfrac{\alpha}{\beta} \cdot \alpha \cdot \alpha\beta = -(-r)$

$\Rightarrow \quad \alpha^3 = r$

But α is a root of equation (1)

$\therefore$ It satisfies (1)

$\therefore \quad \alpha^3 - p\alpha^2 + q\alpha - r = 0$, but $\alpha^3 = r$

$\therefore \quad r - p\alpha^2 + q\alpha - r = 0$

$\Rightarrow \quad p\alpha^2 = q\alpha$, as $(\alpha \neq 0)$

$\therefore \quad p\alpha = q$ Cubing both sides, $p^3\alpha^3 = q^3$

or $\quad p^3 r = q^3$ ...(2)

is the required condition.

To solve the equation $27x^3 + 42x^2 - 28x - 8 = 0$ ...(3)

Divide by 27,

$$\therefore \quad x^3 + \frac{42}{27}x^2 - \frac{28}{27}x - \frac{8}{27} = 0$$

Comparing its co-efficients with those of (1)

$$\therefore \quad p = -\frac{42}{27} = -\frac{14}{9}$$

$$q = -\frac{28}{27}\ ;\ r = \frac{8}{27}$$

Putting values of p, q, r in (2),

$$\therefore \quad \left(-\frac{14}{9}\right)^3 \left(\frac{8}{27}\right) = \left(-\frac{28}{27}\right)^3$$

or $$-\left(\frac{28}{27}\right)^3 = -\left(\frac{28}{27}\right)^3, \text{ which is true.}$$

$\therefore$ The roots of equation (3) are also in G.P.

$\therefore$ One root $\alpha = (r)^{1/3} = \left(\dfrac{8}{27}\right)^{1/3} = \dfrac{2}{3}$

$\Rightarrow \quad x - \frac{2}{3}$ is a factor, dividing by $(x - \frac{2}{3})$

$\frac{2}{3}$)	27	42	− 28	− 8
		18	40	8
	27	60	12	| 0

$\therefore$ The quotient is $27x^2 + 60x + 12$.

Putting $\quad 27x^2 + 60x + 12 = 0$

$\Rightarrow \quad 9x^2 + 20x + 4 = 0$. Solving it

$$\therefore \quad x = \frac{-20 \pm \sqrt{400 - 144}}{18} = \frac{-20 \pm 16}{18}$$

$$\therefore \quad x = -\frac{2}{9}, -2.$$

Hence roots are $-\dfrac{2}{9}, \dfrac{2}{3}, -2$.

Example 3. (*a*) *Find the condition that the sum of two roots of the equation $x^4 - px^3 + qx^2 - rx + s = 0$ may be equal to the sum of the other two.*

(*b*) *Hence solve the equation $x^4 - 8x^3 + 23x^2 - 28x + 12 = 0$.*

Sol. (*a*) Given equation is $x^4 - px^3 + qx^2 - rx + s = 0$ ...(1)

Let its roots be α, β, γ and δ such that $\alpha + \beta = \gamma + \delta$

Now, $\alpha + \beta + \gamma + \delta = p$

$$\therefore \quad \alpha + \beta = \gamma + \delta = \frac{p}{2}$$

Let $\alpha\beta = k$ and $\gamma\delta = k'$

$\therefore$ α and β are the roots of $x^2 - \frac{px}{2} + k = 0$ and γ and δ are the roots of $x^2 - \frac{p}{2}x + k' = 0$

Thus (1) resolves into two quadratic factors

i.e.,
$$x^4 - px^3 + qx^2 - rx + s = \left(x^2 - \frac{p}{2}x + k\right)\left(x^2 - \frac{p}{2}x + k'\right)$$

Equating the co-efficients of x^2, x and constant terms, we have

$$q = k + k' + \frac{p^2}{4} \qquad \therefore \quad k + k' = q - \frac{p^2}{4} \qquad \text{...(2)}$$

$$-r = -\frac{p}{2}(k + k') \qquad \therefore \quad k + k' = \frac{2r}{p} \qquad \text{...(3)}$$

$$s = kk' \qquad kk' = s \qquad \text{...(4)}$$

To find the required condition equating the two values of $(k + k')$ *i.e.*, from (2) and (3), we have

$$q - \frac{p^2}{4} = \frac{2r}{p}$$

or $\quad 4pq - p^3 = 8r \quad$ or $\quad p^3 - 4pq + 8r = 0$

which is the required condition.

(*b*) Given equation is $x^4 - 8x^3 + 23x^2 - 28x + 12 = 0$ ...(1)

Comparing it with $x^4 - px^3 + qx^2 - rx + s = 0$ ...(2)

$\therefore \quad p = 8, \quad q = 23, \quad r = 28, \quad s = 12.$

Putting these values in the condition obtained in part (*a*) *i.e.* in

$$p^3 - 4pq + 8r = 0$$

$$\therefore \quad (8)^3 - 4(8)(23) + 8(28) = 0$$

or $\quad 512 - 736 + 224 = 0$, which is true.

$\therefore$ Roots of equations (1) and (2) are of same type, *i.e.*, sum of two roots of Eq. (2) is equal to the sum of its other two roots.

$\therefore$ Equation (1) reduces into two quadratics

$$x^2 - \frac{p}{2}x + k = 0 \qquad \text{...(3)}$$

$$x^2 - \frac{p}{2}x + k' = 0 \qquad \text{...(4)}$$

where $\quad k + k' = \frac{2r}{p} = \frac{56}{8} = 7, \; kk' = s = 12$

$\Rightarrow \quad k, k'$ are the roots of $t^2 - 7t + 12 = 0 \qquad | \; x^2 - x(s) + p = 0$

$\Rightarrow$ $(t-3)(t-4)=0$

$\Rightarrow$ $t = 3, 4$ $\therefore$ let $k = 3, k' = 4$

Equations (3) and (4) become

$x^2 - 4x + 3 = 0$	$x^2 - 4x + 4 = 0$
$\therefore$ $(x-1)(x-3) = 0$	$(x-2)^2 = 0$
$\therefore$ $x = 1, 3.$	$\therefore$ $x = 2, 2.$

Hence roots are 1, 3, 2, 2.

Example 4. *If $Z^4 + 6HZ^2 + 4GZ + a_0^2I - 3H^2 = 0$ has two pairs of equal roots, prove that $G = 0$ and $a_0^2I - 12H^2 = 0$.*

Sol. Given equation is $Z^4 + 6HZ^2 + 4GZ + a_0^2I - 3H^2 = 0$. Let its roots be $\alpha, \alpha, \beta, \beta$.

$\therefore$ Sum of roots $= 2(\alpha + \beta) = 0$ *i.e.*, $\alpha + \beta = 0$

Let $\alpha\beta = p$

Now α, β are the roots of $Z^2 + p = 0$ $[Z^2 - Z(\text{sum}) + p = 0]$

$\therefore$ $Z^4 + 6HZ^2 + 4GZ + a_0^2I - 3H^2 = (Z^2 + p)^2 = Z^4 + 2pZ^2 + p^2$

$\therefore$ Comparing coeffi. of like powers of Z,

Comparing coeffi. of Z^2, $6H = 2p$ ...(*i*)

Comparing coeffi. of Z, $G = 0$ ...(*ii*)

Comparing constant term, $a_0^2I - 3H^2 = p^2$ ...(*iii*)

Eliminating '*p*' from (*i*) and (*iii*),

from (*i*), $p = 3H$, putting in (*iii*), we get

$a_0^2I - 12H^2 = 0$ and from (*ii*) $G = 0$.

EXERCISE 4

1. Find the relations between the roots and the coefficients of the polynomial equation
$$ax^4 + bx^3 + cx^2 + dx + e = 0.$$
2. Find the condition that the roots of the equation :
   (*i*) $x^3 - ax^2 + p = 0$, be in A.P.
   (*ii*) $px^3 + qx^2 + rx + s = 0$, may be in G.P.
   (*iii*) $x^3 + px^2 + qx + r = 0$, are such that the product of two of its roots is -1.
   (*iv*) $x^3 + ax^2 + bx + c = 0$, are such that one root is reciprocal of the other.
   (*v*) $x^3 - px^2 + qx - r = 0$, are such that sum of two of its roots is zero.
3. Find the condition that the equation $x^4 + px^3 + qx^2 + rx + s = 0$ should have two roots whose sum is zero. Hence or otherwise solve the equation $4x^4 + 8x^3 + 13x^2 + 2x + 3 = 0$.
4. (*a*) Find the condition that the sum of two roots of $f(x) = a_0x^3 + a_1x^2 + a_2x + a_3 = 0$ should also be a root of $f(x) = 0$. Verify the same condition for the equation $8x^3 - 8x^2 + 1 = 0$ and hence solve it.
   (*b*) Find the condition that sum of two roots of $f(x) = ax^3 + bx^2 + cx + d = 0$ should also be a root of $f(x) = 0$. Verify the same condition for $8x^3 - 8x^2 + 1 = 0$.
5. (*a*) Find the condition that the roots $\alpha, \beta, \gamma, \delta$ of the equation $x^4 + px^3 + qx^2 + rx + s = 0$ should be connected by the relation $\alpha\beta = \gamma\delta$.
   (*b*) If the product of two roots of the equation $x^4 + px^3 + qx^2 + rx + s = 0$ be equal in magnitude but opposite in sign to the product of other two, show that $p^2s + r^2 = 4qs$

ANSWERS

1. If $\alpha_1, \alpha_2, \alpha_3, \alpha_4$ are the four roots, then $\Sigma\alpha_1 = -\dfrac{b}{a}$, $\Sigma\alpha_1\alpha_2 = \dfrac{c}{a}$, $\Sigma\alpha_1\alpha_2\alpha_3 = -\dfrac{d}{a}$, $\alpha_1\alpha_2\alpha_3\alpha_4 = \dfrac{e}{a}$

2. (i) $27p = 2a^3$ (ii) $pr^3 = sq^3$ (iii) $r^2 + pr + q + 1 = 0$
(iv) $ac - c^2 - b + 1 = 0$ (v) $pq = r$

3. $pqr = r^2 + sp^2$; $\pm \frac{1}{2}i, -1 \pm i\sqrt{2}$

4. (a) $a_1^{\ 3} - 4a_0a_1a_2 + 8a_3a_0^{\ 2} = 0$; $\frac{1}{2}, \frac{1 \pm \sqrt{5}}{4}$ (b) $b^3 - 4abc + 8da^2 = 0$

5. (a) $p^2s = r^2$.

---

21. COMMON ROOTS OF EQUATIONS

Let $f(x) = 0$ and $g(x) = 0$ be two equations of degree n and m respectively.

Let $\alpha_1, \alpha_2, \alpha_3, \ldots\ldots, \alpha_r$ [$r \le$ min. (m, n)] be common roots of $f(x) = 0$ and $g(x) = 0$.

$\therefore$ $(x - \alpha_1)(x - \alpha_2) \ldots\ldots (x - \alpha_r)$ is the common factor of $f(x)$ and $g(x)$ and this is the highest common factor of $f(x)$ and $g(x)$ as there is no other common root.

Thus $(x - \alpha_1)(x - \alpha_2) \ldots\ldots (x - \alpha_r)$ is the H.C.F. of $f(x)$ and $g(x)$.

Thus to find the common roots of the equations $f(x) = 0$ and $g(x) = 0$, find the H.C.F. of $f(x)$ and $g(x)$. If the H.C.F. is denoted by $\varphi(x)$, then $\varphi(x) = 0$ gives common roots of $f(x) = 0$ and $g(x) = 0$.

Example 1. *Solve the equations $x^4 + 5x^3 - 22x^2 - 50x + 132 = 0$ and $x^4 + x^3 - 20x^2 + 16x + 24 = 0$ being given that they have common roots.*

Sol. Let us find the H.C.F. of the polynomials

$$x^4 + 5x^3 - 22x^2 - 50x + 132$$

and

$$x^4 + x^3 - 20x^2 + 16x + 24$$

	$x^4 + x^3 - 20x^2 + 16x + 24$	$x^4 + 5x^3 - 22x^2 - 50x + 132$	1
	2	$x^4 + x^3 - 20x^2 + 16x + 24$	
x	$2x^4 + 2x^3 - 40x^2 + 32x + 48$	− − + −	
	$2x^4 - x^3 - 33x^2 + 54x$	$4x^3 - 2x^2 - 66x + 108$	
	− + + −	$2(2x^3 - x^2 - 33x + 54)$	$2x$
	$3x^3 - 7x^2 - 22x + 48$	$2x^3 - 10x^2 + 12x$	
	2	− + −	
3	$6x^3 - 14x^2 - 44x + 96$	$9x^2 - 45x + 54$	9
	$6x^3 - 3x^2 - 99x + 162$	$9x^2 - 45x + 54$	
	− + + −	− + −	
	$-11x^2 + 55x - 66$	×	
	$-11(x^2 - 5x + 6)$		

$\therefore$ The H.C.F. of the two polynomials is $x^2 - 5x + 6$. The common roots of the equations are obtained by putting the H.C.F. = 0

i.e., $x^2 - 5x + 6 = 0$ or $(x - 2)(x - 3) = 0$

i.e., $x = 2$ and $x = 3$.

The other roots of the equations can be obtained by dividing the given polynomials by $(x - 2)$ and $(x - 3)$.

2)	1	5	– 22	– 50	132
		2	14	– 16	– 132
3)	1	7	– 8	–66	zero
		3	30	66	
	1	10	22	zero	

The quotient is $x^2 + 10x + 22$

Putting the quotient = 0, we have

$$x^2 + 10x + 22 = 0$$

$$\therefore \quad x = \frac{-10 \pm \sqrt{100-88}}{2} = \frac{-10 \pm 2\sqrt{3}}{2} = -5 \pm \sqrt{3}.$$

Hence all the roots of the first equation are

$$2, 3, -5 + \sqrt{3}, -5 - \sqrt{3}$$

Dividing the second polynomial by $(x - 2)$ and $(x - 3)$

i.e.,

2)	1	1	– 20	16	24
		2	6	– 28	– 24
3)	1	3	– 14	– 12	zero
		3	18	12	
	1	6	4	zero	

$\therefore$ The quotient is $x^2 + 6x + 4$.

Put the quotient = 0

i.e., $\quad x^2 + 6x + 4 = 0$

$$\therefore \quad x = \frac{-6 \pm \sqrt{36-16}}{2} = \frac{-6 \pm 2\sqrt{5}}{2} = -3 \pm \sqrt{5}$$

Hence the roots of the second equation are $2, 3, -3 + \sqrt{5}, -3 - \sqrt{5}$.

22. EQUAL ROOTS OR MULTIPLE ROOTS OR REPEATED ROOTS

Def. A root α is said to be a *repeated* or *multiple* root of an equation $f(x) = 0$ if α occurs more than once among the list of roots of the equation $f(x) = 0$ and if a root 'α' is repeated 'r' times then 'α' is said to be a root of *multiplicity* r. In this case $(x - \alpha)^r$ divides $f(x)$, whereas $(x - \alpha)^{r+1}$ does not divide $f(x)$.

Theorem. If a root 'α' is of multiplicity $r > 1$ of the equation $f(x) = 0$, then 'α' is a root of multiplicity $(r - 1)$ of the equation $f'(x) = 0$, where $f'(x)$ is the derivative of $f(x)$.

Proof. $\because$ α is root of $f(x) = 0$ of multiplicity r.

$\therefore$ $(x - \alpha)$ is a factor repeated r times of $f(x)$

i.e., $(x - \alpha)^r$ is a factor of $f(x)$.

$\therefore$ Divide $f(x)$ by $(x - \alpha)^r$. Let $\varphi(x)$ be the quotient.

$\therefore$ $f(x) = (x - \alpha)^r \varphi(x)$ ...(1), where $\varphi(\alpha) \neq 0$

Differentiating both sides of (1), w.r.t. x, we get

$$f'(x) = (x-\alpha)^r \varphi'(x) + r(x-\alpha)^{r-1}\varphi(x)$$
$$= (x-\alpha)^{r-1}[(x-\alpha)\varphi'(x) + r\varphi(x)] = (x-\alpha)^{r-1}h(x)$$

where $\quad h(x) = (x-\alpha)\varphi'(x) + r\varphi(x)$

Now $\quad h(\alpha) = 0 + r\varphi(\alpha) \neq 0$

Thus $\quad f'(x) = (x-\alpha)^{r-1} h(x)$, where $h(\alpha) \neq 0$.

This proves that α is root of $f'(x) = 0$ repeated $(r-1)$ times.

Cor. 1. By repeated application of above theorem, it follows that

if $\quad \alpha$ is a root repeated r times in $f(x) = 0$

then 'α' is a root repeated $(r-1)$ times in $f'(x) = 0$

and $\quad \alpha$ .... $(r-2)$ times in $f''(x) = 0$ etc.

Important Observations

The equation $f(x) = 0$ has **multiple roots** if $f(x)$ and $f'(x)$ have H.C.F. a polynomial $\varphi(x)$ at least of degree one.

If the H.C.F. of $f(x)$ and $f'(x)$ is a constant polynomial then $f(x) = 0$ does not have equal roots *i.e.,* all roots are distinct.

Method to find equal roots of an equation f(x) = 0

Find $f'(x)$.

Then find H.C.F. of $f(x)$ and $f'(x)$.

Let us call it $\phi(x)$.

Equate $\phi(x)$ to zero and solve it. It gives us **common** roots of $f(x) = 0$ and $f'(x) = 0$.

If α is a root of $\phi(x) = 0$ repeated k times, then α is a root of $f(x) = 0$ repeated $(k+1)$ times.

Cor. 2. Condition for two equal roots. If α is a root repeated twice, in the equation $f(x) = 0$, then α satisfies $f(x) = 0$ and $f'(x) = 0$.

$$\therefore \qquad f(\alpha) = 0 \qquad \text{...(1)}$$

$$\text{and} \qquad f'(\alpha) = 0 \qquad \text{...(2)}$$

The required condition, is then obtained by eliminating α between the equations (1) and (2).

Cor. 3. Condition for three equal roots. If α is a thrice repeated root of $f(x) = 0$, then α satisfies the equations

$$f(x) = 0\ ;\ f'(x) = 0\ ;\ f''(x) = 0$$

$$\therefore \qquad f(\alpha) = 0 \qquad \text{...(1)}$$

$$f'(\alpha) = 0 \qquad \text{...(2)}$$

$$f''(\alpha) = 0 \qquad \text{...(3)}$$

Condition is obtained by eliminating α from the equations (1), (2) and (3).

Example 2. *Solve the equation $x^3 - 3x^2 - 9x + 27 = 0$, which has a multiple root.*

Sol. Here $\qquad f(x) = x^3 - 3x^2 - 9x + 27$

$\therefore \qquad f'(x) = 3x^2 - 6x - 9.$

$\because \quad f(x)$ has a multiple root.

$\therefore \quad$ H.C.F. of $f(x)$ and $f'(x)$ is a polynomial of degree ≥ 1.

Let us find the H.C.F. of $f(x)$ and $f'(x)$

	$3x^2 - 6x - 9$	$x^3 - 3x^2 - 9x + 27$	$x - 1$
$x + 1$	$3(x^2 - 2x - 3)$	$x^3 - 2x^2 - 3x$	
	$x^2 - 3x$	$-\quad+\quad+$	
	$-\quad+$	$-x^2 - 6x + 27$	
	$x - 3$	$-x^2 + 2x + 3$	
	$x - 3$	$+\quad-\quad-$	
	$\times$	$-8x + 24$	
		$-8(x - 3)$	

Thus H.C.F. of $f(x)$ and $f'(x)$ is $\phi(x) = x - 3$.

Putting the H.C.F. = 0, we have $x - 3 = 0$ *i.e.*, $x = 3$.

$\therefore$ $x = 3$ is a root repeated once in $\phi(x) = 0$.

$\therefore$ $x = 3$ is a root repeated twice in $f(x) = 0$.

Now sum of 3 roots of $f(x) = 0$ is 3

and sum of 2 roots of $f(x) = 0$ is 6

$\therefore$ The third root $= 3 - 6 = -3$. Hence roots are $3, 3, -3$.

Example 3. *Solve $2x^3 - 9x^2 + 12x - 4 = 0$ given that two of its roots are equal.*

Sol. Given equation is $f(x) = 2x^3 - 9x^2 + 12x - 4 = 0$ ...(1)

$$\therefore \quad f'(x) = 6x^2 - 18x + 12 = 0$$

or $$x^2 - 3x + 2 = 0 \quad ....(2)$$

Since equation (1) has two equal roots, $\therefore$ the repeated root is also a root of equation (2).

Solving (2), we get $x = 1$ and 2

Putting $x = 1$ in (1), $2 - 9 + 12 - 4 \neq 0$

$\therefore$ $x = 1$ is not a root of given equation

$\therefore$ The twice repeated root of equation (1) is 2

$\therefore$ 3rd root = sum of three roots – sum of two roots $= \dfrac{9}{2} - 4 = \dfrac{1}{2}$

Hence roots are $2, 2, \frac{1}{2}$.

Example 4. *Without solving the equation*

$$x^3 - x^2 - 2x + 2 = 0$$

prove that it has no multiple roots.

Sol. Given equation is $x^3 - x^2 - 2x + 2 = 0$

Hence $$f(x) = x^3 - x^2 - 2x + 2$$

$$\therefore \quad f'(x) = 3x^2 - 2x - 2.$$

Let us find H.C.F. of $f(x)$ and $f'(x)$.

3	$3x^2 - 2x - 2$	$x^3 - x^2 - 2x + 2$	
	$3x^2 + 12x - 18$	3	
	$-\ \ -\ \ +$	$3x^3 - 3x^2 - 6x + 6$	x
	$-14x + 16$	$3x^3 - 2x^2 - 2x$	
	$-2(7x - 8)$	$-\ \ +\ \ +$	
	6	$-x^2 - 4x + 6$	
7	$42x - 48$	$-(x^2 + 4x - 6)$	
	$42x - 49$	7	
	$-\ \ +$	$7x^2 + 28x - 42$	x
	1	$7x^2 - 8x$	
		$-\ \ +$	
		$36x - 42$	
		$6(6x - 7)$	

This shows that H.C.F. of $f(x)$ and $f'(x)$ is not even a polynomial of degree 1.

$f(x) = 0$ has no multiple roots.

Example 5. *Find the condition that the equation $ax^3 + 3bx^2 + 3cx + d = 0$ may have two equal roots and show that each one of them is equal to $\dfrac{bc - ad}{2(ca - b^2)}$ or $\dfrac{2(bd - c^2)}{bc - ad}$.*

Sol. Given equation is $ax^3 + 3bx^2 + 3cx + d = 0$

Hence $\quad f(x) = ax^3 + 3bx^2 + 3cx + d$

$\therefore \quad f'(x) = 3ax^2 + 6bx + 3c$

Let the repeated root of $f(x) = 0$ be α

$\therefore \quad \alpha$ is also a root of $f'(x) = 0$

$$\therefore \quad a\alpha^3 + 3b\alpha^2 + 3c\alpha + d = 0 \qquad ...(1)$$

and $\quad 3a\alpha^2 + 6b\alpha + 3c = 0$

$$\text{or} \quad a\alpha^2 + 2b\alpha + c = 0 \qquad ...(2)$$

Multiplying (2) by α and subtracting from (1), we have

$$(a\alpha^3 + 3b\alpha^2 + 3c\alpha + d) - \alpha(a\alpha^2 + 2b\alpha + c) = 0$$

$$\textit{i.e.,} \quad b\alpha^2 + 2c\alpha + d = 0 \qquad ...(3)$$

(2) and (3) are two quadratics, in α

$\therefore$ From (2) and (3), by cross-multiplication, we have

$$\frac{\alpha^2}{2bd - 2c^2} = \frac{\alpha}{bc - ad} = \frac{1}{2ac - 2b^2}$$

From the first two equalities $\alpha = \dfrac{2(bd - c^2)}{bc - ad}$

From the last two equalities $\alpha = \dfrac{bc - ad}{2(ac - b^2)}$

Equating the two values of α, the required condition is

$$\frac{2(bd - c^2)}{bc - ad} = \frac{bc - ad}{2(ac - b^2)}$$

or $$(bc - ad)^2 = 4(ac - b^2)(bd - c^2)$$

and the common root is

either $$\frac{2(bd - c^2)}{bc - ad} \quad \text{or} \quad \frac{bc - ad}{2(ac - b^2)}.$$

Example 6. *Solve the equation $4x^4 - 4x^3 - 25x^2 + x + 6 = 0$ given that two of its roots differ by 1.*

Sol. The given equation is $4x^4 - 4x^3 - 25x^2 + x + 6 = 0$ ...(1)

Let its two roots be α, $\alpha + 1$

$\therefore$ $$4\alpha^4 - 4\alpha^3 - 25\alpha^2 + \alpha + 6 = 0 \quad ...(2)$$

and $$4(\alpha + 1)^4 - 4(\alpha + 1)^3 - 25(\alpha + 1)^2 + (\alpha + 1) + 6 = 0$$

i.e., $$4\alpha^4 + 12\alpha^3 - 13\alpha^2 - 45\alpha - 18 = 0 \quad ...(3)$$

(on simplification)

Clearly, (2) and (3) have a common root α.

One can verify that H.C.F. of L.H.S.'s of (2) and (3) is $2\alpha + 1$.

$\therefore$ $2\alpha + 1 = 0$ *i.e.,* $\alpha = -\frac{1}{2}$

$\therefore$ two roots of (1) are $-\frac{1}{2}, -\frac{1}{2} + 1$ *i.e.,* $-\frac{1}{2}, \frac{1}{2}$.

$\therefore$ $x + \frac{1}{2}$ and $x - \frac{1}{2}$ are factors of L.H.S. of (1).

$\therefore$ by synthetic division,

$-\frac{1}{2}$	4	– 4	– 25	1	6
		– 2	3	11	– 6
$\frac{1}{2}$	4	– 6	– 22	12	0
		2	– 2	– 12	
	4	– 4	– 24	0	

The depressed equation is $4x^2 - 4x - 24 = 0$ or $x^2 - x - 6 = 0$

$\Rightarrow$ $$(x - 3)(x + 2) = 0$$

$\Rightarrow$ $$x = 3, -2$$

Hence roots of (1) are $-2, -\frac{1}{2}, \frac{1}{2}, 3$.

EXERCISE 5

1. Solve the following equations which have common roots

$$x^3 + 2x^2 - 5x + 2 = 0$$

$$x^3 + x^2 - 8x + 4 = 0.$$

2. Solve the following equations each of which has repeated roots :

(i) $x^3 + x^2 - 16x + 20 = 0$

(ii) $x^4 - 7x^3 + 18x^2 - 20x + 8 = 0$

(iii) $3x^3 - 19x^2 + 33x - 9 = 0.$

3. If 1 is twice repeated root of the equation $ax^3 + bx^2 + bx + d = 0$, show that $a = d = -b$.

4. Solve the equation $x^5 - 15x^3 + 10x^2 + 60x - 72 = 0$ by testing for repeated roots.

5. Find k so that the equation $2x^4 - 3x^2 - 2x + k = 0$ may have a double root and solve the equation.

6. Show that the equation $f(x) = x^4 + 4x^3 + 12x^2 + 24x + 24 = 0$ cannot have multiple roots.

7. Find the condition if two roots of the equation $x^3 - bx^2 + cx - d = 0$ be equal.

8. Find the condition that the equations $x^3 + px + q = 0$ and $x^3 + rx + s = 0$ have a common root. Also show that the equations $x^3 + px + q = 0$ and $x^3 + rx + s = 0$ can have at the most one common root.

[**Hint.** Common root of $x^3 + px + q = 0$ and $x^3 + rx + s = 0$ is also a root of the equation obtained on subtraction *i.e.*, $(p - r)x + q - s = 0$. Since this equation being linear in x, has at the most one root.

$\therefore$ The given equations can have at the most one common root.]

9. If the equation $x^4 + ax^3 + bx^2 + cx + d = 0$ has three equal roots, show that each of them is equal to $\dfrac{6c - ab}{3a^2 - 8b}$, it being given that $3a^2 - 8b \neq 0$.

10. Show that the equation $x^4 + qx^2 + s = 0$ cannot have three equal roots.

[**Hint.** Solve $f'(x) = 0$ and show that no two of the three roots of $f'(x) = 0$ are equal.]

11. Show that the equation $f(x) \equiv x^n + nx^{n-1} + n(n-1)x^{n-2} + \ldots\ldots + n! = 0$ cannot have equal roots.

[**Hint.** $f(x) - f'(x) = x^n$. The given equation has equal roots if $f(x) = 0$ and $f'(x) = 0$ have a common root and that common root must satisfy $f(x) - f'(x) = 0$, *i.e.*, $x^n = 0$ or $x = 0$. But zero is not a root of $f(x) = 0$. Hence the given equation cannot have equal roots.]

ANSWERS

1. $\dfrac{-3 \pm \sqrt{17}}{2}, 1; \dfrac{-3 \pm \sqrt{17}}{2}, 2$

2. (i) 2, 2, – 5 (ii) 2, 2, 2, 1 (iii) 3, 3, $\frac{1}{3}$

4. 2, 2, 2, – 3, – 3

5. $k = 3$; 1, 1, $\dfrac{-2 \pm i\sqrt{2}}{2}$

7. $(bc - 9d)^2 = 4(b^2 - 3c)(c^2 - 3bd)$

8. $(q - s)^3 + p(q - s)(p - r)^2 - q(p - r)^3 = 0.$

MISCELLANEOUS EXERCISE

1. The cubic $2x^3 - 9x^2 + 12x + \lambda = 0$ has two equal roots. Find λ and solve the equation completely.

2. Solve the equation $18x^3 + 81x^2 + 121x + 60 = 0$, given that one of its roots is equal to half the sum of the other two.

3. Find real a and b if the equation $x^3 - 2ax^2 + 9x + b - 5 = 0$ has a root $a + 3i$.

4. If one root of the equation $x^3 + ax + b = 0$ is twice the difference of the other two, prove that one root is $\frac{13b}{3a}$.

5. Determine r so that one root of $x^3 - rx^2 + rx - 4 = 0$ shall be the reciprocal of another and hence find all the roots.

6. Solve the equation $x^5 - 5x^4 - 5x^3 + 25x^2 + 4x - 20 = 0$ whose roots are given to be of the form $\pm \alpha$, $\pm \beta, \gamma$.

7. Solve the equation $x^3 - 9x^2 + 23x - 15 = 0$, it being given that two of its roots are in the ratio $3 : 5$.

8. If $\alpha, \beta, \gamma, \delta$ be the roots of $x^4 + px^3 + qx^2 + rx + s = 0$, show that
$(1 + \alpha^2)(1 + \beta^2)(1 + \gamma^2)(1 + \delta^2) = (1 - q + s)^2 + (p - r)^2$.
[**Hint.** Because $\alpha, \beta, \gamma, \delta$ are roots of $x^4 + px^3 + qx^2 + rx + s = 0$,
therefore $x^4 + px^3 + qx^2 + rx + s = (x - \alpha)(x - \beta)(x - \gamma)(x - \delta)$
Put $x = i$ and then $x = -i$ on both sides and multiply the two equations.]

9. Prove that the equation $ax^5 + bx^3 + cx^2 + d = 0$ will have three equal roots if $-\frac{c}{b} = \frac{b^2}{5ac} = \frac{5bd}{c^2}$, each of the quantities being equal to the repeated root.

10. Find x, y, z from the equations

$$a^3 + a^2x + ay + z = 0$$
$$b^3 + b^2x + by + z = 0$$
$$c^3 + c^2x + cy + z = 0.$$

11. Solve for x, y, z the equations

$$x + y + z = 6$$
$$yz + zx + xy = 11$$
$$xyz = 6.$$

[**Hint.** From the given results it follows that x, y, z are the roots of the cubic
$$t^3 - t^2(x + y + z) + t(yz + zx + xy) - xyz = 0$$
i.e., $t^3 - 6t^2 + 11t - 6 = 0$, $t = 1$ satisfies it
$\therefore$ One root of the cubic in t is 1, find the other two roots. These are values of x, y, z.]

12. The distances of three points, A, B, C on a straight line from a fixed origin O on the line are the roots of the equation $ax^3 + 3bx^2 + 3cx + d = 0$
Find the condition that one of the points should bisect the distance between the other two.

ANSWERS

1. $\lambda = -5$, roots are 1, 1, 5/2 or $\lambda = -4$, roots are 2, 2, 1/2

2. $-3/2, -4/3, -5/3$

3. $a = 0, b = 5$

5. $r = 5$; $\frac{1 \pm \sqrt{-3}}{2}$, 4

6. $\pm 1, \pm 2, 5$

7. 1, 3, 5

10. $x = -(a + b + c)$; $y = ab + bc + ca$; $z = -abc$

11. 1, 2, 3

12. $2b^3 - 3abc + a^2d = 0$.

7

Transformations of Equations

1. Transformations of Equations

Here we shall transform a given equation into another equation whose roots bear a given relation to the roots of the original equation. This process of transformation sometimes helps us in solving certain equations, which are otherwise difficult to solve.

2. To Transform the Equation

$$\mathbf{f(x) = a_0x^n + a_1x^{n-1} + a_2x^{n-2} + ... + a_{n-1}x + a_n = 0}$$

into another equation whose roots are the roots of f(x) = 0 with their signs changed.

Sol. Given equation is $f(x) = a_0x^n + a_1x^{n-1} + a_2x^{n-2} + ... + a_{n-1}x + a_n = 0$ ...(1)

If y denotes the general root of required equation and x that of the given equation ; then the relation is $\mathbf{y = -x}$ ...(2)

Eliminate x from (1) and (2),

From (2), $x = -y$, putting in (1), we have

$$f(-y) = a_0(-y)^n + a_1(-y)^{n-1} + a_2(-y)^{n-2} + ... + a_{n-1}(-y) + a_n = 0$$

or expressing it in x, we have

$$f(-x) = a_0(-x)^n + a_1(-x)^{n-1} + a_2(-x)^{n-2} + ... + a_{n-1}(-x) + a_n = 0$$

which is the required transformed equation.

Thus Rule is Change x → to – x and simplify.

Example 1. *Transform the equation $x^7 - 7x^6 - 3x^4 + 4x^2 - 3x - 2 = 0$ into another whose roots shall be equal in magnitude but opposite in sign to those of this equation.*

Sol. Given equation is $f(x) = x^7 - 7x^6 - 3x^4 + 4x^2 - 3x - 2 = 0$

Changing $x \to -x$, the required equation is

$$f(-x) = (-x)^7 - 7(-x)^6 - 3(-x)^4 + 4(-x)^2 - 3(-x) - 2 = 0$$

or $$= -x^7 - 7x^6 - 3x^4 + 4x^2 + 3x - 2 = 0$$

or $$x^7 + 7x^6 + 3x^4 - 4x^2 - 3x + 2 = 0$$

which is the required transformed equation.

3. To multiply the roots by a given quantity

To transform a given equation f(x) = 0 into another whose roots are m times the roots of f(x) = 0.

Sol. Let the given equation be $f(x) = a_0x^n + a_1x^{n-1} + a_2x^{n-2} + ... + a_{n-1}x + a_n = 0$ ...(1)

if y denotes any general root of required equation and x that of given equation, the **relation** is

$$\mathbf{y = mx} \quad ...(2)$$

Eliminate x from (1) and (2)

From (2) ; $x = \frac{y}{m}$, putting in (1), we get

$$f\left(\frac{y}{m}\right) = a_0\left(\frac{y}{m}\right)^n + a_1\left(\frac{y}{m}\right)^{n-1} + a_2\left(\frac{y}{m}\right)^{n-2} + ... + a_{n-1}\left(\frac{y}{m}\right) + a_n = 0$$

i.e., $$a_0y^n + ma_1y^{n-1} + m^2a_2y^{n-2} + ... + m^{n-1}a_{n-1}y + m^na_n = 0$$

or expressing it in x, the *required* equation is

$$a_0x^n + ma_1x^{n-1} + m^2a_2x^{n-2} + ... + m^{n-1}a_{n-1}x + m^na_n = 0$$

Thus,

To multiply the roots of given equation by m ; the Rule is : Change $x \to$ to $\frac{x}{m}$ and simplify

or alternately (*i*) Complete the given equation by supplying zero co-efficients to the missing terms and then

(*ii*) Multiply successive terms by m^0, m^1, m^2, m^3 .... so on.

Note. To divide the roots by $m \neq 0$ means to multiply the roots by $\frac{1}{m}$.

Example 2. *Find an equation whose roots are three times the roots of equation*

$$x^5 - \frac{1}{3}x^4 + \frac{25}{27}x^2 + \frac{14}{81}x - \frac{8}{81} = 0.$$

Sol. Given equation is $f(x) = x^5 - \frac{1}{3}x^4 + \frac{25}{27}x^2 + \frac{14}{81}x - \frac{8}{81} = 0$

Complete the equation by supplying zero co-efficients to the missing terms

i.e., $$f(x) = x^5 - \frac{1}{3}x^4 + 0x^3 + \frac{25}{27}x^2 + \frac{14}{81}x - \frac{8}{81} = 0$$

Multiplying the terms respectively by $3^0, 3^1, 3^2, 3^3, 3^4, 3^5$, we have

$$3^0.x^5 + 3^1\left(-\frac{1}{3}x^4\right) + 3^2(0x^3) + 3^3\left(\frac{25}{27}x^2\right) + 3^4\left(\frac{14}{81}x\right) + 3^5\left(-\frac{8}{81}\right) = 0$$

or $$x^5 - x^4 + 25x^2 + 14x - 24 = 0$$

which is the required equation.

Example 3. *Remove the fractional coefficients, keeping the co-efficient of highest degree term unity of the equation* $x^4 - \frac{1}{2}x^3 + \frac{1}{9}x^2 - \frac{1}{8}x + \frac{1}{12} = 0.$

Sol. L.C.M. of the denominators of the fractional co-efficients *i.e.*, L.C.M. of 2, 9, 8 and 12 is 72.

∴ If we multiply the roots of the given equation by 72 the equation shall transform into the desired form, but we shall have to multiply the terms by $(72)^0, (72)^1, (72)^2, (72)^3, (72)^4$. In this process all the coefficients, in particular the constant term will be a very big number and the equation will be in a clumsy form. This can be **avoided** by choosing a much smaller number, known as **Least Multiplier.**

To find the least multiplier, the method is as follows :

Re-write the equation in the form :

$$x^4 - \frac{x^3}{2} + \frac{x^2}{3^2} - \frac{x}{2^3} + \frac{1}{2^2.3} = 0$$

[Factorising the denominators into product of primes]

or $$x^4 - \frac{x^3}{2^1 . 3^0} + \frac{x^2}{2^0 . 3^2} - \frac{x}{2^3 . 3^0} + \frac{1}{2^2 . 3^1} = 0$$

[Writing proper powers of each prime occurring in the denominator]

Write the indices of 2, *i.e.*, 1, 0, 3, 2.

Divide these respectively by 1, 2, 3 and 4 getting $\frac{1}{1}, \frac{0}{2}, \frac{3}{3}, \frac{2}{4}$.

Choose the least integer next or equal to the above numbers, *i.e.*, 1 here.

$\therefore$ *2^1 is a factor of the least multiplier.*

Similarly, writing the indices of 3, *i.e.*, 0, 2, 0, 1.

Dividing by 1, 2, 3, 4 respectively, we get $\frac{0}{1}, \frac{2}{2}, \frac{0}{3}, \frac{1}{4}$.

$\therefore$ 1 is an integer greater than or equal to all of these

$\therefore$ 3^1 is a factor of the least multiplier.

Hence if m is the least multiplier, then $m = 2^1 \times 3^1 = 6$

$\therefore$ Multiplying the root of the given equation

$$f(x) = x^4 - \tfrac{1}{2}x^3 + \tfrac{1}{9}x^2 - \tfrac{1}{8}x + \tfrac{1}{12} = 0 \text{ by } 6.$$

$\therefore$ The required transformed equation is

$$6^0(x^4) + 6^1(-\tfrac{1}{2}x^3) + 6^2(\tfrac{1}{9}x^2) + 6^3(-\tfrac{1}{8}x) + 6^4(\tfrac{1}{12}) = 0$$

or $$x^4 - 3x^3 + 4x^2 - 27x + 108 = 0].$$

4. Reciprocal Roots

Transform the equation into another whose roots are the reciprocals of the roots of given equation.

Proof. Let the given equation be

$$f(x) = a_0x^n + a_1x^{n-1} + a_2x^{n-2} + \dots + a_{n-1}x + a_n = 0 \qquad \text{...(1)}$$

If a root of the required equation be denoted by y and that of given equation by x, the relation between the roots is $y = \dfrac{1}{x}$ ...(2)

Eliminate x between (1) and (2)

From (2), $x = \dfrac{1}{y}$ put in (1),

$$\therefore \quad f\left(\frac{1}{y}\right) = a_0\left(\frac{1}{y}\right)^n + a_1\left(\frac{1}{y}\right)^{n-1} + a_2\left(\frac{1}{y}\right)^{n-2} + \dots\dots + a_{n-1}\left(\frac{1}{y}\right) + a_n = 0$$

or $$a_0 + a_1y + a_2y^2 + a_3y^3 + \dots + a_{n-1}y^{n-1} + a_ny^n = 0$$

or expressing it in the variable x, we get

$$a_0 + a_1x + a_2x^2 + \dots\dots + a_{n-1}x^{n-1} + a_nx^n = 0$$

or $$a_nx^n + a_{n-1}x^{n-1} + a_{n-2}x^{n-2} + \dots\dots + a_1x + a_0 = 0$$

this is the required equation.

The Rule is : Change $x \to \dfrac{1}{x}$ and multiply the equation by the L.C.M. or when the equation is complete, the co-efficients are simply reversed in their order.

Example 4. *(a) Find the condition that the roots of the equation $x^3 + px^2 + qx + r = 0$ may be in H.P. ($r \neq 0$).*

(b) Solve the cubic $28x^3 - 39x^2 + 12x - 1 = 0$ it being given that the roots are in H.P.

Sol. (*a*) The given equation is $x^3 + px^2 + qx + r = 0$ ...(1)

The equation whose roots are reciprocals of the roots of (1) is

$$rx^3 + qx^2 + px + 1 = 0 \left(\text{Changing } x \to \frac{1}{x}\right) \quad ...(2)$$

As roots of (1) are in H.P., so roots of (2) are in A.P.

$\therefore$ Let the roots of (2) be $\alpha - \beta, \alpha, \alpha + \beta$.

$\therefore$ Sum of roots $= (\alpha - \beta) + \alpha + (\alpha + \beta) = -\frac{q}{r}$ $(\neq 0)$

i.e., $$3\alpha = -\frac{q}{r} \quad \text{or} \quad \alpha = \frac{-q}{3r}$$

But $\alpha = -\frac{q}{3r}$ is a root of (2). $\therefore$ It satisfies equation (2)

Putting the value of α in (2), we get

$$r\left(\frac{-q}{3r}\right)^3 + q\left(\frac{-q}{3r}\right)^2 + p\left(\frac{-q}{3r}\right) + 1 = 0$$

or $$-r \cdot \frac{q^3}{27r^3} + q\frac{q^2}{9r^2} - \frac{pq}{3r} + 1 = 0$$

i.e., $$\frac{q^3}{27r^2} - \frac{q^3}{9r^2} + \frac{pq}{3r} - 1 = 0$$

or Multiplying by $27r^2$, we get

$$q^3 - 3q^3 + 9pqr - 27r^2 = 0 \quad \text{or} \quad 2q^3 - 9pqr + 27r^2 = 0$$

this is the required condition.

(*b*) Roots of the equation $28x^3 - 39x^2 + 12x - 1 = 0$ ...(1)

are in H.P.

$\therefore$ The roots of the equation $-x^3 + 12x^2 - 39x + 28 = 0$

or of $x^3 - 12x^2 + 39x - 28 = 0$ ...(2)

are in A.P.

Let the roots in A.P. of (2) be $\alpha - \beta, \alpha, \alpha + \beta$

$\therefore$ Sum of roots $(\alpha - \beta) + \alpha + (\alpha + \beta) = 12$ $\therefore$ $3\alpha = 12$ or $\alpha = 4$

i.e., $\alpha = 4$ is one root of (2)

$\therefore$ $\frac{1}{4}$ is a root of given equation (1)

$\therefore$ $(x - \frac{1}{4})$ is a factor of L.H.S. of (1)

Dividing (1) by $(x - \frac{1}{4})$,

$(\frac{1}{4})$	28	−39	12	−1
		7	−8	1
	28	−32	4	zero

$\therefore$ The quotient is $28x^2 - 32x + 4$

Put $28x^2 - 32x + 4 = 0$

or $7x^2 - 8x + 1 = 0$

$$\therefore \quad x = \frac{8 \pm \sqrt{64-28}}{14} = \frac{8 \pm 6}{14} = 1, \frac{1}{7}.$$

Hence roots of given equation are $\frac{1}{7}, \frac{1}{4}, 1$.

5. Formation of the equation whose roots are squares, cubes of roots of the given equation

Example 5. *Find the equation whose roots are the squares of the roots of the equation*

$$x^3 + ax^2 + bx + c = 0.$$

Sol. Given equation is $x^3 + ax^2 + bx + c = 0$ ...(1)

If y is any root of the required equation and x that of the given equation, then

$$y = x^2 \quad ...(2)$$

Eliminate x between (1) and (2)

Putting the value of x^2 from (2) in (1), we have

$$xy + ay + bx + c = 0 \qquad [\because \quad x^3 = x.x^2 = xy]$$

$$x(y + b) = -(ay + c) \qquad \text{(Keeping terms containing } x \text{ on one side)}$$

Squaring both sides and replacing x^2 by y, we get $y(y + b)^2 = (ay + c)^2$

i.e., $y(y^2 + 2by + b^2) = a^2y^2 + 2acy + c^2$

i.e., $y^3 + (2b - a^2)y^2 + (b^2 - 2ac)y - c^2 = 0$

which is the required equation.

Example 6. *Find the equation whose roots are the cubes of the roots of*

$$x^3 + px^2 + qx + r = 0.$$

Hence evaluate $\Sigma\alpha^3\beta^3$ *if* α, β, γ *are roots of given equation.*

Sol. Given equation is $x^3 + px^2 + qx + r = 0$ ...(1)

If the transformed equation is in y, then $y = x^3$ ...(2)

Eliminate x between (1) and (2)

Substituting the value of $x^3 (= y)$ from (2) in (1)

$$\therefore \quad y + px^2 + qx + r = 0 \quad \text{or} \quad (px^2 + qx) = -(y + r) \quad ...(3)$$

[Keeping terms containing x on one side]

Cubing both sides, we get $(px^2 + qx)^3 = -(y + r)^3$

or $p^3x^6 + q^3x^3 + 3px^2.qx(px^2 + qx) = -(y^3 + 3y^2r + 3yr^2 + r^3)$

But $px^2 + qx = -(y + r)$ from (3) and $x^3 = y$ from (2)

$\therefore$ Using these, we have $p^3y^2 + q^3y + 3pqy\,[-(y + r)] = -(y^3 + 3y^2r + 3yr^2 + r^3)$

or $y^3 + (p^3 - 3pq + 3r)y^2 + (q^3 - 3pqr + 3r^2)y + r^3 = 0$ ...(4)

which is the required equation whose roots are cubes of the roots of (1).

Now $\alpha^3, \beta^3, \gamma^3$ are roots of equation (4).

$$\therefore \quad \Sigma\alpha^3\beta^3 = q^3 - 3pqr + 3r^2.$$

EXERCISE 1

1. Obtain the equation whose roots are the roots of the following equation with their signs changed:
$$x^5 + 8x^4 - 9x^3 + 13x^2 - 12x + 9 = 0.$$
2. Multiply the roots of the equation by the number indicated and find the transformed equation
(*i*) $3x^3 - 10x^2 + 9x - 2 = 0$ by 3
(*ii*) $x^4 + 3x^3 - 5x^2 + 2x - 6 = 0$ by $\frac{1}{2}$.

3. Remove the fractional co-efficients from the equation so that the co-efficient of leading term remains unity $x^4 - \frac{1}{2}x^3 + \frac{2}{9}x^2 - \frac{3}{8}x + \frac{5}{12} = 0$.

4. Use the least multiplier to remove the fractional co-efficients of the equations :

(i) $x^5 + \frac{4}{3}x^4 + \frac{2}{9}x^3 + \frac{1}{12}x^2 + \frac{1}{36} = 0$

(ii) $x^4 + \frac{1}{ab}x^3 + \frac{1}{a^2b}x^2 + \frac{1}{a^4b^2}x + \frac{1}{a^3b^3} = 0$.

5. Find the equation whose roots are four times and with their signs changed of the roots of the equation $4x^4 - x^3 + x + 1 = 0$.

6. Solve $6x^3 - 11x^2 + 6x - 1 = 0$, given that its roots are in H.P.

7. Find the condition that the roots of the equation $x^3 - px^2 + qx - r = 0$ may be in H.P.

8. (i) Solve the equation $40x^4 - 22x^3 - 21x^2 + 2x + 1 = 0$, whose roots are in H.P.

(ii) Solve $15x^4 - 8x^3 - 14x^2 + 8x - 1 = 0$ given that its roots are in H.P.

9. Form equations whose roots shall be squares of the roots of the following equations :

(i) $x^3 + qx + r = 0$ (ii) $x^3 + 3x^2 + 6x + 1 = 0$

(iii) $x^4 - x^2 + x - 3 = 0$.

10. Form equations whose roots are cubes of the roots of

(i) $x^3 + x + 1 = 0$ (ii) $2x^3 - x^2 - 3 = 0$

(iii) $x^4 - x^3 + 2x^2 + 3x + 1 = 0$.

11. Form an equation whose roots are squares of roots of the equation $x^3 - x^2 - 2x + 2 = 0$ and prove without solving that the new equation has a multiple root.

12. Show that the cubes of the roots of the equation $x^3 + ax^2 + bx + ab = 0$ are given by the equation

$$x^3 + a^3x^2 + b^3x + a^3b^3 = 0.$$

ANSWERS

1. $x^5 - 8x^4 - 9x^3 - 13x^2 - 12x - 9 = 0$

2. (i) $x^3 - 10x^2 + 27x - 18 = 0$ (ii) $x^4 + \frac{3}{2}x^3 - \frac{5}{4}x^2 + \frac{1}{4}x - \frac{3}{8} = 0$

3. $x^4 - 3x^3 + 8x^2 - 81x + 540 = 0$

4. (i) $x^5 + 8x^4 + 8x^3 + 18x^2 + 216 = 0$ (ii) $x^4 + ax^3 + a^2bx^2 + a^2bx + a^5b = 0$

5. $x^4 + x^3 - 16x + 64 = 0$ **6.** $1, \frac{1}{2}, \frac{1}{3}$

7. $2q^3 - 9pqr + 27r^2 = 0$ **8.** (i) $-\frac{1}{5}, -\frac{1}{2}, 1, \frac{1}{4}$ (ii) $-1, 1, \frac{1}{3}, \frac{1}{5}$

9. (i) $y^3 + 2qy^2 + q^2y - r^2 = 0$ (ii) $y^3 + 3y^2 + 30y - 1 = 0$ (iii) $y^4 - 2y^3 - 5y^2 + 5y + 9 = 0$

10. (i) $y^3 + 3y^2 + 4y + 1 = 0$ (ii) $8y^3 - 37y^2 + 54y - 27 = 0$ (iii) $y^4 + 14y^3 + 50y^2 + 6y + 1 = 0$

11. $(y - 1)(y - 2)^2 = 0$.

---

6. To Form an equation whose roots are less by 'h' than the roots of a given equation

We shall explain the process by considering a cubic equation only. However, the method holds for equations of any degree.

Let the cubic be $f(x) = a_0x^3 + a_1x^2 + a_2x + a_3 = 0$ ...(1)

Expressing $f(x)$ as a polynomial in $(x - h)$, we can write (1) as

$$f(x) = a_0(x - h)^3 + A_1(x - h)^2 + A_2(x - h) + A_3 \quad ...(2)$$

where A_1, A_2, A_3 are to be determined.

If the required equation be in y, then

$$y = x - h \text{ i.e., } x = y + h \qquad ...(3)$$

Eliminate 'x' between (2) and (3), we have

$$F(y) = A_0y^3 + A_1y^2 + A_2y + A_3 \; (A_0 = a_0) \qquad ...(4)$$

which is the required equation.

From (2), it is clear, if we divide (2),

successively by $(x - h)$, the remainders obtained are respectively,

A_3, A_2, A_1 and A_0 and (4) is the required equation.

Hence rule to decrease the roots by h :

(1) Change x to $(x + h)$ in the given equation and simplify.

Or

(2) Divide the given polynomial equation $f(x)$ successively by $(x - h)$ and get the coefficients of the required equation in the reverse order.

Note. To increase the roots by h means to decrease by $(- h)$.

Example 1. *Find the equation whose roots are less than the roots of equation*

$$3x^4 - 2x^3 + 4x^2 - 3x + 16 = 0 \text{ by } 3.$$

Sol. Given equation is $3x^4 - 2x^3 + 4x^2 - 3x + 16 = 0$ ...(1)

Let the transformed equation be

$$A_0x^4 + A_1x^3 + A_2x^2 + A_3x + A_4 = 0 \qquad ...(2)$$

where A_0, A_1, A_2, A_3, A_4 are determined by dividing L.H.S. of (1) successively by $(x - 3)$ using synthetic division

3)	3	– 2	4	– 3	16
		9	21	75	216
	3	7	25	72	232 = A_4
		9	48	219	
	3	16	73	291 = A_3	
		9	75		
	3	25	148 = A_2		
		9			
	3	34 = A_1			
	3 = A_0				

Hence from (2), the required equation is

$$3x^4 + 34x^3 + 148x^2 + 291x + 232 = 0.$$

Example 2. *Increase by 4 the roots of the equation*

$$3x^5 - 5x^3 + 7 = 0.$$

Sol. Completing the given equation by supplying zero co-efficients, we get

$$3x^5 + 0x^4 - 5x^3 + 0x^2 + 0x + 7 = 0$$

Let the transformed equation be

$$A_0x^5 + A_1x^4 + A_2x^3 + A_3x^2 + A_4x + A_5 = 0$$

To increase the roots by 4 means to decrease the roots by – 4.

Dividing the L.H.S. of given equation successively by $x - (-4)$, we have

– 4)	3	0	– 5	0	0	7
		– 12	48	– 172	688	– 2752
	3	– 12	43	– 172	688	– 2745 = A_5
		– 12	96	– 556	2912	
	3	– 24	139	– 728	3600 = A_4	
		– 12	144	– 1132		
	3	– 36	283	– 1860 = A_3		
		– 12	192			
	3	– 48	475 = A_2			
		– 12				
	3	– 60 = A_1				
	3 = A_0					

Hence the required transformed equation is

$$3x^5 - 60x^4 + 475x^3 - 1860x^2 + 3600x - 2745 = 0.$$

Example 3. *The difference between two roots of the equation $2x^3 + x^2 - 7x - 6 = 0$ is 3. Solve it diminishing the roots by 3.*

Sol. Let roots of the equation $f(x) \equiv 2x^3 + x^2 - 7x - 6 = 0$ ...(1)

be $\alpha, \alpha + 3, \beta$.

Let us diminish the roots of (1) by 3

Using successively synthetic division, we have

3	2	1	– 7	– 6
		6	21	42
	2	7	14	36
		6	39	
	2	13	53	
		6		
	2	19		
	2			

The transformed equation is $g(x) \equiv 2x^3 + 19x^2 + 53x + 36 = 0$ ...(2)

whose roots are $\alpha - 3, \alpha, \beta - 3$.

Clearly, equation (1) and (2) have a common root α.

But, H.C.F. of $f(x)$ and $g(x)$ is $x + 1$ (Verify !)

$\therefore$ Common root α is given by $x + 1 = 0$ or $x = -1$ $\therefore$ $\alpha = -1$

$\therefore$ two roots of (1) are $-1, 2$.

Since sum of 3 roots of (1) is $-\frac{1}{2}$

$$\therefore \quad -1 + 2 + \beta = -\frac{1}{2} \quad \text{or} \quad \beta = -\frac{3}{2}$$

Hence roots of (1) are $-1, 2, -\dfrac{3}{2}$.

EXERCISE 2

1. Diminish the roots of the equation $x^4 + 3x^2 - 4x + 9 = 0$ by 5.
2. (*a*) Find the equation whose roots exceed by 2, the roots of the equation $4x^4 + 32x^3 + 83x^2 + 76x + 21 = 0$. Hence solve the equation.

   (*b*) Transform the equation $x^4 + 4x^3 - 19x^2 - 46x + 120 = 0$ into another whose roots are increased by one than the roots of this equation. Hence find the roots of the given equation.
3. (*a*) The difference of two roots of the equation $x^3 - 13x^2 + 15x + 189 = 0$ is 2. Solve it by decreasing the roots by 2.

   (*b*) The difference of two roots of the equation $x^3 - 13x^2 + 15x + 189 = 0$ is 2. Solve it by increasing the roots by 2.
4. If α, β, γ are the roots of the equation $x^3 - 6x^2 + 12x - 8 = 0$,
   find the equation whose roots are $\alpha - 2, \beta - 2, \gamma - 2$. Hence find the roots of the given equation.

ANSWERS

1. $x^4 + 20x^3 + 153x^2 + 526x + 689 = 0$
2. (*a*) $4x^4 - 13x^2 + 9 = 0$; $-\frac{7}{2}, -3, -1, -\frac{1}{2}$ (*b*) $x^4 - 25x^2 + 144 = 0$; $-5, -4, 2, 3$
3. (*a*) $-3, 7, 9$ (*b*) $-3, 7, 9$ **4.** $x^3 = 0$; 2, 2, 2.

7. Removal of Terms, in General

One or more terms of an equation can be removed by decreasing the roots of the given equation by a suitable number say h.

Let the given equation be $a_0x^n + a_1x^{n-1} + a_2x^{n-2} + \ldots + a_{n-1}x + a_n = 0$ ...(1)

decrease the roots by '*h*'

i.e., Changing x to $(x + h)$, we get [Rule Art. 5]

$$a_0(x + h)^n + a_1(x + h)^{n-1} + a_2(x + h)^{n-2} + \ldots\ldots + a_{n-1}(x + h) + a_n = 0$$

Expanding by binomial theorem,

$$a_0(x^n + {}^nC_1x^{n-1}h + {}^nC_2x^{n-2}h^2 + \ldots) + a_1(x^{n-1} + {}^{n-1}C_1x^{n-2}h + \ldots)$$
$$+ a_2(x^{n-2} + \ldots) + \ldots + a_{n-1}(x + h) + a_n = 0$$

i.e., $$a_0\left(x^n + nx^{n-1}h + \frac{n(n-1)}{2!}x^{n-2}h^2 + \ldots\right) + a_1(x^{n-1} + (n-1)x^{n-2}h + \ldots)$$
$$+ a_2(x^{n-2} + \ldots) + \ldots + a_{n-1}(x + h) + a_n = 0$$

or $$a_0x^n + (na_0h + a_1)x^{n-1} + \left(\frac{n(n-1)}{2!}a_0h^2 + (n-1)a_1h + a_2\right)x^{n-2} + \ldots + \ldots = 0$$

(*i*) **For removing the second term**

Put co-efficient of $x^{n-1} = 0$.

i.e., Put $na_0h + a_1 = 0 \quad \therefore \quad \mathbf{h} = \dfrac{-\mathbf{a_1}}{\mathbf{na_0}}$.

Thus, if we diminish the roots of the equation $f(x) = 0$ by $h = \dfrac{-a_1}{na_0}$, the transformed equation will not have the *second* term.

(*ii*) **For removing the third term**

Put co-efficient of $x^{n-2} = 0$

i.e., Put $\dfrac{n(n-1)}{2!} a_0h^2 + (n-1)a_1h + a_2 = 0$

This is a quadratic in h, giving in general two values of h say h_1 and h_2 and if we decrease the roots either by h_1 or h_2, the transformed equation will have no *third term*.

(*iii*) Similarly to remove any other term, put the co-efficient of that term = 0. A polynomial in h will be obtained giving you values of h, by which, if the roots are decreased, the transformed equation will be *devoid* (wanting or missing) of the desired term.

Remember

(*i*) To remove **second term,** decrease the roots by

$$\mathbf{h} = \frac{-\mathbf{a_1}}{\mathbf{na_0}} = \frac{-\text{(Co-efficient of 2nd term)}}{\text{(Degree of the equation)(Co-efficient of 1st term)}}$$

(*ii*) To remove **third term,** decrease the roots by h, where 'h' is given by

$$\frac{\mathbf{n(n-1)}}{\mathbf{2}}\mathbf{a_0h^2 + (n-1)a_1h + a_2 = 0.}$$

(iii) To find the condition for removing the second and the third terms simultaneously eliminate 'h' from (i) and (ii) above.

8. Equations with Binomial Co-efficients and the Notation used

An nth degree equation in x written in the form

$${}^nC_0a_0x^n + {}^nC_1a_1x^{n-1} + {}^nC_2a_2x^{n-2} + \ldots\ldots + {}^nC_{n-r}a_{n-r}x^r \ldots + {}^nC_{n-1}\, a_{n-1}\, x + {}^nC_na_n = 0$$

or in the form

$$a_0x^n + na_1x^{n-1} + \frac{n(n-1)}{2!}a_2x^{n-2} + \ldots\ldots + \frac{n(n-1)}{2!}a_{n-2}x^2 + na_{n-1}x + a_n = 0$$

is called an *equation with co-efficients in the* **binomial form.**

In particular

The **cubic** equation with binomial co-efficients is $a_0x^3 + 3a_1x^2 + 3a_2x + a_3 = 0$

The **bi-quadratic** equation with binomial co-efficients is

$$a_0x^4 + 4a_1x^3 + 6a_2x^2 + 4a_3x + a_4 = 0$$

Now some functions of the co-efficients of these equations in the standard notations are given below

(*i*) $\mathrm{H} = a_0a_2 - a_1^2 = \begin{vmatrix} a_0 & a_1 \\ a_1 & a_2 \end{vmatrix}$

(*ii*) $\mathrm{G} = a_0^2a_3 - 3a_0a_1a_2 + 2a_1^3$

$\mathrm{I} = a_0a_4 - 4a_1a_3 + 3a_2^2$

$$J = 2a_1a_2a_3 + a_0a_2a_4 - a_1^2a_4 - a_0a_3^2 - a_2^3 = \begin{vmatrix} a_0 & a_1 & a_2 \\ a_1 & a_2 & a_3 \\ a_2 & a_3 & a_4 \end{vmatrix}.$$

Example 1. *Find the condition that the second and third terms of the equation $a_0x^3 + 3a_1x^2 + 3a_2x + a_3 = 0$ may be removed by the same transformation.*

Sol. Given equation is $\quad a_0x^3 + 3a_1x^2 + 3a_2x + a_3 = 0 \quad$ ...(1)

Decrease the roots by 'h'

i.e., Change x to $(x + h)$

$\therefore$ Equation (1) transforms into

$$a_0(x + h)^3 + 3a_1(x + h)^2 + 3a_2(x + h) + a_3 = 0$$

i.e., $$a_0(x^3 + 3x^2h + 3xh^2 + h^3) + 3a_1(x^2 + 2xh + h^2) + 3a_2(x + h) + a_3 = 0$$

or $$a_0x^3 + 3(a_0h + a_1)x^2 + 3(a_0h^2 + 2a_1h + a_2)x + (a_0h^3 + 3a_1h^2 + 3a_2h + a_3) = 0$$

The second and the third terms will be removed simultaneously if

$$a_0h + a_1 = 0 \quad \text{...(2)}$$

and $$a_0h^2 + 2a_1h + a_2 = 0 \quad \text{...(3)}$$

To find the condition, eliminate 'h' between (2) and (3)

From (2), $$h = -\frac{a_1}{a_0}$$

Putting in (3), we get

$$a_0\left(\frac{a_1^2}{a_0^2}\right) + 2a_1\left(-\frac{a_1}{a_0}\right) + a_2 = 0$$

i.e., $$\frac{a_1^2}{a_0} - \frac{2a_1^2}{a_0} + a_2 = 0$$

or $$\mathbf{a_0a_2 - a_1^2 = 0} \quad \text{or} \quad \mathbf{H = 0}.$$

Hence the required condition when second and third terms can be removed simultaneously is **H = 0** and this is done, if we decrease the roots by

$$h = -\frac{a_1}{a_0}.$$

Note 1. This condition remains the same even when the equation is of higher degree with coefficients in binomial form.

2. By following the above method, we can find the condition when any of two terms are to be removed by the same transformation.

Example 2. *Solve the equation $x^4 + 8x^3 + 19x^2 + 12x - 5 = 0$ by removing the second term by a transformation.*

Sol. Given equation is $x^4 + 8x^3 + 19x^2 + 12x - 5 = 0$ ...(1)

For removing the second term, decrease the roots by h, where

$$h = -\frac{\text{Co-efficient of 2nd term}}{(\text{Degree of the equation}) \times (\text{Co-effi. of 1st term})}$$

i.e., $$h = \frac{-8}{4} = -2$$

−2)	1	8	19	12	−5
		−2	−12	−14	4
	1	6	7	−2	−1
		−2	−8	2	
	1	4	−1	0	
		−2	−4		
	1	2	−5		
		−2			
	1	0			

$\therefore$ **The transformed equation is $x^4 - 5x^2 - 1 = 0$** ...(2)

It is a quadratic equation in x^2.

$$\therefore \qquad x^2 = \frac{5 \pm \sqrt{25+4}}{2} = \frac{5 \pm \sqrt{29}}{2}$$

$$\therefore \qquad x = \pm \{\tfrac{1}{2}(5 \pm \sqrt{(29)}\}^{1/2}$$

Hence the roots of the given equation are $\pm \{\frac{1}{2}(5 \pm \sqrt{(29)}\}^{1/2}, -2$.

EXERCISE 3

1. Remove the second term from the equations :
   (i) $3x^3 + 5x^2 + 7x + 2 = 0$ (ii) $x^4 + 8x^3 + x - 5 = 0$.
2. Transform the equation $2x^3 - 9x^2 + 13x - 6 = 0$ into one in which the second term is missing and hence solve the given equation.
3. Reduce the equation $2x^3 - 3x^2 + 10x - 4 = 0$ to the form $z^3 + \text{H}z + \text{G} = 0$, where G and H are integers. Also find the relation between z and x.
4. Show that the same transformation can remove both second and fourth terms of the equation $x^4 + 16x^3 + 83x^2 + 152x + 84 = 0$ and hence solve it.
5. Transform the equation $x^4 - 4x^3 - 18x^2 - 3x + 2 = 0$ into another in which the third term is missing.
6. Find the condition that the same transformation may remove second and the fourth terms from the equation $a_0x^4 + 4a_1x^3 + 6a_2x^2 + 4a_3x + a_4 = 0$.

ANSWERS

1. (i) $3x^3 + \dfrac{114}{27}x - \dfrac{209}{243} = 0$ (ii) $x^4 - 24x^2 + 65x - 55 = 0$
2. $4x^3 - x = 0$; $\frac{3}{2}, 2, 1$
3. $z^3 + 17z + 2 = 0$; $z = 2x - 1$
4. $x^4 - 13x^2 + 36 = 0$; $-7, -6, -2, -1$
5. $h = 3$ or -1 ; $x^4 + 8x^3 - 111x - 196 = 0$ or $x^4 - 8x^3 + 17x - 8 = 0$
6. $a_0{}^2a_3 - 3a_0a_1a_2 + 2a_1{}^3 = 0$ *i.e.,* $\text{G} = 0$.

9. The Cubic

(a) To transform the cubic $a_0x^3 + 3a_1x^2 + 3a_2x + a_3 = 0$ into the cubic in which the second term is missing and the co-efficient of highest degree term unity and the other co-efficients integral numbers, where a_0, a_1, a_2, a_3 are given to be integers.

(b) Find the relation between the roots of the transformed and the given equation.

Sol. (*a*) Let the given equation be $a_0x^3 + 3a_1x^2 + 3a_2x + a_3 = 0$ ...(1)

Diminish the roots by h, if the new equation is in y, then $y = x - h$ ...(2)

Eliminate x from (1) and (2),

From (2), $x = y + h$, put in (1)

$$\therefore \quad a_0(y + h)^3 + 3a_1(y + h)^2 + 3a_2(y + h) + a_3 = 0$$

or $$a_0(y^3 + 3y^2h + 3yh^2 + h^3) + 3a_1(y^2 + 2yh + h^2) + 3a_2(y + h) + a_3 = 0$$

or $$a_0y^3 + 3(a_0h + a_1)y^2 + 3(a_0h^2 + 2a_1h + a_2)y + a_0h^3 + 3a_1h^2 + 3a_2h + a_3 = 0 \quad ...(3)$$

To remove the second term, put the co-efficient of $y^2 = 0$

i.e., $$a_0h + a_1 = 0 \quad \text{or} \quad h = \frac{-a_1}{a_0}.$$

Putting the value of h in the co-efficients of equation (3), we get

(*i*) $$a_0h^2 + 2a_1h + a_2 = a_0\left(\frac{-a_1}{a_0}\right)^2 + 2a_1\left(\frac{-a_1}{a_0}\right) + a_2$$

$$= a_0\frac{a_1^2}{a_0^2} - \frac{2a_1^2}{a_0} + a_2 = \frac{-a_1^2}{a_0} + a_2$$

$$= \frac{a_0a_2 - a_1^2}{a_0} = \frac{H}{a_0}, \text{ where } H = a_0a_2 - a_1^2.$$

(*ii*) $a_0h^3 + 3a_1h^2 + 3a_2h + a_3$

$$= a_0\left(\frac{-a_1}{a_0}\right)^3 + 3a_1\left(\frac{-a_1}{a_0}\right)^2 + 3a_2\left(\frac{-a_1}{a_0}\right) + a_3$$

$$= \frac{-a_0a_1^3}{a_0^3} + \frac{3a_1^3}{a_0^2} - \frac{3a_1a_2}{a_0} + a_3 = \frac{2a_1^3}{a_0^2} - \frac{3a_1a_2}{a_0} + a_3$$

$$= \frac{2a_1^3 - 3a_0a_1a_2 + a_0^2a_3}{a_0^2}$$

$$= \frac{G}{a_0^2}, \text{ where } G = a_0^2a_3 - 3a_0a_1a_2 + 2a_1^3$$

Substituting (*i*) and (*ii*) in (3), the equation (3) becomes

$$a_0y^3 + \frac{3.H}{a_0}y + \frac{G}{a_0^2} = 0$$

or dividing by $a_0 (\neq 0)$, we have $y^3 + \dfrac{3H}{a_0^2}y + \dfrac{G}{a_0^3} = 0$ ...(4)

Multiply the roots of this equation by a_0 and if the transformed equation is in Z, we get

$$Z^3 + 3HZ + G = 0 \quad ...(5)$$

where $\mathbf{Z = a_0y}$ ...(6)

This is the required transformed equation.

(*b*) If Z denotes the root of the transformed equation (5) and x that of the given equation (1), then

$$Z = a_0 y \qquad \text{[By (6)]}$$
$$= a_0(x - h) \qquad \text{[By (2) ; } y = x - h]$$
$$= a_0\left(x + \frac{a_1}{a_0}\right) \qquad \left[\because\ h = \frac{-a_1}{a_0}\right]$$

i.e., $$\mathbf{Z = a_0x + a_1}.$$

Thus, if the roots of the given equation are x_1, x_2, x_3 and that of the transformed equation Z_1, Z_2, Z_3, then $Z_1 = a_0x_1 + a_1$; $Z_2 = a_0x_2 + a_1$; $Z_3 = a_0x_3 + a_1$.

Cor. *To express a root of the transformed equation in terms of the roots of the given equation.*

Since $$Z_1 = a_0x_1 + a_1 = a_0\left(x_1 + \frac{a_1}{a_0}\right)$$

From the given equation (1),

Sum of roots $$= x_1 + x_2 + x_3 = \frac{-3a_1}{a_0}$$

$$\Rightarrow \quad -\frac{1}{3}(x_1 + x_2 + x_3) = \frac{a_1}{a_0}$$

Putting the value of $\frac{a_1}{a_0}$, we get

$$Z_1 = a_0\left(x_1 - \frac{x_1 + x_2 + x_3}{3}\right) = \frac{a_0}{3}(3x_1 - x_1 - x_2 - x_3)$$

Hence $$Z_1 = \frac{a_0}{3}(2x_1 - x_2 - x_3)$$

Similarly, $$Z_2 = \frac{a_0}{3}(2x_2 - x_3 - x_1)$$

and $$Z_3 = \frac{a_0}{3}(2x_3 - x_1 - x_2).$$

Note. The equation $Z^3 + 3HZ + G = 0$ can be obtained from the equation $a_0x^3 + 3a_1x^2 + 3a_2x + a_3 = 0$ directly using the transformation $\mathbf{Z = a_0x + a_1}$.

10. The Bi-quadratic

(*a*) *Transform the bi-quadratic* $a_0x^4 + 4a_1x^3 + 6a_2x^2 + 4a_3x + a_4 = 0$ *into another equation in which the second term is wanting, the coefficient of highest degree term unity, and the coefficient of other terms integral numbers, when* a_0, a_1, a_2, a_3 *and* a_4 *are integers.*

(*b*) *Express the roots of the transformed equation in terms of the roots of the given equation.*

Sol. (*a*) Let the equation be $\quad a_0x^4 + 4a_1x^3 + 6a_2x^2 + 4a_3x + a_4 = 0 \quad$ ...(1)

Diminish the roots by h and if the new equation is in y, then $y = x - h$ ...(2)

Eliminate x between (1) and (2)

From (2) $x = y + h$, put in (1)

$$\therefore \quad a_0(y + h)^4 + 4a_1(y + h)^3 + 6a_2(y + h)^2 + 4a_3(y + h) + a_4 = 0$$

i.e., $$a_0(y^4 + 4y^3h + 6y^2h^2 + 4yh^3 + h^4) + 4a_1(y^3 + 3y^2h + 3h^2y + h^3) + 6a_2(y^2 + 2yh + h^2) + 4a_3(y + h) + a_4 = 0$$

i.e., $a_0y^4 + 4(a_0h + a_1)y^3 + 6(a_0h^2 + 2a_1h + a_2)y^2 + 4(a_0h^3 + 3a_1h^2 + 3a_2h + a_3)y + (a_0h^4 + 4a_1h^3 + 6a_2h^2 + 4a_3h + a_4) = 0$ ...(3)

To remove the second term, put the co-efficient of $y^3 = 0$

i.e., put $$a_0h + a_1 = 0 \quad \therefore \quad \mathbf{h} = \frac{-\mathbf{a_1}}{\mathbf{a_0}}$$

substituting this value of h in the other co-efficients of (3), we get as in *Art. 9.*

(*i*) $a_0h^2 + 2a_1h + a_2 = \dfrac{\mathrm{H}}{a_0}$, where $\mathrm{H} = a_0a_2 - a_1^2$

(*ii*) $a_0h^3 + 3a_1h^2 + 3a_2h + a_3 = \dfrac{\mathrm{G}}{a_0^2}$

where $\mathrm{G} = a_0^2a_3 - 3a_0a_1a_2 + 2a_1^3$

(*iii*) $a_0h^4 + 4a_1h^3 + 6a_2h^2 + 4a_3h + a_4$

$$= a_0\frac{a_1^4}{a_0^4} + 4a_1\left(\frac{-a_1^3}{a_0^3}\right) + 6a_2\left(\frac{a_1^2}{a_0^2}\right) + 4a_3\left(\frac{-a_1}{a_0}\right) + a_4$$

$$= \frac{a_1^4}{a_0^3} - \frac{4a_1^4}{a_0^3} + \frac{6a_1^2a_2}{a_0^2} - \frac{4a_1a_3}{a_0} + a_4$$

$$= \frac{a_0^3a_4 - 4a_0^2a_1a_3 + 6a_0a_1^2a_2 - 3a_1^4}{a_0^3}$$

$$= \frac{(a_0^3a_4 - 4a_0^2a_1a_3 + 3a_0^2a_2^2) - 3(a_0^2a_2^2 - 2a_0a_2a_1^2 + a_1^4)}{a_0^3}$$

$$= \frac{a_0^2(a_0a_4 - 4a_1a_3 + 3a_2^2) - 3(a_0a_2 - a_1^2)^2}{a_0^3}$$

$$= \frac{a_0^2\,\mathrm{I} - 3\mathrm{H}^2}{a_0^3}, \text{ where } \mathrm{I} = a_0a_4 - 4a_1a_3 + 3a_2^2$$

Substituting, these values in (3), the equation becomes

$$a_0y^4 + \frac{6\mathrm{H}}{a_0}y^2 + \frac{4\mathrm{G}}{a_0^2}y + \frac{a_0^2\mathrm{I} - 3\mathrm{H}^2}{a_0^3} = 0$$

or $$y^4 + \frac{6\mathrm{H}}{a_0^2}y^2 + \frac{4\mathrm{G}}{a_0^3}y + \frac{a_0^2\mathrm{I} - 3\mathrm{H}^2}{a_0^4} = 0 \qquad \text{...(4)}$$

Multiplying the roots of equation (4) by a_0 and let the new equation be in z.

$\therefore$ Equation (4) transforms into $z^4 + 6\mathrm{H}z^2 + 4\mathrm{G}z + a_0^2\mathrm{I} - 3\mathrm{H}^2 = 0$

which is the required transformed equation. It is also known as the *Standard form.*

(*b*) *To find the relation between z and x.*

z is any general root of the transformed equation and x that of the given equation

$$z = a_0y = a_0(x - h) = a_0\left(x + \frac{a_1}{a_0}\right) = \mathbf{a_0x + a_1}$$

Thus the relation $z = a_0x + a_1$ will directly transform equation (1) into equation (5).

Cor. *To find a root of the transformed equation in terms of the roots of the given equation.*

$$z_1 = a_0x_1 + a_1, \text{ where } z_1 \text{ is any one root of the transformed equation}$$

$$= a_0\left(x_1 + \frac{a_1}{a_0}\right)$$

From equation (1),

$$x_1 + x_2 + x_3 + x_4 = -\frac{4a_1}{a_0}$$

or
$$-\frac{1}{4}(x_1 + x_2 + x_3 + x_4) = \frac{a_1}{a_0}$$

$$\therefore \quad z_1 = a_0\left(x_1 - \frac{x_1 + x_2 + x_3 + x_4}{4}\right)$$

$$\therefore \quad z_1 = \frac{a_0}{4}(3x_1 - x_2 - x_3 - x_4)$$

which expresses z_1 in terms of the roots of the given equation.

11. Transformation in General

Given an equation $f(x) = 0$, to transform the equation $f(x) = 0$ into another whose roots are some function of the roots of given equation. The method consists of the following steps :

Step 1. If α, β, γ are the roots of the given equation $f(x) = 0$, write down the values of $\Sigma\alpha$, $\Sigma\alpha\beta$, $\Sigma\alpha\beta\gamma$ etc.

Step 2. Put y = one of the roots of the required equation and simplify it with the help of equations obtained in step 1, so that it contains only one of the roots say α of $f(x) = 0$. Replace this root α by the variable x, we obtain a relation between x and y.

Step 3. Eliminate x between the relation obtained in step 2 and the given equation $f(x) = 0$, we shall obtain the required equation in the variable y.

The method is *illustrated* with the help of following examples.

Example 1. *If α, β, γ are the roots of the equation $x^3 + px^2 + qx + r = 0$, find an equation whose roots are $\beta + \gamma - 2\alpha$; $\gamma + \alpha - 2\beta$; $\alpha + \beta - 2\gamma$.*

Sol. The given equation is $x^3 + px^2 + qx + r = 0$ ...(1)

As α, β, γ are its roots

$$\therefore \quad \left.\begin{aligned} \sigma_1 &= \alpha + \beta + \gamma = -p \\ \sigma_2 &= \alpha\beta + \beta\gamma + \gamma\alpha = q \\ \sigma_3 &= \alpha\beta\gamma = -r \end{aligned}\right\} \quad ...(2)$$

If the required equation be in the variable y, then

$$y = \beta + \gamma - 2\alpha = (\alpha + \beta + \gamma) - 3\alpha = -p - 3\alpha \quad \text{[Using (2)]}$$

or
$$y = -p - 3x \quad ...(3)$$

Eliminate x between (1) and (3)

From (3), $x = -\dfrac{p+y}{3}$, putting in (1), we get

$$-\tfrac{1}{27}(p + y)^3 + p \cdot \tfrac{1}{9}(p + y)^2 - q \cdot \tfrac{1}{3}(p + y) + r = 0$$

or
$$(p + y)^3 - 3p(p + y)^2 + 9q(p + y) - 27r = 0$$

or $(p^3 + 3p^2y + 3py^2 + y^3) - 3p(p^2 + 2py + y^2) + 9pq + 9qy - 27r = 0$

or $y^3 + y(3p^2 - 6p^2 + 9q) + p^3 - 3p^3 + 9pq - 27r = 0$

or $y^3 + (9q - 3p^2)y + 9pq - 2p^3 - 27r = 0.$

This is the required equation.

Example 2. *If α, β, γ are the roots of the cubic equation $x^3 + ax^2 + bx + c = 0$, find the equation whose roots are*

$$\frac{\alpha}{\beta+\gamma}, \frac{\beta}{\gamma+\alpha}, \frac{\gamma}{\alpha+\beta}.$$

Sol. The given equation is $x^3 + ax^2 + bx + c = 0$ ...(1)

Let the required equation be in y

$$\therefore \quad y = \frac{\alpha}{\beta+\gamma} = \frac{\alpha}{\alpha+\beta+\gamma-\alpha} = \frac{\alpha}{-a-\alpha} \qquad [\because \ \alpha+\beta+\gamma = -a]$$

Replacing α by x, we have $y = -\left(\frac{x}{a+x}\right)$ ...(2)

Eliminate x between (1) and (2)

From (2), $ay + xy + x = 0$

i.e., $ay + x(y + 1) = 0$

or $x = -\frac{ay}{y+1},$

$$\therefore \quad (1) \Rightarrow \quad -\frac{a^3y^3}{(y+1)^3} + a \cdot \frac{a^2y^2}{(y+1)^2} - b \cdot \frac{ay}{y+1} + c = 0$$

Multiplying by $(y + 1)^3$ and dividing by -1, we get

$$a^3y^3 - a^3y^2(y + 1) + aby(y + 1)^2 - c(y + 1)^3 = 0$$

i.e., $a^3y^3 - a^3y^3 - a^3y^2 + aby(y^2 + 2y + 1) - c(y^3 + 3y^2 + 3y + 1) = 0$

i.e., $(ab - c)y^3 + (2ab - a^3 - 3c)y^2 + (ab - 3c)y - c = 0$

This is the required transformed equation.

Example 3. *If α, β, γ are the roots of the equation $2x^3 + x^2 + x + 1 = 0$, find the equation whose roots are*

$$\frac{1}{\beta^2} + \frac{1}{\gamma^2} - \frac{1}{\alpha^2}; \frac{1}{\gamma^2} + \frac{1}{\alpha^2} - \frac{1}{\beta^2}; \frac{1}{\alpha^2} + \frac{1}{\beta^2} - \frac{1}{\gamma^2}.$$

Sol. Given equation is $2x^3 + x^2 + x + 1 = 0$ ...(1)

Let us first find an equation whose roots are $\frac{1}{\alpha^2}, \frac{1}{\beta^2}, \frac{1}{\gamma^2}$

If this equation is in y, then $y = \frac{1}{\alpha^2}$ or $y = \frac{1}{x^2}$ ...(2)

We eliminate x between (1) and (2).

Putting the value of $x^2 = \frac{1}{y}$ from (2) in (1), we get

$$2x \cdot \frac{1}{y} + \frac{1}{y} + x + 1 = 0$$

$$\Rightarrow \quad x\left(\frac{2}{y} + 1\right) = -\left(\frac{1}{y} + 1\right)$$

Square both sides and put the value of x^2, we get

$$\frac{1}{y}\left(\frac{2}{y}+1\right)^2 = \left(\frac{1}{y}+1\right)^2$$

or $$\frac{1}{y}\left(\frac{4}{y^2}+\frac{4}{y}+1\right) = \left(\frac{1}{y^2}+\frac{2}{y}+1\right)$$

or $$\frac{4}{y^3}+\frac{4}{y^2}+\frac{1}{y} = \frac{1}{y^2}+\frac{2}{y}+1$$

or $$4 + 4y + y^2 = y + 2y^2 + y^3$$

or $$y^3 + y^2 - 3y - 4 = 0 \qquad ...(3)$$

Roots of (3) are $\frac{1}{\alpha^2}, \frac{1}{\beta^2}, \frac{1}{\gamma^2}$ which we denote by α', β', γ' respectively.

Now let the required equation be in z

$$\therefore \quad z = \frac{1}{\beta^2}+\frac{1}{\gamma^2}-\frac{1}{\alpha^2} = \beta' + \gamma' - \alpha'$$

$$= (\alpha' + \beta' + \gamma') - 2\alpha'$$

$$= -1 - 2\alpha' \qquad [\because \text{ from (3) ; } \alpha' + \beta' + \gamma' = -1]$$

$$\therefore \quad z = -(1 + 2y) \qquad ...(4)$$

[As α' is a root of (3) an equation in y replace α' by y]

Eliminate y from (3) and (4)

From (4), $y = -\left(\frac{z+1}{2}\right)$, put in (3)

$$\therefore \quad -\frac{(z+1)^3}{8} + \frac{(z+1)^2}{4} + \frac{3(z+1)}{2} - 4 = 0$$

i.e., $$(z + 1)^3 - 2(z + 1)^2 - 12(z + 1) + 32 = 0$$

i.e., $$(z^3 + 3z^2 + 3z + 1) - 2(z^2 + 2z + 1) - 12z - 12 + 32 = 0$$

i.e., $$z^3 + z^2 - 13z + 19 = 0$$

which is the required equation.

Example 4. *If α, β, γ are roots of the cubic $x^3 + 3x + 2 = 0$, find the equation whose roots are $(\alpha - \beta)(\alpha - \gamma), (\beta - \gamma)(\beta - \alpha), (\gamma - \alpha)(\gamma - \beta)$. Hence show that the given equation has two imaginary roots.*

Sol. Since α, β, γ are roots of the equation $x^3 + 3x + 2 = 0$ ...(1)

$$\therefore \quad \left.\begin{aligned} \alpha + \beta + \gamma &= 0 \\ \alpha\beta + \beta\gamma + \gamma\alpha &= 3 \\ \alpha\beta\gamma &= -2 \end{aligned}\right\} \qquad ...(2)$$

If y denotes the general root of the transformed equation, then

$$y = (\alpha - \beta)(\alpha - \gamma) = \alpha^2 - \alpha\beta - \alpha\gamma + \beta\gamma = \alpha^2 - \alpha(\beta + \gamma) + \beta\gamma$$

$$= \alpha^2 - \alpha\,[(\alpha + \beta + \gamma) - \alpha] + \frac{\alpha\beta\gamma}{\alpha} = \alpha^2 + \alpha^2 - \frac{2}{\alpha} \qquad \text{[By (2)]}$$

$$= 2x^2 - \frac{2}{x} \qquad (\text{Replacing } \alpha \text{ by } x)$$

or $$2x^3 - xy - 2 = 0 \qquad ...(3)$$

To have the transformed equation, we shall eliminate x between (1) and (3).

(1) – (3) gives, $$6x + xy + 6 = 0$$

$$\Rightarrow \quad x = -\frac{6}{6+y}$$

Substituting this value of x in (1), we have

$$-\frac{216}{(6+y)^3} - \frac{18}{6+y} + 2 = 0$$

$$\Rightarrow \quad y^3 + 9y^2 - 216 = 0 \qquad \text{...(4)}$$

(on simplification)

This is the required equation.

Now, product of roots of equation (4) is 216.

$$\Rightarrow \quad (\alpha - \beta)(\alpha - \gamma)(\beta - \gamma)(\beta - \alpha)(\gamma - \alpha)(\gamma - \beta) = 216$$

$$\Rightarrow \quad (\alpha - \beta)^2 (\beta - \gamma)^2 (\gamma - \alpha)^2 = -216$$

$\Rightarrow$ Either one of $(\alpha - \beta)^2, (\beta - \gamma)^2, (\gamma - \alpha)^2$ is negative or all the three are negative.

If one of $(\alpha - \beta)^2, (\beta - \gamma)^2, (\gamma - \alpha)^2$, namely $(\alpha - \beta)^2$ is negative, then α, β are conjugate complex numbers *i.e.*, the equation (1) has two imaginary roots.

All of $(\alpha - \beta)^2, (\beta - \gamma)^2, (\gamma - \alpha)^2$ cannot be negative because the equation (1) is a cubic and so it has one root real.

Hence, (1) has two imaginary roots.

EXERCISE 4

1. If α, β, γ be the roots of the equation $x^3 + qx + r = 0$, form an equation whose roots are :

(*i*) $\frac{1}{\alpha^2}, \frac{1}{\beta^2}, \frac{1}{\gamma^2}$ (*ii*) $\beta\gamma + \frac{1}{\alpha}, \gamma\alpha + \frac{1}{\beta}, \alpha\beta + \frac{1}{\gamma}$.

2. If α, β, γ are the roots of the equation $x^3 + px + q = 0$, form an equation whose roots are :

(*i*) $\frac{\beta+\gamma}{\alpha^2}, \frac{\gamma+\alpha}{\beta^2}, \frac{\alpha+\beta}{\gamma^2}$

(*ii*) $\gamma^2 + \gamma\alpha + \alpha^2 ; \alpha^2 + \alpha\beta + \beta^2 ; \beta^2 + \beta\gamma + \gamma^2$.

3. If α, β, γ are the roots of the equation $x^3 - ax^2 + bx - c = 0$, find an equation whose roots are :

(*i*) $\beta + \gamma, \gamma + \alpha, \alpha + \beta$ (*ii*) $\frac{\alpha}{\beta+\gamma}, \frac{\beta}{\gamma+\alpha}, \frac{\gamma}{\alpha+\beta}$.

4. If α, β, γ are the roots of the equation $x^3 + px^2 + qx + r = 0$, form the equation whose roots are :

(*i*) $\alpha - \frac{1}{\beta\gamma}, \beta - \frac{1}{\gamma\alpha}, \gamma - \frac{1}{\alpha\beta}$ (*ii*) $\beta\gamma + \frac{1}{\alpha}, \gamma\alpha + \frac{1}{\beta}, \alpha\beta + \frac{1}{\gamma}$.

5. (*a*) If α, β, γ are the roots of the equation $x^3 + x^2 + 2x + 3 = 0$, form the equation whose roots are : $\beta + \gamma - \alpha ; \gamma + \alpha - \beta ; \alpha + \beta - \gamma$.

(*b*) If α, β, γ are roots of $x^3 - 3x - 2 = 0$, find an equation whose roots are $(\beta + \gamma)^2, (\gamma + \alpha)^2, (\alpha + \beta)^2$. Hence find an equation whose roots are α^2, β^2 and γ^2.

6. If α, β, γ are the roots of the equation $x^3 - 6x^2 + 11x + k = 0$, find an equation whose roots are $\alpha - 2\beta + \gamma, \beta - 2\gamma + \alpha, \gamma - 2\alpha + \beta$.

Hence find the condition that roots of the given equation are in A.P.

7. If α, β, γ are the roots of the equation $z^3 + 3Hz + G = 0$, find the equation with roots

$$\frac{\alpha+1}{\beta+\gamma-\alpha}, \frac{\beta+1}{\gamma+\alpha-\beta}, \frac{\gamma+1}{\alpha+\beta-\gamma}.$$

8. (*a*) If x_1, x_2, x_3 are the roots of $x^3 - x^2 + 4 = 0$, form the equation whose roots are

$$x_1 + x_2^2 + x_3^2, x_2 + x_3^2 + x_1^2, x_3 + x_1^2 + x_2^2.$$

(*b*) If α, β, γ are roots of $x^3 - x^2 + 1 = 0$, form the equation whose roots are

$$\beta^2 + \gamma^2 + \alpha, \gamma^2 + \alpha^2 + \beta, \alpha^2 + \beta^2 + \gamma.$$

9. If α, β, γ are the roots of the equation $x^3 + px^2 + qx + r = 0$ find the equation whose roots are

$$\alpha^2 - \beta\gamma, \beta^2 - \gamma\alpha, \gamma^2 - \alpha\beta.$$

10. If α, β, γ are roots of $x^3 - x - 1 = 0$, then prove that

$$\frac{1+\alpha}{1-\alpha} + \frac{1+\beta}{1-\beta} + \frac{1+\gamma}{1-\gamma} = -7.$$

11. If α, β, γ are roots of $ax^3 + 3bx^2 + 3cx + d = 0$, find the equation whose roots are $\alpha(\beta + \gamma), \beta(\gamma + \alpha), \gamma(\alpha + \beta)$.

[**Hint.** Let $y = \alpha(\beta + \gamma) = \alpha\beta + \alpha\gamma = \alpha\beta + \alpha\gamma + \beta\gamma - \beta\gamma = \frac{3c}{a} - \frac{\alpha\beta\gamma}{\alpha} = \frac{3c}{a} - \frac{\left(-\frac{d}{a}\right)}{x}$. Eliminate x]

12. If α, β, γ are roots of $x^3 + px^2 + qx + r = 0$, find the value of $(\beta + \gamma)(\gamma + \alpha)(\alpha + \beta)$ in terms of p, q, r.

[**Hint.** $\beta + \gamma = \alpha + \beta + \gamma - \alpha = -p - \alpha = -(p + \alpha)$ etc]

13. If the roots of the equation $x^3 - 6x^2 - 11x - 6 = 0$ be α, β, γ, form the equation whose roots are

$$\alpha^2 + \beta^2, \beta^2 + \gamma^2, \gamma^2 + \alpha^2.$$

ANSWERS

1. (*i*) $r^2y^3 - q^2y^2 - 2qy - 1 = 0$ (*ii*) $ry^3 + q(1 - r)y^2 + (1 - r)^3 = 0$

2. (*i*) $qy^3 - py^2 - 1 = 0$ (*ii*) $(y + p)^3 = 0$

3. (*i*) $y^3 - 2ay^2 + (a^2 + b)y + c - ab = 0$ (*ii*) $(ab - c)y^3 - (a^3 - 2ab + 3c)y^2 + (ab - 3c)y - c = 0$

4. (*i*) $r^2y^3 + pr(r + 1)y^2 + q(r + 1)^2y + (r + 1)^3 = 0$ (*ii*) $ry^3 - q(r - 1)y^2 + p(r - 1)^2y - (r - 1)^3 = 0$

5. (*a*) $y^3 + y^2 + 7y - 17 = 0$ (*b*) $y^3 - 6y^2 + 9y - 4 = 0$, $y^3 - 6y^2 + 9y - 4 = 0$

6. $y^3 - 9y - 27(k + 6) = 0$; $k = -6$

7. $8Gt^3 + 12(G - H)t^2 + 6(G - 2H)t + (G - 3H - 1) = 0$

8. (*a*) $y^3 - 3y^2 - y + 19 = 0$ (*b*) $y^3 - 3y^2 + 2y + 1 = 0$

9. $y^3 + (3q - p^2)y^2 + (3q^2 - p^2q)y + (q^3 - p^3r) = 0$

11. $a^3y^3 - 6a^2cy^2 + 3a(3c^2 + bd)y + (ad^2 - 9bcd) = 0$

12. $r - pq$ **13.** $z^3 - 116z^2 + 3413z - 2806 = 0$.

12. Equation of Squared Differences of a Cubic

If α, β, γ are the roots of $y^3 + 3Hy + G = 0$, find an equation whose roots are

$$(\beta - \gamma)^2, (\gamma - \alpha)^2, (\alpha - \beta)^2.$$

Sol. α, β, γ are the roots of the given equation

$$y^3 + 3Hy + G = 0 \quad ...(1)$$

$$\therefore \left.\begin{aligned} \alpha + \beta + \gamma &= 0 \\ \alpha\beta\gamma &= -G \end{aligned}\right\} \quad ...(2)$$

To form an equation whose roots are $(\beta - \gamma)^2, (\gamma - \alpha)^2, (\alpha - \beta)^2$, let the transformed equation be in the variable z.

$$\therefore \quad z = (\beta - \gamma)^2 \quad [z = \text{any one of the roots of transformed equation}]$$

$$= (\beta + \gamma)^2 - 4\beta\gamma$$

$$= (-\alpha)^2 - 4\left(\frac{-G}{\alpha}\right) \quad \Big| \because \ \beta + \gamma = -\alpha, \beta\gamma = \frac{-G}{\alpha}$$

$$= \alpha^2 + \frac{4G}{\alpha} \quad [\text{replace } \alpha \text{ by } y]$$

i.e.,

$$z = y^2 + \frac{4G}{y} \quad \text{or} \quad y^3 - yz + 4G = 0 \quad \text{...(3)}$$

Eliminate y from (1) and (3)

Subtracting (3) from (1), we have

$$(3H + z)y - 3G = 0 \quad \text{or} \quad y = \frac{3G}{3H + z}$$

Put in (1),

$$\therefore \quad \frac{27G^3}{(3H + z)^3} + 3H\left(\frac{3G}{3H + z}\right) + G = 0$$

or

$$27G^2 + 9H(3H + z)^2 + (3H + z)^3 = 0 \quad \text{(cancelling G)}$$

or

$$27G^2 + 9H(9H^2 + 6Hz + z^2) + 27H^3 + 27H^2z + 9Hz^2 + z^3 = 0$$

or

$$z^3 + 18Hz^2 + 81H^2z + 27(G^2 + 4H^3) = 0$$

This is the required equation of the **squared differences** of the roots of equation (1).

13. Find the equation of the Squared Differences of the roots of the equation

$$\mathbf{a_0x^3 + 3a_1x^2 + 3a_2x + a_3 = 0.}$$

Sol. Let x_1, x_2, x_3 be the roots of the equation $a_0x^3 + 3a_1x^2 + 3a_2x + a_3 = 0$ ...(1)

Transform the equation (1) into another equation in which the second term is missing, this can be done by making the substitution $y = a_0x + a_1$ ...(2)

in (1)

The transformed equation is $y^3 + 3Hy + G = 0$ ...(3)

where $H = a_0a_2 - a_1^2$ and $G = a_0^2a_3 - 3a_0a_1a_2 + 2a_1^3$

If α, β, γ are the roots of equation (3), we have from (2)

$$\alpha = a_0x_1 + a_1 \ ; \ \beta = a_0x_2 + a_1 \ ; \ \gamma = a_0x_3 + a_1$$

Let us form an equation whose roots are $(\beta - \gamma)^2, (\gamma - \alpha)^2, (\alpha - \beta)^2$

If this equation is in the variable Z, then

$$Z = (\beta - \gamma)^2 = (\beta + \gamma)^2 - 4\beta\gamma \quad [\because \ \alpha + \beta + \gamma = 0, \alpha\beta\gamma = -G]$$

$$= \alpha^2 - 4\left(\frac{-G}{\alpha}\right) = \frac{\alpha^3 + 4G}{\alpha}$$

$$Z = \frac{y^3 + 4G}{y} \quad \text{or} \quad y^3 - yZ + 4G = 0 \quad \text{...(4)}$$

To eliminate y from (3) and (4), we have by subtracting (4) from (3), $y(3H + Z) = 3G$

or

$$y = \frac{3G}{3H + Z}$$

Put in (3),

$$\therefore \quad \frac{27G^3}{(3H+Z)^3} + 3H \cdot \frac{3G}{3H+Z} + G = 0$$

or $$27G^2 + 9H(3H + Z)^2 + (3H + Z)^3 = 0$$

or $$Z^3 + 18HZ^2 + 81H^2Z + 27(G^2 + 4H^3) = 0 \quad ...(5)$$

Now we are required to form an equation whose roots are

$$(x_2 - x_3)^2 \,;\, (x_3 - x_1)^2 \,;\, (x_1 - x_2)^2$$

if p is any one of these roots, then

$$p = (x_2 - x_3)^2 = \frac{1}{a_0^{\,2}} (\beta - \gamma)^2 = \frac{1}{a_0^{\,2}} \cdot Z$$

or $$\mathbf{Z = a_0^{\,2}p} \quad ...(6)$$

Eliminate 'Z' from (5) and (6)

$$\therefore \quad a_0^{\,6}p^3 + 18Ha_0^{\,4}p^2 + 81H^2a_0^{\,2}p + 27(G^2 + 4H^3) = 0 \quad ...(7)$$

which is the required equation of the squared differences of the roots of equation (1).

Cor. If $a_0 = 1$

Then from (6), $Z = p$

$\therefore$ Equation (7) becomes the same as equation (6).

Hence, if the co-efficient of leading term is unity, the equation of squared differences of the roots of equations (1) and (3) is same.

Note. For numerical problems, first remove the second term and then proceed as in Art. 12.

Example 1. *Find the equation of squared differences of the roots of the equation $x^3 + 3x + 2 = 0$.*

Hence show that the equation has a pair of imaginary roots.

Sol. Let α, β, γ be the roots of the given equation $x^3 + 3x + 2 = 0$ ...(1)

To form an equation, whose roots are $(\beta - \gamma)^2$; $(\gamma - \alpha)^2$; $(\alpha - \beta)^2$ if the new equation is in the variable y, then let

$$y = (\beta - \gamma)^2 = (\beta + \gamma)^2 - 4\beta\gamma = (-\alpha)^2 - 4\left(-\frac{2}{\alpha}\right)$$

$$\left[\because \ \alpha + \beta + \gamma = 0 \ \therefore \ \beta + \gamma = (-\alpha), \alpha\beta\gamma = -2 \ \therefore \ \beta\gamma = -\frac{2}{\alpha}\right]$$

$$= \alpha^2 + \frac{8}{\alpha} = \frac{\alpha^3 + 8}{\alpha} \text{ or } y = \frac{x^3 + 8}{x} \quad \text{[replacing } x \text{ by } \alpha]$$

i.e., $$x^3 - xy + 8 = 0 \quad ...(2)$$

Eliminate x between (1) and (2) by subtracting

$$x(3 + y) - 6 = 0$$

$$\therefore \quad x = \frac{6}{3 + y}, \text{ put in (1)}$$

$$\therefore \quad \left(\frac{6}{3+y}\right)^3 + 3\left(\frac{6}{3+y}\right) + 2 = 0$$

or $$216 + 18(3 + y)^2 + 2(3 + y)^3 = 0$$

or $$216 + 18(9 + 6y + y^2) + 2(27 + y^3 + 27y + 9y^2) = 0$$

or $$2y^3 + 36y^2 + 162y + 432 = 0$$

or $$\mathbf{y^3 + 18y^2 + 81y + 216 = 0}$$

This is the required equation of the squared differences of the roots of (1).

$\therefore$ **Product of all its roots = – 216**

i.e., $$(\alpha - \beta)^2(\beta - \gamma)^2(\gamma - \alpha)^2 = -216$$

(*i*) Either one of the three factors is negative or (*ii*) all the three factors are negative.

Let one of the factors *i.e.*, $(\alpha - \beta)^2$ be negative

$\Rightarrow$ **$(\alpha - \beta)$ is purely imaginary**

$\Rightarrow$ **α and β are conjugate complex numbers**

$\therefore$ **Two roots are conjugate complex.**

Again, all the three factors cannot be negative as a cubic has at least one real root.

Hence two roots are imaginary.

Example 2. *Find the equation of the squared differences of the roots of*

$$x^3 + 6x^2 + 9x + 4 = 0.$$

What conclusion with respect to the roots of the given cubic can be drawn from the form of the resulting cubic ?

Sol. **Let α, β, γ be the roots of the given equation $x^3 + 6x^2 + 9x + 4 = 0$ ...(1)**

We remove the second term by decreasing the roots by

$$h = -\frac{a_1}{na_0} = -\frac{6}{3(1)} = -2.$$

$$\therefore \quad \begin{array}{r|rrrr}
-2) & 1 & 6 & 9 & 4 \\
 & & -2 & -8 & -2 \\ \hline
 & 1 & 4 & 1 & 2 \\
 & & -2 & -4 & \\ \hline
 & 1 & 2 & -3 & \\
 & & -2 & & \\ \hline
 & 1 & 0 & & \\ \hline
 & 1 & & &
\end{array}$$

$\therefore$ **The transformed equation is $y^3 - 3y + 2 = 0$ ...(2)**

If its roots be $\alpha_1, \beta_1, \gamma_1$, then $\alpha_1 = \alpha + 2$, $\beta_1 = \beta + 2$, $\gamma_1 = \gamma + 2$

[Decreasing by – 2 means we have increased by 2]

Hence, $$(\alpha_1 - \beta_1)^2 = (\alpha - \beta)^2, (\beta_1 - \gamma_1)^2 = (\beta - \gamma)^2$$

and $$(\gamma_1 - \alpha_1)^2 = (\gamma - \alpha)^2$$

$\therefore$ **Equation of the squared differences of the roots of (1) and (2) will be exactly the same.**

$\therefore$ **Let** $$z = (\beta - \gamma)^2 = (\beta_1 - \gamma_1)^2$$
$$= (\beta_1 + \gamma_1)^2 - 4\beta_1\gamma_1$$

$$= (-\alpha_1)^2 - 4\left(-\frac{2}{\alpha_1}\right)$$

$$= \alpha_1^2 + \frac{8}{\alpha_1} = \frac{\alpha_1^3 + 8}{\alpha_1} = \frac{y^3 + 8}{y}$$

[Replacing α_1 by y, as α_1 is a root of equation in y]

or $\quad y^3 - yz + 8 = 0 \quad$ ...(3)

Eliminate y from (2) and (3) by subtracting (3) from (2)

$\therefore \quad -y(3 - z) - 6 = 0$

or $\quad y = \dfrac{6}{z-3}$

$\therefore$ (2) $\Rightarrow \quad \dfrac{216}{(z-3)^3} - 3 \cdot \dfrac{6}{z-3} + 2 = 0$

or $\quad 216 - 18(z-3)^2 + 2(z-3)^3 = 0$

or $\quad 216 - 18(z^2 - 6z + 9) + 2(z^3 - 9z^2 + 27z - 27) = 0$

$\quad 2z^3 - 36z^2 + 162z = 0$

or $\quad 2z(z^2 - 18z + 81) = 0 \quad$ or $\quad 2z(z-9)^2 = 0$

or $\quad z(z-9)^2 = 0 \quad$ ...(4)

This is the required equation of the squared differences of the roots of (1).

From (4), value of one root is 0.

But roots of Eqn. (4) are $(\alpha - \beta)^2, (\beta - \gamma)^2, (\gamma - \alpha)^2$

So let $(\alpha - \beta)^2 = 0 \quad \therefore \quad \alpha = \beta$

$\therefore$ Two roots of (1) are equal.

EXERCISE 5

1. Find the equation whose roots are squared differences of roots of the equations :

(i) $x^3 - 3x + 1 = 0$ (ii) $x^3 - 7x + 6 = 0$

(iii) $x^3 + 6x^2 + 2 = 0$ [**Hint.** First remove second term.]

(iv) $x^3 + 3x^2 - 3x + 1 = 0$ (v) $x^3 - 9x^2 + 23x - 15 = 0$.

2. Find the equation whose roots are the squared differences of the roots of the cubic

$x^3 - 6x + 4\sqrt{2} = 0.$

Hence prove that the given equation has a multiple root.

ANSWERS

1. (i) $y^3 - 18y^2 + 81y - 81 = 0$ (ii) $y^3 - 42y^2 + 441y - 400 = 0$

(iii) $z^3 - 72z^2 + 1296z + 1836 = 0$ (iv) $z^3 - 36z^2 + 324z + 108 = 0$

(v) $z^3 - 24z^2 + 144z - 256 = 0$ **2.** $y^3 - 36y^2 + 324y = 0$.

8

Descarte's Rule of Signs

1. Definitions

(*a*) **Continuation or Permanence of Sign.** In a polynomial (terms being written in descending order), if a + sign follows a + sign or a – sign follows a – sign, a continuation or a permanence is said to occur, *e.g.*, in the polynomial $x^6 + 2x^5 - x^4 - 3x^3 + x^2 + 2x + 1$ ...(1) continuation of sign occurs at $2x^5$, $-3x^3$, $2x$ and 1.

(*b*) **Variation or Change of Sign.** If a + sign follows a – sign or a – sign follows a + sign in a polynomial, then a variation or change of sign is said to occur, *e.g.*, in polynomial (1), the change of sign occurs at $-x^4$ and x^2.

It may be noted here that the sum of continuations and variations of signs is equal to degree of the polynomial. But it is not so, in case of incomplete polynomials (in which one or more terms are missing), *e.g.*, In polynomial (1), the number of continuations of sign is 4 and the number of veriations of signs is 2. Thus, the sum of continuations and variations of signs is six which is equal to the degree of the polynomial. In the incomplete polynomial $x^6 + 1$, there is only one continuation of sign and no variation of sign occurs. The sum of continuation and variation of signs is one which is less then the degree of the polynomial.

2. A lemma

If a real polynomial f(x), complete or incomplete, be multiplied by x – h, (h > 0) the product will contain atleast one more change of sign than the original.

Proof. Let the polynomial $f(x)$ be

$$(a_0x^n + \ldots\ldots + a_{n_1-1}x^{n-n_1+1}) - (a_{n_1}x^{n-n_1} + \ldots\ldots + a_{n_2-1}x^{n-n_2+1})$$
$$+ (a_{n_2}x^{n-n_2} + \ldots\ldots + a_{n_3-1}x^{n-n_3+1}) \ldots\ldots + (-1)^m(a_{n_m}x^{n-n_m} + \ldots\ldots + a_n)$$

in which the terms inside each parenthesis are positive and $0 < n_1 < n_2 < n_3 < \ldots\ldots < n$. The number of variations of signs is evidently m.

Multiplying $f(x)$ by $(x - h)$, we get

$$(x - h)f(x) = x\,f(x) - hf(x)$$
$$= [(a_0x^{n+1} + \ldots\ldots a_{n_1-1}x^{n-n_1+2}) - (a_{n_1}x^{n-n_1+1} + \ldots\ldots + a_{n_2-1}x^{n-n_2+2})$$
$$+ (a_{n_2}x^{n-n_2+1} + \ldots\ldots + a_{n_3-1}x^{n-n_3+2}) - \ldots\ldots + (-1)^m(a_{n_m}x^{n-n_m+1} + \ldots\ldots + a_nx)]$$
$$- [(a_0hx^n + \ldots\ldots + a_{n_1-1}hx^{n-n_1+1}) - (a_{n_1}hx^{n-n_1} + \ldots\ldots + a_{n_2-1}hx^{n-n_2-1})$$
$$+ (a_{n_2}hx^{n-n_2} + \ldots\ldots + a_{n_3-1}hx^{n-n_3+1}) - \ldots\ldots + (-1)^m(a_{n_m}hx^{n-n_m} + \ldots\ldots + a_nh)]$$
$$= [(a_0x^{n+1} \pm \ldots\ldots] - [(a_{n_1-1}h + a_{n_1})x^{n-n_1+1} \pm \ldots\ldots] + [(a_{n_2-1}h + a_{n_2})x^{n-n_2+1} \pm \ldots\ldots]$$
$$- \ldots\ldots + (-1)^m[(a_{n_m-1}h + a_{n_m})x^{n-n_m+1} \pm \ldots\ldots - (a_{n-1}h - a_n)x] + (-1)^{m+1}a_nh.$$

In product $(x - h)f(x)$, the terms of the degree $n + 1, n - n_1 + 1, n - n_2 + 1, \ldots\ldots, n - n_m + 1$ are of the same sign as in $xf(x)$, and therefore, the number of variations of signs is m in both the cases. But in $(x - h)f(x)$ the last term is $(-1`)^{m+1} a_nh$ whose sign is opposite to the sign placed before the last parenthesis. Thus, there is atleast one additional variation of sign in $(x - h)f(x)$, as compared to $xf(x)$ or $f(x)$.

Hence irrespective of the number of variations of signs in each parenthesis of $(x-h)\,f(x)$, the number of variations of signs in it is atleast on more than that of $f(x)$.

3. Descarte's Rule of Sign

In a polynomial equation $f(x) = 0$, the number of positive roots cannot exceed the number of variations of signs of the coefficients and the number of negative roots cannot exceed the number of variations of signs of the coefficients in $f(-x)$.

Proof. If $\alpha_1, \alpha_2, \ldots\ldots, \alpha_p$ are the p positive roots of the equation $f(x) = 0$, then $f(x) = (x-\alpha_1)(x-\alpha_2) \ldots\ldots (x-\alpha_p)\,\phi(x)$, where the roots of $\phi(x) = 0$ are either negative or imaginary or both.

When $\phi(x)$ is multiplied by $(x-\alpha_1)(x-\alpha_2) \ldots\ldots (x-\alpha_p)$ atleast p variation of signs are introduced since each factor introduces atleast one variation of sign on multiplication. (See Section 2). This shows that if there are p positive roots of the equation $f(x) = 0$, then the number of variation of signs in $f(x)$ is p or more. Hence the number of positive roots of $f(x) = 0$ cannot exceed the number of variation of signs in $f(x)$.

If $-\beta_1, -\beta_2, \ldots\ldots, -\beta_q$ are the q negative roots of $f(x) = 0$, then $\beta_1, \beta_2, \ldots\ldots, \beta_q$ are the positive roots of $f(-x) = 0$. This equation has atleast q variations of signs, *i.e.*, then number of negative roots cannot exceed the number of variations of sign in $f(-x)$.

It should be noted that we do not get exact number of real roots by Descartes' Rule of Sign but we can find the maximum number of real roots, the equation $f(x) = 0$ may have. This rule helps us to know about the existence of imaginary roots. If there are atmost r real roots, the least number of imaginary roots is $n - r$.

Remark. Every real number is either zero, positive or negative.

Example 1. *Show that the equation $2x^7 - 5x^4 + 3x^3 - 1 = 0$ has atleast four imaginary roots.*

Sol. Let
$$f(x) = 2x^7 - 5x^4 + 3x^3 - 1$$
$$\therefore \quad f(-x) = -2x^7 - 5x^4 - 3x^3 - 1$$

Here number of changes of sign in $f(x)$ is 3.

$\therefore$ By Descarte's rule of signs, maximum number of positive real roots of the given equation is 3.

Again there is no change of sign in $f(-x)$.

$\therefore$ The given equation has no negative real root.

$\therefore$ Maximum number of real roots of eqn. (1) is 3.

But the given equation is of degree 7 and hence the total number of roots is 7.

$\therefore$ Least number of imaginary roots of eqn. (1) $= 7 - 3 = 4$.

Example 2. *Show that for all values of k, the equation $x^5 + 2x^2 + 3x + k = 0$ has at least two imagnary roots.*

Sol. The given equation is $x^5 + 2x^2 + 3x + k = 0$ ...(1)

Let $f(x) = x^5 + 2x^2 + 3x + k$ ...(2)

Case I. k > 0

Here there is no change of sign in $f(x)$.

$\therefore$ $f(x) = 0$ *i.e.* eqn. (1) has no positive root. $f(-x) = -x^5 + 2x^2 - 3x + k$

There are three changes of sign in $f(-x)$.

$\therefore$ Maximum number of negative real roots of eqn. (1) is 3.

But $f(x) = 0$ being of degree 5 has 5 roots.

$\therefore$ Least number of imaginary roots of eqn. (1) $= 5 - 3 = 2$.

Case II. k < 0

Let $k = -k'$, where $k' > 0$.

$\therefore$ From (2), $f(x) = x^5 + 2x^2 + 3x - k'$

There is only one change of sign in $f(x)$.

$\therefore$ Maximum number of positive real roots of eqn. (1) is 1, by Descarte's rule of signs.

Also $f(-x) = -x^5 + 2x^2 - 3x - k'$

There are two changes of sign in $f(-x)$.

$\therefore$ Maximum number of negative real roots of eqn. (1) is 2.

$\therefore$ Maximum number of real roots = 2 + 1 = 3

$\therefore$ Least number of imaginary roots of eqn. (1) is 5 – 3 = 2.

Case III. k = 0

$\therefore$ Eqn. (1) becomes $x^5 + 2x^2 + 3x = 0$ or $x(x^4 + 2x + 3) = 0$

$\therefore$ Either $x = 0$ or $x^4 + 2x + 3 = 0$

$x = 0 \Rightarrow$ one root is zero.

Let $\phi(x) = x^4 + 2x + 3$

There is no change of sign in $\phi(x)$.

$\therefore$ $\phi(x) = 0$ has no positive real root.

Also $\phi(-x) = x^4 - 2x + 3$

There are two changes of sign in $\phi(-x)$.

$\therefore$ Maximum number of negative real roots of $\phi(x) = 0$ is 2.

$\therefore$ Maximum number of real roots of eqn. (1) is 1 + 2 = 3.

$\therefore$ Least number of imaginary roots of eqn. (1) is 5 – 3 = 2.

EXERCISE

1. Apply Descarte's Rule of signs to discuss the nature of the roots of the equations :
   (*i*) $x^6 - 3x^2 - x + 1 = 0$ (*ii*) $3x^4 + 12x^2 + 5x - 4 = 0$ (*iii*) $x^4 + 15x^2 + 7x - 11 = 0$
2. Show that the equation whose terms are all positive, cannot have positive roots.
3. Show that the equation $x^8 - x^3 + x^2 - x + 1 = 0$ must have atleast 4 non-real roots.
4. Show that the equation $x^7 - 3x^4 + 2x^2 - 1 = 0$ has atleast two complex roots.
5. Show that the equation $x^{12} - 2x^4 + x^3 - 3x^2 + 12 = 0$ must have atleast six imaginary roots.
6. Prove that equation $x^5 + 2x^3 + 3x - 1 = 0$ has no negative root. Also find the number of positive and imaginary roots.
7. State Descarte's rule of signs. Prove that if n is even, the equation $x^n - 1 = 0$ has two and only two real roots, one positive and one negative and the rest complex and when n is odd the real root is unity and the rest are complex.
8. (*a*) Show that the equation $x^{2n} - 1 = 0$ has only two real roots and all other roots are non-real complex roots.
   (*b*) Prove that the equation $x^{2n} + 1 = 0$ has no real root and the equation $x^{2n+1} + 1 = 0$ has only one real root.
9. Prove that the equation $x^3 + 3Hx + G = 0$, where $H > 0$ has two non-real roots.
   [**Hint.** Do as Example 2 and also use that non-real roots of an equation always occur in conjugate pairs.]
10. Prove that the equation $2x^7 + 3x^4 + 3x + k = 0$ has atleast four imaginary roots for all real values of k.
11. Prove that the equation whose terms are alternately positive and negative can't have negative roots.

ANSWER

1. (*i*) At most two positive roots, at most two negative roots and at least two imaginary roots.
   (*ii*) At most one positive root, at most one negative root and at least two imaginary roots.
   (*iii*) At most one positive root, at most one negative root and at least two imaginary roots.

9

Solutions of General Cubic and Bi-quadratic Equations

1. Cardan's Method of Solving a Cubic equation

The general cubic equation is $a_0x^3 + 3a_1x^2 + 3a_2x + a_3 = 0$ ...(1)

First remove the second term, keeping the co-efficient of highest degree term unity. This can be done (see Art. 9, Chapter 7) by the substitution $z = a_0x + a_1$

The equation (1) transforms into $z^3 + 3Hz + G = 0$ ...(2)

where
$$H = a_0a_2 - a_1^2$$
$$G = a_0^2a_3 - 3a_0a_1a_2 + 2a_1^3$$

Let $z = u + v$ ...(3)

where u and v are numbers to be determined so that z becomes a root of (2).

Cubing it, we get
$$z^3 = u^3 + v^3 + 3uv(u + v)$$
or
$$z^3 = u^3 + v^3 + 3uv(z)$$
i.e.,
$$z^3 - 3uvz - (u^3 + v^3) = 0 \quad ...(4)$$

Identifying (2) and (4), we get
$$uv = -H \quad \text{or} \quad u^3v^3 = -H^3 \quad \text{and} \quad u^3 + v^3 = -G$$

$\therefore$ u^3, v^3 are the roots of the quadratic equation
$$t^2 + Gt - H^3 = 0 \ [t^2 - t \text{ (sum of roots)} + \text{product of roots} = 0]$$

i.e.,
$$t = \frac{-G \pm \sqrt{G^2 + 4H^3}}{2}$$

$\therefore$ Let
$$u^3 = \frac{-G + \sqrt{G^2 + 4H^3}}{2} \quad ...(5)$$
and
$$v^3 = \frac{-G - \sqrt{G^2 + 4H^3}}{2} \quad ...(6)$$

From equation (5), we get three values of u, namely u, ωu, $\omega^2 u$, where ω is a complex root of unity.

Substituting these values of u in the relation $uv = -H$ the corresponding three values of v are
$$v = -\frac{H}{u}, -\frac{H}{\omega u}, -\frac{H}{\omega^2 u}$$
or
$$v = -\frac{H}{u}, -\omega^2 \cdot \frac{H}{u}, -\frac{H}{u}\,\omega$$

Substituting the values of u and v in (3), the three roots of (2) are

$$z = u - \frac{H}{u}, \left(\omega u - \frac{H}{u}\omega^2\right); \left(\omega^2 u - \frac{H}{u}\omega\right)$$

Having known the solution of equation (2), the solution of equation (1) can be known by the relation $z = a_0 x + a_1$.

Note. The solutions of cubic $z^3 + 3Hz + G = 0$ are

$$z = u - \frac{H}{u}, u\omega - \frac{H}{u}\omega^2, u\omega^2 - \frac{H}{u}\omega$$

But $$-\frac{H}{u} = v$$

$\therefore$ These solutions may be written as

$$z = u + v, u\omega + v\omega^2, u\omega^2 + v\omega.$$

2. To Discuss the nature of Roots of the Cubic $z^3 + 3Hz + G = 0$

Given equation is $z^3 + 3Hz + G = 0$ ...(1)

Let $z = u + v$ be a root of (1) ; cubing,

$$z^3 = u^3 + v^3 + 3uv(u + v)$$

or $$z^3 - 3uvz - (u^3 + v^3) = 0 \quad ...(2)$$

$\therefore$ (1) and (2) have the same roots,

$\therefore$ $$uv = -H \Rightarrow u^3v^3 = -H^3 \text{ and } u^3 + v^3 = -G$$

$\therefore$ u^3, v^3 are the roots of $t^2 - t(-G) - H^3 = 0$

i.e., of $$t^2 + Gt - H^3 = 0 \quad ...(3)$$

$\therefore$ $$t = \frac{-G \pm \sqrt{G^2 + 4H^3}}{2}$$

$\therefore$ Let $$u^3 = \frac{-G + \sqrt{G^2 + 4H^3}}{2} \quad ...(4)$$

$$v^3 = \frac{-G - \sqrt{G^2 + 4H^3}}{2} \quad ...(5)$$

From (4) and (5), we get three values of u and v *i.e.,* $u, u\omega, u\omega^2$ and $v, v\omega, v\omega^2$

The three roots of (1), knowing that $uv = -H$ (Real), are

$$z_1 = u + v, z_2 = u\omega + v\omega^2, z_3 = u\omega^2 + v\omega$$

But $$\omega = \frac{-1 + i\sqrt{3}}{2}, \omega^2 = \frac{-1 - i\sqrt{3}}{2}$$

$\therefore$ $$z_1 = u + v, z_2 = u\left(\frac{-1 + i\sqrt{3}}{2}\right) + v\left(\frac{-1 - i\sqrt{3}}{2}\right)$$

$$= -\frac{1}{2}(u + v) + i\frac{\sqrt{3}}{2}(u - v)$$

$$z_3 = u\left(\frac{-1 - i\sqrt{3}}{2}\right) + v\left(\frac{-1 + i\sqrt{3}}{2}\right)$$

$$= -\frac{1}{2}(u + v) - \frac{i\sqrt{3}}{2}(u - v)$$

Now nature of the roots of equation (1) depends upon the nature of the values of u^3 and v^3, which depend upon $\mathbf{G^2 + 4H^3}$, which is called the **Discriminant** of the cubic.

Case 1. $G^2 + 4H^3 > 0$

$\Rightarrow$ u^3 and v^3 are both real $(u^3 \neq v^3)$

$\Rightarrow$ u and v are both real

$\Rightarrow$ $u + v$ and $(u - v)$ are both real.

$\Rightarrow$ z_1 is real and z_2 and z_3 are conjugate complex numbers

$\therefore$ *ONE root of (1) is real and the two roots are imaginary.*

Case 2. $G^2 + 4H^3 < 0$

$\Rightarrow$ u^3 and v^3 are imaginary and conjugate complex

$\Rightarrow$ u and v are conjugate complex numbers

$\Rightarrow$ $(u + v)$ is real and $(u - v)$ is purely imaginary

$\Rightarrow$ All the three roots, z_1, z_2, z_3 are real.

Case 3. $G^2 + 4H^3 = 0$

$\Rightarrow$ u^3 and v^3 are equal and real

$\Rightarrow$ u and v are equal and real

$\Rightarrow$ $u + v$ is real and $(u - v) = 0$

$\Rightarrow$ $z_2 = z_3$.

$\therefore$ All the roots are real and two roots are equal.

Case 4. If $G = 0, H = 0$

then given cubic becomes $z^3 = 0$

$\Rightarrow$ All the roots of (1) are equal and each equal to zero.

Remember. For the cubic $z^3 + 3Hz + G = 0$

$$D = \text{Discriminant} = G^2 + 4H^3.$$

(*i*) If $D > 0 \Rightarrow$ one root real and two imaginary

(*ii*) If $D < 0 \Rightarrow$ All roots real.

(*iii*) $D = 0$, with $G \neq 0, H \neq 0$.

All roots real and two equal roots

(*iv*) If $G = 0, H = 0$, their all roots equal.

3. Irreducible case of Cardon's method

The case when the Discriminant $G^2 + 4H^3 < 0$ is called an irreducible case of Cardon's method. Here u^3 and v^3 are conjugate complex numbers and we find cube roots of the complex numbers by using DeMovire's theorem given in chapter 1, section IV.

Example 1. *Solve the equation $x^3 - 15x - 126 = 0$ by Cardon's method.*

Sol. Given equation is
$$x^3 - 15x - 126 = 0 \quad ...(1)$$

Let
$$x = u + v$$

Cubing both sides, we have
$$x^3 = u^3 + v^3 + 3uv(u + v)$$

or
$$x^3 = u^3 + v^3 + 3uv(x)$$

i.e.,
$$x^3 - 3uvx - (u^3 + v^3) = 0 \quad ...(2)$$

Comparing (1) and (2), we have

$$uv = 5 \quad \text{and} \quad u^3 + v^3 = 126$$

i.e., $$u^3v^3 = 125$$

$\therefore$ u^3, v^3 are the roots of

$$t^2 - t(126) + 125 = 0$$

i.e., $$t^2 - 126t + 125 = 0$$

or $$(t - 125)(t - 1) = 0$$

$\therefore$ $$t = 1 \quad \text{and} \quad t = 125$$

Let $$u^3 = 1 \quad \text{and} \quad v^3 = 125$$

$\therefore$ $$u = 1 \quad \text{and} \quad v = 5$$

are one of the values of u and v respectively.

$x = u + v = 1 + 5 = 6$ is one of the required roots.

Dividing (1) by $(x - 6)$,

6)	1	0	– 15	– 126
		6	36	126
	1	6	21	zero

The quotient is $x^2 + 6x + 21$

Put the quotient = 0 *i.e.,* $x^2 + 6x + 21 = 0$

$\therefore$ $$x = \frac{-6 \pm \sqrt{36 - 84}}{2} = \frac{-6 \pm 4\sqrt{-3}}{2}$$

or $$x = -3 \pm 2\sqrt{-3}$$

Hence the three roots of the cubic are $6, -3 + 2\sqrt{-3}, -3 - 2\sqrt{-3}$.

Note. In a cubic equation, whenever, we get one root $x = u + v$, as here $x = 6$, the other two roots can be found by equating to zero, the quotient obtained by dividing the L.H.S. of the given cubic by the factor corresponding to the known root.

Example 2. *Solve by Cardon's method, the equation*

$$x^3 + 6x^2 + 9x + 4 = 0.$$

Sol. Given equation is $x^3 + 6x^2 + 9x + 4 = 0$ ...(1)

First remove the second term by decreasing the roots by h,

where $$h = -\tfrac{6}{3} = -2$$

i.e.,

– 2)	1	6	9	4
		– 2	– 8	– 2
	1	4	1	2
		– 2	– 4	
	1	2	– 3	
		– 2		
	1	0		
	1			

$\therefore$ The transformed equation in y obtained by increasing the roots of the given equation by 2 is $\quad y^3 - 3y + 2 = 0 \quad$ ...(2)

Let $\quad y = u + v$

$\therefore \quad y^3 = u^3 + v^3 + 3uvy$

or $\quad y^3 - 3uvy - (u^3 + v^3) = 0 \quad$ ...(3)

Identifying (2) and (3), we have $uv = 1$

$\therefore \quad u^3v^3 = 1$

and $\quad u^3 + v^3 = -2$

$\therefore \quad u^3, v^3$ are the roots of $t^2 + 2t + 1 = 0$ *i.e.*, $(t + 1)^2 = 0$

$(t + 1)^2 = 0 \Rightarrow t = -1, -1$

$\therefore \quad u^3 = v^3 = -1$

$\therefore \quad u = -1, v = -1.$

are the real cube roots of u^3 and v^3.

$\therefore \quad y = u + v = -2$ is one of the roots of cubic (2).

$\therefore$ Dividing (2) by $(y + 2)$, we have

$-2)$	1	0	-3	2
		-2	4	-2
	1	-2	1	zero

$\therefore$ The reduced equation is

$$y^2 - 2y + 1 = 0$$

i.e., $\quad (y - 1)^2 = 0$

$\Rightarrow \quad y = 1, 1.$

$\therefore$ The roots of the equation (2) are $1, 1, -2$.

Hence the roots of the given equation (1) are $= 1 - 2, 1 - 2, -2 - 2$ or $-1, -1, -4$.

Example 3. *Solve the cubic $2x^3 - 7x^2 + 8x - 3 = 0$ by Cardon's method.*

Sol. Given equation is $2x^3 - 7x^2 + 8x - 3 = 0 \quad$ ...(1)

To remove 2nd term, we have to decrease the roots by $h = \dfrac{7}{3(2)} = \dfrac{7}{6}$.

To avoid, decreasing the roots by a fraction, we first multiply the roots by 6. If the changed equation be in y, then $y = 6x$ and transformed equation is

$$2y^3 - 42y^2 + 288y - 648 = 0$$

or $\quad y^3 - 21y^2 + 144y - 324 = 0$

Now decreasing the roots by $h = 7$, we have $\quad$ ...(2)

7)	1	-21	144	-324
		7	-98	322
	1	-14	46	-2
		7	-49	
	1	-7	-3	
		7		
	1	0		
	1			

If transformed equation be in Z, then $Z^3 - 3Z - 2 = 0$, ...(3)

where $$Z = y - 7 = 6x - 7$$

Let $Z = u + v$ be a solution

$\therefore$ Cubing $$Z^3 = u^3 + v^3 + 3uvZ$$

or $$Z^3 - 3uvZ - (u^3 + v^3) = 0 \quad ...(4)$$

Comparing co-efficients of (3) and (4)

$$uv = 1 \quad \therefore \quad u^3v^3 = 1 \quad \text{and} \quad u^3 + v^3 = 2$$

$\therefore$ u^3, v^3 are the roots of $t^2 - 2t + 1 = 0$ or $(t - 1)^2 = 0$ $\therefore$ $t = 1, 1$

$\therefore$ $$u^3 = 1 \quad \text{and} \quad v^3 = 1.$$

$\therefore$ One real value each of u and v is 1

$\therefore$ Z = 2 is a solution of (3)

Dividing (3) by Z – 2,

2)	1	0	– 3	– 2
		2	4	2
	1	2	1	0

Quotient is $$Z^2 + 2Z + 1$$

Put $$Z^2 + 2Z + 1 = 0$$

or $$(Z + 1)^2 = 0 \quad \text{or} \quad Z = -1, -1$$

Hence roots of equation (3) are $2, -1, -1$

$\therefore$ Roots of (1) are given by $$x = \frac{Z + 7}{6}$$

i.e., $$x = \frac{9}{6}, \frac{6}{6}, \frac{6}{6}$$

Hence roots of given equation are $\frac{3}{2}$, 1, 1.

Example 4. *Solve the cubic* $28x^3 - 9x^2 + 1 = 0$ *by Cardon's method.*

Sol. The given cubic is $$28x^3 - 9x^2 + 1 = 0 \quad ...(1)$$

Since the constant term is 1 and second term from the end is missing, so the equation can be reduced to the standard form by the transformation

$$y = \frac{1}{x} \quad i.e., \quad x = \frac{1}{y}.$$

Putting $x = \frac{1}{y}$ in (1), we have

$$\frac{28}{y^3} - \frac{9}{y^2} + 1 = 0$$

or $$y^3 - 9y + 28 = 0 \quad ...(2)$$

Let $y = u + v$ be a root of (2)

Cubing, we have $$y^3 = u^3 + v^3 + 3uv\,(u + v)$$

$\Rightarrow$ $$y^3 - 3uvy - (u^3 + v^3) = 0 \quad ...(3)$$

Comparing (2) and (3), we have

$$uv = 3 \quad \Rightarrow \quad u^3v^3 = 27$$

and $$u^3 + v^3 = -28.$$

$\therefore$ u^3, v^3 are roots of $t^2 + 28t + 27 = 0.$

$$\Rightarrow \quad (t + 1)(t + 27) = 0$$

$$\Rightarrow \quad t = -1, -27$$

Let $u^3 = -1, v^3 = -27$

$$\Rightarrow \quad u = -1, v = -3$$

$$\Rightarrow \quad y = u + v = -4 \text{ is a root of (2)}$$

$\therefore$ $-\frac{1}{4}$ is a root of (1).

By synthetic division, we have

$-\frac{1}{4}$	28	-9	0	1
		-7	4	-1
	28	-16	4	0

$\therefore$ The depressed equation is $28x^2 - 16x + 4 = 0$

$$\Rightarrow \quad 7x^2 - 4x + 1 = 0$$

$$\Rightarrow \quad x = \frac{4 \pm \sqrt{16 - 28}}{14} = \frac{2 \pm i\sqrt{3}}{7}$$

Hence roots of (1) are $-\frac{1}{4}, \frac{2 \pm i\sqrt{3}}{7}$.

Example 5. *Solve the cubic equation $x^3 + 3x - 14 = 0$ by Cardon's method.*

Sol. Given equation is $x^3 + 3x - 14 = 0$ ...(1)

Let $x = u + v$

$$\therefore \quad x^3 = u^3 + v^3 + 3uv(u + v) = u^3 + v^3 + 3uvx$$

i.e., $x^3 - 3uvx - (u^3 + v^3) = 0$ ...(2)

Identifying (1) and (2), we have

$$uv = -1 \quad \therefore \quad u^3v^3 = -1$$

and $u^3 + v^3 = 14$

$\therefore$ u^3, v^3 are the roots of $t^2 - 14t - 1 = 0$

$$\therefore \quad t = \frac{14 \pm \sqrt{196 + 4}}{2} = \frac{14 \pm 10\sqrt{2}}{2} = 7 \pm 5\sqrt{2}$$

$\therefore$ Let $u^3 = 7 + 5\sqrt{2}$ and $v^3 = 7 - 5\sqrt{2}$.

To find u and v from these equations ; u and v will be of the form $p \pm q\sqrt{2}$. Let $u = p + q\sqrt{2}$.

$$\therefore \quad (p + q\sqrt{2})^3 = 7 + 5\sqrt{2}$$

$$\therefore \quad p^3 + 2\sqrt{2}q^3 + 3p.q.\sqrt{2}(p + q\sqrt{2}) = 7 + 5\sqrt{2}$$

i.e., $p^3 + 2\sqrt{2}q^3 + 3p^2q\sqrt{2} + 6pq^2 = 7 + 5\sqrt{2}$

i.e., $(p^3 + 6pq^2) + \sqrt{2}(2q^3 + 3p^2q) = 7 + 5\sqrt{2}$.

$\therefore$ Equating rational and irrational parts, we get

$p^3 + 6pq^2 = 7$ *i.e.,* $p(p^2 + 6q^2) = 7$ ...(*i*)

$2q^3 + 3p^2q = 5$ *i.e.,* $q(2q^2 + 3p^2) = 5$ ...(*ii*)

(*i*) and (*ii*) show that p is a factor of 7 and q is a factor of 5.

But it is easily seen that $p = 1$, $q = 1$, satisfy (*i*) and (*ii*).

$\therefore$ $\left.\begin{array}{l} u = p + q\sqrt{2} = 1 + \sqrt{2} \\ v = p - q\sqrt{2} = 1 - \sqrt{2} \end{array}\right\}$ are the real cube root of u^3 and v^3,

and satisfying $uv = -1$.

$\therefore$ One root of the given equation (1) is $x = u + v = 2$.

$\therefore$ Dividing (1) by $x - 2$,

2)	1	0	3	– 14
		2	4	14
	1	2	7	zero

$\therefore$ Reduced equation is $x^2 + 2x + 7 = 0$

$$x = \frac{-2 \pm \sqrt{4-28}}{2} = \frac{-2 \pm 2\sqrt{-6}}{2}$$

$$= -1 \pm \sqrt{-6}.$$

Hence the roots are $2, -1 + \sqrt{-6}, -1 - \sqrt{-6}$.

Note. We may also say that the roots are $u + v$, $u\omega^2 + v\omega$, $u\omega + v\omega^2$, where u, v have been obtained.

Example 6. *Using Cardon's method, show that the roots of the equation*

$$x^3 - 3x + 1 = 0 \text{ are } 2\cos\frac{2\pi}{9}, 2\cos\frac{8\pi}{9}, 2\cos\frac{14\pi}{9}.$$

Sol. Given equation is $x^3 - 3x + 1 = 0$ ...(1)

Let $x = u + v$

Cubing $x^3 = u^3 + v^3 + 3uv.x$

or $x^3 - 3uvx - (u^3 + v^3) = 0$ ...(2)

Comparing co-efficients of (1) and (2),

$$uv = 1 \text{ or } u^3v^3 = 1 \text{ and } u^3 + v^3 = -1$$

$\therefore$ u^3 and v^3 are the roots of

$$t^2 - t(-1) + 1 = 0 \text{ or } t^2 + t + 1 = 0$$

$\therefore$ $$t = \frac{-1 \pm \sqrt{1-4}}{2} = \frac{-1 \pm \sqrt{-3}}{2}$$

Let $u^3 = -\frac{1}{2} + i\frac{\sqrt{3}}{2}$ and $v^3 = -\frac{1}{2} - i\frac{\sqrt{3}}{2}$.

To find u from u^3 which is non-real

$\therefore$ $$u = \left(-\frac{1}{2} + i\frac{\sqrt{3}}{2}\right)^{1/3}$$

Put $\frac{-1}{2} = r\cos\theta$, $\frac{\sqrt{3}}{2} = r\sin\theta$

$$1 = r^2$$

$\therefore$ $r = 1$ (Taking +ve value of r)

$\therefore$ $\cos\theta = -\frac{1}{2}$ and $\sin\theta = \frac{\sqrt{3}}{2}$

$\therefore$ The value of θ satisfying both the equations is

$$\theta = 120^\circ = \frac{2\pi}{3}$$

$$\therefore \quad u = (r \cos \theta + i\, r \sin \theta)^{1/3} = r^{1/3} [\cos (\theta + 2k\pi) + i \sin (\theta + 2k\pi)]^{1/3}$$

$$= \left[\cos\left(\frac{2\pi}{3} + 2k\pi\right) + i \sin\left(\frac{2\pi}{3} + 2k\pi\right)\right]^{1/3}$$

$$= \cos (3k + 1)\frac{2\pi}{9} + i \sin (3k + 1)\frac{2\pi}{9}, k = 0, 1, 2$$

Putting $k = 0, 1, 2$, the three values of u are

$$u = \cos \frac{2\pi}{9} + i \sin \frac{2\pi}{9}, \cos \frac{8\pi}{9} + i \sin \frac{8\pi}{9}, \cos \frac{14\pi}{9} + i \sin \frac{14\pi}{9}$$

Similarly, the values of v are

$$v = \cos \frac{2\pi}{9} - i \sin \frac{2\pi}{9}, \cos \frac{8\pi}{9} - i \sin \frac{8\pi}{9}, \cos \frac{14\pi}{9} - i \sin \frac{14\pi}{9}$$

We have $\quad x = u + v.$

Hence the three required roots are $x = 2 \cos \frac{2\pi}{9}, 2 \cos \frac{8\pi}{9}, 2 \cos \frac{14\pi}{9}$.

Example 7. *Find the nature of roots of the equation* $x^3 - 6x^2 - 6x - 14 = 0$.

Sol. The given equation is $x^3 - 6x^2 - 6x - 14 = 0$

Comparing it with $a_0x^3 + 3a_1x^2 + 3a_2x + a_3 = 0$, we have

$$a_0 = 1, 3a_1 = -6, 3a_2 = -6 \text{ and } a_3 = -14$$

or $\quad a_0 = 1, a_1 = -2, a_2 = -2 \text{ and } a_3 = -14$

$$\therefore \quad H = a_0a_2 - a_1^2 = 1(-2) - (-2)^2 = -2 - 4 = -6$$

and $\quad G = a_0^2a_3 - 3a_0a_1a_2 + 2a_1^3$

$$= (1)^2(-14) - 3(1)(-2)(-2) + 2(-2)^3 = -14 - 12 - 16 = -42$$

$$\therefore \quad G^2 + 4H^3 = (-42)^2 + 4(-6)^3 = 1764 - 864 = 900 > 0$$

$\therefore$ By Case I, Art. 2, Chatper 9 ; one root of the given equation is real and other two are conjugate imaginary.

EXERCISE 1

1. Solve the following equations by Cardon's method :
   (*i*) $x^3 - 18x - 35 = 0$ (*ii*) $x^3 - 36x - 91 = 0$
   (*iii*) $x^3 - 147x + 686 = 0$ (*iv*) $x^3 - 6x + 9 = 0$.
2. Solve the equation $9x^3 + 6x^2 - 1 = 0$ by Cardon's method.
3. Solve the following equations using Cardan's method of solving a cubic :
   (*i*) $x^3 - 12x^2 - 60x + 800 = 0$ (*ii*) $x^3 + 6x^2 - 12x - 112 = 0$
   (*iii*) $x^3 - 3x^2 + 12x + 16 = 0$ (*iv*) $x^3 - 6x^2 - 6x - 7 = 0$
   (*v*) $x^3 - 3x^2 - 9x - 54 = 0$ (*vi*) $x^3 - 15x^2 - 33x + 847 = 0$
4. Solve the following equations by Cardon's method :
   (*i*) $2x^3 + 3x^2 + 3x + 1 = 0$ (*ii*) $x^3 + x^2 - 16x + 20 = 0$
   (*iii*) $27x^3 + 54x^2 + 198x - 73 = 0$.
5. Determine the roots of $x^3 - 12x + 8 = 0$ by Cardon's method.
   [**Hint.** Proceed as in Solved Example 6]

6. Discuss the nature of roots of the equation $p_0x^3 + 3p_1x^2 + 3p_2x + p_3 = 0$.

[**Hint.** It is Art. 2.]

7. Show that the equation $x^3 + x^2 - 2x - 1 = 0$ has all three roots real.

8. Discuss the nature of roots of the equation $x^3 + 6x + 3 = 0$.

9. Solve $x^3 - 3x^2 - 10x + 24 = 0$ by Cardon's method

[**Hint.** To remove the second term, decrease roots by $h = \dfrac{-a_1}{na_0} = 1$.

The reduced equation is $y^3 - 13y + 12 = 0$ where $y = x - 1$.

Put $$y = u + v$$

Applying Cardon's Method, $uv = \dfrac{13}{3}$ and $u^3 + v^3 = -12$

$$\therefore \quad u^3v^3 = \left(\frac{13}{3}\right)^3$$

$\therefore$ u^3 and v^3 are roots of $t^2 + 12t + \left(\dfrac{13}{3}\right)^3 = 0$.

$$\therefore \quad t = \frac{-12 \pm \sqrt{144 - \dfrac{4 \times 169 \times 13}{27}}}{2}$$

or $$t = \frac{-12 \pm \sqrt{-\dfrac{4900}{27}}}{2} = \frac{-12 \pm \dfrac{70i}{3\sqrt{3}}}{2} = -6 \pm \frac{35i}{3\sqrt{3}} = -6 \pm \frac{35\sqrt{3}i}{9}$$

Let $$u^3 = -6 + \frac{35\sqrt{3}i}{9} \text{ and } v^3 = -6 - \frac{35\sqrt{3}i}{9}$$

Let $$u = p + i\sqrt{3}\,q$$

$$\therefore \quad u^3 = -6 + \frac{35\sqrt{3}}{9}i = (p + i\sqrt{3}\,q)^3 = p^3 + i^3(\sqrt{3}\,q)^3 + 3pqi\sqrt{3}\,(p + i\sqrt{3}\,q)$$

$$= p^3 - 3\sqrt{3}\,q^3i + 3p^2q\sqrt{3}\,i - 9pq^2 = p(p^2 - 9q^2) + 3\sqrt{3}\,qi\,(p^2 - q^2)$$

Equating Real and Imaginary parts on both sides, we have

$$p(p^2 - 9q^2) = -6 \quad \text{...(1)}$$

and $$3\sqrt{3}\,q(p^2 - q^2) = \frac{35\sqrt{3}}{9}$$

or $$q(p^2 - q^2) = \frac{35}{27} \quad \text{...(2)}$$

Using Trial Method $p = -2$ and $q = \dfrac{1}{3}$ satisfy Eqns. (1) and (2)

$$\therefore \quad u = p + i\sqrt{3}\,q = -2 + \frac{i\sqrt{3}}{3}$$

$$\therefore \quad v = \text{Conjugate of } u = -2 - \frac{i\sqrt{3}}{3}$$

[$\because$ u^3 and v^3 are conjugate of each other]

$$\therefore \quad y = u + v = -4$$

$$\therefore \quad x = y + 1 = -3 \text{ is a root of the given equation.]}$$

ANSWERS

1. (*i*) 5, $\dfrac{-5 \pm \sqrt{-3}}{2}$ (*ii*) 7, $\dfrac{-7 \pm \sqrt{-3}}{2}$ (*iii*) – 14, 7, 7 (*iv*) – 3, $\dfrac{3 \pm i\sqrt{3}}{2}$

2. $\frac{1}{3}$, $\dfrac{-3 \pm i\sqrt{3}}{2}$

3. (*i*) – 8, 10, 10 (*ii*) 4, $-5 \pm \sqrt{-3}$ (*iii*) – 1, $2 \pm 2\sqrt{-3}$ (*iv*) 7, $\dfrac{-1 \pm i\sqrt{3}}{2}$

(*v*) 6, $\dfrac{3(-1 \pm i\sqrt{3})}{2}$ (*vi*) – 7, 11, 11

4. (*i*) $\dfrac{-1}{2}$, $\dfrac{-1 \pm i\sqrt{3}}{2}$ (*ii*) – 5, 2, 2 (*iii*) $\frac{1}{3}$, $\dfrac{-7 \pm i9\sqrt{3}}{6}$

5. $4\cos\dfrac{2\pi}{9}$, $4\cos\dfrac{4\pi}{9}$, $4\cos\dfrac{8\pi}{9}$ **6.** Art. 1. and 2

8. One real root and two are conjugate imaginary. **9.** 2, 4, – 3.

---

4. Solution of Bi-quadratic Equation by resolution into quadratic factors

Descarte's Method. To Solve the bi-quadratic equation

$$\mathbf{a_0x^4 + 4a_1x^3 + 6a_2x^2 + 4a_3x + a_4 = 0}$$

by resolving into quadratic factors.

Sol. Given equation is $a_0x^4 + 4a_1x^3 + 6a_2x^2 + 4a_3x + a_4 = 0$ ...(1)

Remove the second term by using the transformation $z = a_0x + a_1$ (see Art. 10, Chapter 7)

$\therefore$ The equation (1) transforms into

$$z^4 + 6Hz^2 + 4Gz + (a_0^2 I - 3H^2) = 0 \quad ...(2)$$

where $H = a_0a_2 - a_1^2$, $G = a_0^2 a_3 - 3a_0a_1a_2 + 2a_1^3$, $I = a_0a_4 - 4a_1a_3 + 3a_2^2$

Let L.H.S. of (2) resolves into two quadratic factors

$$\mathbf{z^2 + pz + q,\ z^2 - pz + q'}$$

$\therefore$ $z^4 + 6Hz^2 + 4Gz + (a_0^2I - 3H^2) = (z^2 + pz + q)(z^2 - pz + q')$ equating the co-efficients of like powers of z, we get

$$q + q' - p^2 = 6H \quad \therefore \quad q + q' = p^2 + 6H \quad ...(3)$$

$$pq' - pq = 4G \quad i.e., \quad q' - q = \frac{4G}{p} \quad ...(4)$$

$$qq' = a_0^2I - 3H^2 \quad ...(5)$$

Eliminate q and q' from (3), (4) and (5).

Using the identity $(q' + q)^2 - (q' - q)^2 = 4qq'$

Substituting from (3), (4) and (5), we have

$$(p^2 + 6H)^2 - \left(\frac{4G}{p}\right)^2 = 4(a_0^2I - 3H^2)$$

i.e.,
$$p^4 + 12Hp^2 + 36H^2 - \frac{16G^2}{p^2} = 4(a_0^2I - 2H^2)$$

or
$$p^6 + 12Hp^4 + 4(12H^2 - a_0^2I)p^2 - 16G^2 = 0 \quad ...(6)$$

It is a cubic in 'p^2' and is known as **Descorte's Resolvant.** From (6), we can find a value of p^2 either by *inspection method* or by Cardon's method of solving a cubic.

If $p = 0$, then q and q' can be determined from (3) and (5).

If $p \neq 0$, then q and q' can be determined from equations (3) and (4). Thus, p, q, q' are determined.

Thus, the bi-quadratic reduces to two quadratic equations $z^2 + pz + q = 0$ and $z^2 - pz + q' = 0$ which gives four values of z.

Having determined z, we can find the roots of (1) by the relation $z = a_0x + a_1$.

Note. In Descarte's resolvant, find p^2 by inspection method by putting $p^2 = 1^2, 2^2, 3^2, \ldots$ etc.

Example 1. *Solve the bi-quadratic equation $x^4 + 12x - 5 = 0$ by resolving into quadratic factors.*

Sol. Given equation is $x^4 + 12x - 5 = 0$

Let it be resolved into two quadratic equations

$$x^2 + px + q = 0 \text{ and } x^2 - px + q' = 0$$

$$\therefore \quad x^4 + 12x - 5 = (x^2 + px + q)(x^2 - px + q') \qquad \ldots(1)$$

Equating co-efficients of like powers of x, we have

$$0 = q + q' - p^2 \text{ (equation co-efficients of } x^2) \qquad \ldots(2)$$

$$12 = pq' - pq \qquad \ldots(3)$$

$$-5 = qq' \qquad \ldots(4)$$

From (2), (3) and (4), we have $q + q' = p^2,\ q' - q = \dfrac{12}{p},\ qq' = -5.$

Eliminate, q and q', by using the identity

$$(q + q')^2 - (q - q')^2 = 4qq'$$

$$\therefore \quad p^4 - \frac{144}{p^2} = -20 \quad \text{or} \quad p^6 + 20p^2 - 144 = 0 \qquad \ldots(5)$$

[Descarte's Resolvant]

It is a cubic in p^2

put $p^2 = 1^2, \quad 1 + 20 - 144 \neq 0$

$p^2 = 2^2, \quad 64 + 80 - 144 = 0$

$\therefore \quad p^2 = 4$ is root of (5) by inspection $\therefore \ p = \pm 2$. Let us take $\mathbf{p = 2}$

$\therefore \quad q' + q = 4$ and $q' - q = 6$

which gives $q' = 5$ and $q = -1$, they satisfy $qq' = -5$

Hence the two quadratics are

$$x^2 + 2x - 1 = 0 \quad \text{and} \quad x^2 - 2x + 5 = 0$$

$$\therefore \quad x = \frac{-2 \pm \sqrt{4+4}}{2} \quad \text{and} \quad x = \frac{2 \pm \sqrt{4-20}}{2}$$

i.e., $$x = \frac{-2 \pm 2\sqrt{2}}{2} \quad \text{and} \quad x = \frac{2 \pm 4i}{2}$$

i.e., $$x = -1 \pm \sqrt{2} \quad \text{and} \quad x = 1 \pm 2i$$

Hence the four roots of the bi-quadratic are $-1 + \sqrt{2}$; $-1 - \sqrt{2}$; $1 + 2i,\ 1 - 2i$.

Example 2. *Solve the bi-quadratic $x^4 - 4x^3 + 9x^2 - 12x + 18 = 0$ by resolving it into quadratic factors.*

Sol. Given equation is $x^4 - 4x^3 + 9x^2 - 12x + 18 = 0$ ...(1)

First remove the second term by diminishing the roots by $h = -\left(\frac{-4}{4.1}\right) = 1$

$$\begin{array}{r|rrrrr}
1) & 1 & -4 & 9 & -12 & 18 \\
 & & 1 & -3 & 6 & -6 \\
\hline
 & 1 & -3 & 6 & -6 & 12 \\
 & & 1 & -2 & 4 & \\
\hline
 & 1 & -2 & 4 & -2 & \\
 & & 1 & -1 & & \\
\hline
 & 1 & -1 & 3 & & \\
 & & 1 & & & \\
\hline
 & 1 & 0 & & & \\
\hline
 & 1 & & & &
\end{array}$$

$\therefore$ The transformed equation is $y^4 + 3y^2 - 2y + 12 = 0$, where $y = x - 1$ ...(2)

Let it resolve into two quadratic equations

$$y^2 + py + q = 0$$

and
$$y^2 - py + q' = 0$$

$$\therefore \quad y^4 + 3y^2 - 2y + 12 = (y^2 + py + q)(y^2 - py + q')$$

Equating the co-efficients of y^2, y and constant term, we have

$$3 = q + q' - p^2 \qquad \therefore \quad q + q' = 3 + p^2 \qquad \text{...}(i)$$

$$-2 = pq' - pq \qquad \therefore \quad q' - q = \frac{-2}{p} \qquad \text{...}(ii)$$

$$12 = qq' \qquad \therefore \quad qq' = 12 \qquad \text{...}(iii)$$

Using the identity $(q' + q)^2 - (q' - q)^2 = 4qq'$

we obtain
$$(3 + p^2)^2 - \left(\frac{-2}{p}\right)^2 = 48$$

i.e.,
$$9 + 6p^2 + p^4 - \frac{4}{p^2} = 48$$

or
$$p^6 + 6p^4 - 39p^2 - 4 = 0 \qquad \text{...}(3)$$

which is a cubic in p^2. Let us find a value of p^2 by inspection method.

Put $p^2 = 1^2$; $1 + 6 - 39 - 4 \neq 0$

$p^2 = 2^2$; $64 + 96 - 156 - 4 = 0$

$\therefore$ $p^2 = 4$ satisfies the cubic in p^2

$\therefore$ $p = \pm 2$

Take $p = 2$

$\therefore$ $q' + q = 7$

$q' - q = -1$

which gives $q' = 3$ and $q = 4$ satisfying (*iii*)

$\therefore$ The two quadratic equations are $y^2 + 2y + 4 = 0$ ...(4)

and $$y^2 - 2y + 3 = 0 \qquad ...(5)$$

Solving (4), $y = \dfrac{-2 \pm \sqrt{4-16}}{2} = \dfrac{-2 \pm 2\sqrt{-3}}{2} = -1 \pm \sqrt{-3}$

Solving (5), $y = \dfrac{2 \pm \sqrt{4-12}}{2} = \dfrac{2 \pm \sqrt{-8}}{2} = \dfrac{2 \pm 2\sqrt{-2}}{2} = 1 \pm \sqrt{-2}$

$\therefore$ The roots of (2) are, $-1 \pm \sqrt{-3}$ and $1 \pm \sqrt{-2}$.

The roots of given equation (1) are obtained by adding 1 to the roots of (2).

$\therefore$ Roots of (1) are $\pm\sqrt{-3}$, $2 \pm \sqrt{-2}$.

Example 3. *Solve the biquadratic equation $x^4 - 10x^3 + 35x^2 - 50x + 24 = 0$ by Descarte's method.*

Sol. The given equation is $x^4 - 10x^3 + 35x^2 - 50x + 24 = 0$ ...(1)

To remove the second term, we diminish the roots by $h = \dfrac{10}{4} = \dfrac{5}{2}$.

To avoid fractions, we multiply the roots of (1) by 2 and the equation transforms to

$$y^4 - 10.2y^3 + 35.2^2y^2 - 50.2^3y + 24.2^4 = 0$$

or $$y^4 - 20y^3 + 140y^2 - 400y + 384 = 0 \qquad ...(2)$$

where $$y = 2x.$$

Now, we diminish the roots of (2) by 5, using synthetic division,

$$\begin{array}{c|ccccc}
5) & 1 & -20 & 140 & -400 & 384 \\
 & & 5 & -75 & 325 & -375 \\ \hline
 & 1 & -15 & 65 & -75 & 9 \\
 & & 5 & -50 & 75 & \\ \hline
 & 1 & -10 & 15 & 0 & \\
 & & 5 & -25 & & \\ \hline
 & 1 & -5 & -10 & & \\
 & & 5 & & & \\ \hline
 & 1 & 0 & & & \\ \hline
 & 1 & & & &
\end{array}$$

The transformed equation is $z^4 - 10z^2 + 9 = 0$ ...(3)

where $$z = y - 5 = 2x - 5 \quad \text{or} \quad x = \frac{z+5}{2} \qquad ...(4)$$

From (3), $$(z^2 - 1)(z^2 - 9) = 0$$

$\Rightarrow$ $$z = \pm 1, \pm 3.$$

$\therefore$ from (4), $$x = \frac{1+5}{2}, \frac{-1+5}{2}, \frac{3+5}{2}, \frac{-3+5}{2} = 3, 2, 4, 1.$$

Hence roots of given equation are 1, 2, 3, 4.

Example 4. *Solve the biquadratic equation $12x^4 - 2x^3 + 3x^2 + 1 = 0$ by resolving it into quadratic factors.*

Sol. The given biquadratic equation is $12x^4 - 2x^3 + 3x^2 + 1 = 0$ ...(1)

Since second term from the end is missing, so it can be reduced to the standard form by using the transformation

$$y = \frac{1}{x} \quad \text{or} \quad x = \frac{1}{y}$$

Substituting this value of x in (1), we get $y^4 + 3y^2 - 2y + 12 = 0$ ...(2)

Now proceed after equation (2) as we did in solved example 2.

EXERCISE 2

1. Solve $x^4 + qx^2 + rx + s = 0$ by Descarte's method.
2. Solve the following equations by Descarte's method :
   (*i*) $x^4 - 10x^2 - 20x - 16 = 0$
   (*ii*) $x^4 - 8x^2 - 24x + 7 = 0$
   (*iii*) $x^4 - 4x^3 + 5x + 2 = 0$
   (*iv*) $x^4 - 3x^2 - 42x - 40 = 0$
   (*v*) $x^4 - 15x^2 - 10x + 24 = 0$
   (*vi*) $x^4 - 3x^2 - 6x - 2 = 0$.
3. (*i*) Solve the equation $x^4 - 15x^2 + 20x - 6 = 0$ by the method of resolution into quadratic factors
   (*ii*) Solve the equation $x^4 - 6x^3 + 8x^2 + 2x - 1 = 0$ by resolving the left-hand side into quadratic factors.
4. Solve (*i*) $2x^4 + 6x^3 - 3x^2 + 2 = 0$
   (*ii*) $40x^4 + 42x^3 + 3x^2 - 1 = 0$, by Descarte's method.

[**Hint. 4** (*i*) and (*ii*), put $x = \frac{1}{y}$.]

ANSWERS

2. (*i*) $4, -2, -1 \pm i$ (*ii*) $2 \pm \sqrt{3}, -2 \pm \sqrt{-3}$ (*iii*) $\frac{1 \pm \sqrt{5}}{2}; \frac{3 \pm \sqrt{17}}{2}$
   (*iv*) $-1, 4, \frac{-3 \pm i\sqrt{31}}{2}$ (*v*) $1, -2, -3, 4$ (*vi*) $-1 \pm i, 1 \pm \sqrt{2}$
3. (*i*) $1, 3, -2 \pm \sqrt{6}$ (*ii*) $1 \pm \sqrt{2}, 2 \pm \sqrt{3}$
4. (*i*) $\frac{1}{2}(1 \pm i), -2 \pm \sqrt{2}$ (*ii*) $-1, \frac{1}{4}, \frac{-3 \pm i\sqrt{31}}{20}$.

5. Ferrari's Method of Solving a Biquadratic Equation

This *Method* expresses the biquadratic as the difference of two squares, which is further written as product of two quadratic factors.

Let the given quadratic be $a_0x^4 + 4a_1x^3 + 6a_2x^2 + 4a_3x + a_4 = 0$ ...(1) $(a_0 \neq 0)$

Multiply by a_0, $a_0^2x^4 + 4a_0a_1x^3 + 6a_0a_2x^2 + 4a_0a_3x + a_0a_4 = 0$

$$a_0^2x^4 + 4a_0a_1x^3 = -6a_0a_2x^2 - 4a_0a_3x - a_0a_4$$

Add $(2a_1x)^2 = 4a_1^2x^2$ to both sides to make L.H.S. a perfect square

$$\therefore \quad a_0^2x^4 + 4a_0a_1x^2 + 4a_1^2x^2 = (4a_1^2 - 6a_0a_2)x^2 - 4a_0a_3x - a_0a_4$$

or $$(a_0x^2 + 2a_1x)^2 = (4a_1^2 - 6a_0a_2)x^2 - 4a_0a_3x - a_0a_4 \quad ...(3)$$

Let us introduce a real number λ, so that both sides of (3) become perfect squares.

$$\therefore\ (a_0x^2 + 2a_1x + \lambda)^2 = (a_0x^2 + 2a_1x)^2 + 2\lambda(a_0x^2 + 2a_1x) + \lambda^2$$
$$= (4a_1^2 - 6a_0a_2)\,x^2 - 4a_0a_3x - a_0a_4 + 2\lambda(a_0x^2 + 2a_1x) + \lambda^2$$
$$= (4a_1^2 - 6a_0a_2 + 2\lambda a_0)x^2 + 4(a_1\lambda - a_0a_3)x + \lambda^2 - a_0a_4 \quad \text{...(4)}$$

R.H.S. of (4) will be a perfect square,

if discriminant of R.H.S. of (4) is zero

if $16(a_1\lambda - a_0a_3)^2 - 4(4a_1^2 - 6a_0a_2 + 2\lambda a_0)(\lambda^2 - a_0a_4) = 0$

if $4(a_1^2\lambda^2 - 2a_0a_1a_3\lambda + a_0^2a_3^2) - (4a_1^2\lambda^2 - 6a_0a_2\lambda^2 + 2a_0\lambda^3 - 4a_0a_1^2a_4 + 6a_0^2a_2a_4 - 2\lambda a_0^2a_4) = 0$

if $-2a_0\lambda^3 + 6a_0a_2\lambda^2 - 2a_0\lambda(4a_1a_3 - a_0a_4) + 4a_0^2a_3^2 + 4a_0a_1^2a_4 - 6a_0^2a_2a_4 = 0$

On dividing by $(-2a_0)$,

$$\lambda^3 - 3a_2\lambda^2 + (4a_1a_3 - a_0a_4)\lambda + 3a_0a_2a_4 - 2a_1^2a_4 - 2a_0a_3^2 = 0 \quad \text{...(5)}$$

It is cubic in λ and is known as **Resolvent cubic.**

Find a value of λ and put in (4), then both sides of (4) will become perfect squares, and we shall get equation of type

$$(a_0x^2 + 2a_1x + \lambda)^2 = (px + q)^2 \quad \text{or} \quad a_0x^2 + 2a_1x + \lambda = \pm (px + q).$$

Taking +ve sign and –ve signs, Roots of given biquadratic equation are given by two quadratic equations

$$a_0x^2 + (2a_1 - p)x + \lambda - q = 0$$
$$a_0x^2 + (2a_1 + p)x + \lambda + q = 0.$$

Remark 1. Ferrari's method is better than Descarte's method in the sense that it is not necessary to remove the second term and make the coefficient of highest degree term unity.

2. Usually we find λ by inspection, considering those values of λ which shall make coefficient of x^2 equal to 1, 4, 9, 16 or $\frac{1}{4}, \frac{1}{9}, \frac{1}{16}$, ... etc. and choosing that value of λ which makes the constant term also a perfect square.

Example 1. *Solve the equation $x^4 - 10x^3 + 35x^2 - 50x + 24 = 0$ by Ferrari's method.*

Sol. Given equation is $x^4 - 10x^3 + 35x^2 - 50x + 24 = 0$

$$x^4 - 10x^3 = -35x^2 + 50x - 24$$

Completing square of L.H.S. by adding $25x^2$ (*i.e.*, $(\frac{1}{2}$ coeff. of $x^3)^2x^2$) on both sides,

$$x^4 - 10x^3 + 25x^2 = -10x^2 + 50x - 24$$

or
$$(x^2 - 5x)^2 = -10x^2 + 50x - 24 \quad \text{...(1)}$$

$$\therefore\ (x^2 - 5x + \lambda)^2 = (x^2 - 5x)^2 + 2\lambda(x^2 - 5x) + \lambda^2$$

Putting the value of $(x^2 - 5x)^2$ from (1) on the R.H.S.

$$\therefore\ (x^2 - 5x + \lambda)^2 = -10x^2 + 50x - 24 + 2\lambda(x^2 - 5x) + \lambda^2$$
$$= (2\lambda - 10)x^2 + (50 - 10\lambda)x + \lambda^2 - 24 \quad \text{....(2)}$$

Choosing λ, so that R.H.S. of (2) is a perfect square. This requires,

$(50 - 10\lambda)^2 - 4(2\lambda - 10)(\lambda^2 - 24) = 0$

$\Rightarrow\ 100(5 - \lambda)^2 - 8(\lambda - 5)(\lambda^2 - 24) = 0$

$\Rightarrow\ 4(5 - \lambda)[25(5 - \lambda) + 2(\lambda^2 - 24)] = 0$

Let $\lambda - 5 = 0$ *i.e.,* $\lambda = 5$. Putting $\lambda = 5$ in (2), we get

$$(x^2 - 5x + 5)^2 = 1$$

$$\therefore \quad x^2 - 5x + 5 = \pm 1$$

$$\therefore \quad x^2 - 5x + 4 = 0 \quad \text{or} \quad x^2 - 5x + 6 = 0$$

$$\therefore \quad (x-4)(x-1) = 0 \quad \text{or} \quad (x-2)(x-3) = 0$$

$$\therefore \quad x = 1, 4, 2, 3.$$

Hence roots are 1, 2, 3 and 4.

Example 2. *Solve the equation $x^4 - 4x^3 - 4x^2 - 24x + 15 = 0$ by Ferrari's method.*

Sol. The given equation is $\quad x^4 - 4x^3 - 4x^2 - 24x + 15 = 0 \qquad ...(1)$

$$\Rightarrow \quad x^4 - 4x^3 = 4x^2 + 24x - 15$$

Completing square of L.H.S., by adding $4x^2$ on both sides, we get

$$x^4 - 4x^3 + 4x^2 = 8x^2 + 24x - 15$$

$$\Rightarrow \quad (x^2 - 2x)^2 = 8x^2 + 24x - 15 \qquad ...(2)$$

Now $\quad (x^2 - 2x + \lambda)^2 = (x^2 - 2x)^2 + 2(x^2 - 2x)\lambda + \lambda^2$

Putting the value of $(x^2 - 2x)^2$ from (2) on R.H.S., we have

$$(x^2 - 2x + \lambda)^2 = 8x^2 + 24x - 15 + 2(x^2 - 2x)\lambda + \lambda^2$$

$$= (8 + 2\lambda)x^2 + (24 - 4\lambda)x + \lambda^2 - 15 \qquad ...(3)$$

Choosing λ so that R.H.S. of (3) is a perfect square, this requires discriminant of R.H.S. zero

i.e., $\quad (24 - 4\lambda)^2 - 4(8 + 2\lambda)(\lambda^2 - 15) = 0$

$$\Rightarrow \quad \lambda^3 + 2\lambda^2 + 9\lambda - 132 = 0 \qquad ...(4)$$

By inspection $\lambda = 4$ is a root of (4)

$\therefore$ Putting $\quad \lambda = 4$ in (3), we get

$$(x^2 - 2x + 4)^2 = 16x^2 + 8x + 1$$

or $\quad (x^2 - 2x + 4)^2 = (4x + 1)^2$

$$\Rightarrow \quad x^2 - 2x + 4 = \pm (4x + 1)$$

$$\Rightarrow \quad x^2 - 6x + 3 = 0 \quad \text{or} \quad x^2 + 2x + 5 = 0$$

$$\Rightarrow \quad x = \frac{6 \pm \sqrt{36 - 12}}{2}, \frac{-2 \pm \sqrt{4 - 20}}{2}$$

$$\Rightarrow \quad x = 3 \pm \sqrt{6}, -1 \pm 2i.$$

Example 3. *Solve the equation $2x^4 + 6x^3 - 3x^2 + 2 = 0$ by Ferrari's method.*

Sol. Given equation is $\quad 2x^4 + 6x^3 - 3x^2 + 2 = 0$

i.e., $\quad x^4 + 3x^3 - \frac{3}{2}x^2 + 0x + 1 = 0 \qquad ...(1)$

Multiply the roots of (1) by 2, by multiplying the terms of (1) by $2^0, 2^1, 2^2, 2^3, 2^4$.

$$\therefore \quad x^4 + 6x^3 - 6x^2 + 0x + 16 = 0 \qquad ...(2)$$

$$\therefore \quad x^4 + 6x^3 = 6x^2 - 16$$

Making L.H.S. a perfect square

$$x^4 + 6x^3 + 9x^2 = 15x^2 - 16$$

or $\quad (x^2 + 3x)^2 = 15x^2 - 16$

$$\therefore \quad (x^2 + 3x + \lambda)^2 = (x^2 + 3x)^2 + 2\lambda(x^2 + 3x) + \lambda^2$$
$$= 15x^2 - 16 + 2\lambda(x^2 + 3x) + \lambda^2$$
$$= (15 + 2\lambda)x^2 + 6\lambda x + \lambda^2 - 16 \qquad ...(3)$$

Choosing λ so that R.H.S. also becomes a perfect square, this requires D = 0

$$\therefore \quad 36\lambda^2 - 4(15 + 2\lambda)(\lambda^2 - 16) = 0$$

or $$9\lambda^2 - (15\lambda^2 - 240 + 2\lambda^3 - 32\lambda) = 0$$

or $$2\lambda^3 + 6\lambda^2 - 32\lambda - 240 = 0$$

or $$\lambda^3 + 3\lambda^2 - 16\lambda - 120 = 0$$

By inspection $\lambda = 5$ is a root.

Put in (3) $\therefore$ (3) becomes

$$(x^2 + 3x + 5)^2 = 25x^2 + 30x + 9 = (5x + 3)^2$$

$$\therefore \quad x^2 + 3x + 5 = \pm (5x + 3)$$

Taking positive sign, $x^2 - 2x + 2 = 0$

$$\therefore \quad x = \frac{2 \pm \sqrt{4-8}}{2} = \frac{2 \pm 2i}{2} = 1 \pm i$$

Taking negative sign, $x^2 + 8x + 8 = 0$

$$\therefore \quad x = \frac{-8 \pm \sqrt{64 - 32}}{2} = \frac{-8 \pm 4\sqrt{2}}{2} = -4 \pm 2\sqrt{2}$$

Dividing each root by 2, roots of given equation are $\frac{1}{2}(1 \pm i)$, $-2 \pm \sqrt{2}$.

EXERCISE 3

Solve the following equations by Ferrari's method :

1. $x^4 + 2x^3 - 7x^2 - 8x + 12 = 0.$
2. $x^4 - 8x^3 + 9x^2 + 8x - 10 = 0.$
   [**Hint.** $2\lambda^3 - 9\lambda^2 - 12\lambda + 54 = 0$ or $\lambda^2(2\lambda - 9) - 6(2\lambda - 9) = 0$ or $(2\lambda - 9)(\lambda^2 - 6) = 0$
   $\therefore$ one value of λ is $\frac{9}{2}$.]
3. $x^4 + 3x^3 - 2x^2 - 7x + 3 = 0.$
4. $x^4 - 10x^3 + 26x^2 - 10x + 1 = 0.$
5. $x^4 + 4x^3 + 12x^2 - 8x + 95 = 0.$
   [**Hint.** Cubic in λ is $\lambda^3 - 6\lambda^2 - 103\lambda + 372 = 0.$
   $\lambda = 12$ is a root by trial method.]

ANSWERS

1. $-3, -2, 1, 2$
2. $-1, 1, 4 \pm \sqrt{6}$
3. $(-1 \pm \sqrt{2}), \frac{1}{2}(-1 \pm \sqrt{13})$
4. $2 \pm \sqrt{3}, 3 \pm 2\sqrt{2}$
5. $1 \pm 2i, -3 \pm \sqrt{10}\, i$

SECTION III

10

Groups

1. BINARY OPERATION

Let S be a non-empty set. A function from S × S into S is called a **binary operation** on S. Thus, if '$*$' is a binary operation on S, then it means that '$*$' is a function from S × S into S.

If $(a, b) \in S \times S$, then the image $*(a, b)$ of (a, b) under the binary operation '$*$' is written $a * b$.

For example :

(*i*) Ordinary addition '+' is a binary operation on **Z**, because '+' is a function from **Z** × **Z** into **Z**. We write $+ (4, 7) = 4 + 7 = 11 \in \mathbf{Z}$.

(*ii*) Ordinary subtraction '–' is not a binary operation on **N**, because difference of two natural numbers may not be a natural number. We have $-(5, 8) = 5 - 8 = -3 \notin \mathbf{N}$.

2. ALGEBRAIC STRUCTURE

A non-empty set with one or more binary operations defined on it is called an **algebraic structure.**

In an algebraic structure, we can combine two elements of it to obtain another element of the algebraic structure.

For example, (**Z**, +), (**Q**, +), (**R**, –) are algebraic structures.

3. SEMI GROUP

A non-empty set G with a binary operation, denoted by $*$, is called a **semi group** if

$$x * (y * z) = (x * y) * z \quad \forall\, x, y, z \in G$$

Equivalently, an algebraic structure (G, $*$) is a semi group if

$$x * (y * z) = (x * y) * z \quad \forall\, x, y, z \in G.$$

If the binary operation on the semi group G is denoted by ' . ' then we have

$$x \,.\, (y \,.\, z) = (x \,.\, y) \,.\, z \quad \forall\, x, y, z \in G.$$

4. GROUP

A non-empty set G with a binary operation, denoted by '.', is called a **group** if :

(*i*) $x \,.\, (y \,.\, z) = (x \,.\, y) \,.\, z \quad \forall\, x, y, z \in G$

(*ii*) There exists an element $e \in G$ such that :

$$x \,.\, e = x = e \,.\, x \quad \forall\, x \in G$$

e is called an **identity** of G.

(*iii*) For $x \in G$, there exists an element $y \in G$ such that :

$$x \cdot y = e = y \cdot x$$

y is called an **inverse** of x. An inverse of x is written as x^{-1}.

If in addition, we have

(*iv*) $x \cdot y = y \cdot x \quad \forall\, x, y \in G$,

then G is called a **commutative group.** A commutative group is also called an **abelian group** after the famous Norwegian mathematician **N.H. Abel.** A group is called **non-commutative** or **non-abelian** if it is not commutative.

If the binary operation of a group is written as '+', then we write

(*i*) $x + (y + z) = (x + y) + z \quad \forall\, x, y, z \in G$

(*ii*) $\exists\, e \in G : x + e = x = e + x \quad \forall\, x \in G$

(*iii*) For $x \in G, \exists\, y \in G : x + y = e = y + x$

And for commutativity, $x + y = y + x \quad \forall\, x, y \in G$.

Remark 1. The binary operation '+' used above has got nothing to do with the ordinary addition of numbers. In fact $*$, +, . are just symbols representing binary operations.

Remark 2. In order to show a non-empty set G with a given binary operation to be a group, we should verify all the three conditions given above.

5. ORDER OF A GROUP

If the number of elements in a group is finite, the group is said to be a **finite group,** otherwise an **infinite group.** The number of elements in a finite group is called the **order of the group.**

If a group G is finite and has n elements, then we write $o(G) = n$.

Examples :

1. (**Z**, +) is an abelian group, where '+' represents the usual addition of numbers. This is true because, we have

(*i*) $x + (y + z) = (x + y) + z \quad \forall\, x, y, z \in \mathbf{Z}$

(*ii*) $0 \in \mathbf{Z}$ and $x + 0 = x = 0 + x \quad \forall\, x \in \mathbf{Z}$

(*iii*) For $x \in \mathbf{Z}$, we have $-x \in \mathbf{Z}$ and $x + (-x) = 0 = (-x) + x$

(*iv*) $x + y = y + x \quad \forall\, x, y \in \mathbf{Z}$.

2. (**Q**, +), (**R**, +), (**C**, +), (**Q**°, .), (**R**°, .), (**C**°, .) are all abelian groups, where $\mathbf{Q}° = \mathbf{Q} - \{0\}$, $\mathbf{R}° = \mathbf{R} - \{0\}$, $\mathbf{C}° = \mathbf{C} - \{0\}$.

All these groups are infinite groups.

3. The set of all $m \times n$ matrices whose elements are complex numbers forms an infinite abelian group with respect to matrix addition.

4. The set of all non-singular $n \times n$ square matrices whose elements are complex numbers forms an infinite non-abelian group with respect to matrix multiplication.

5. The set of all integers which are multiple of m (an integer) *i.e.*, the set $\{0, \pm m, \pm 2m, \pm 3m, ...\}$ forms an infinite abelian group w.r.t. usual addition. Note that 0 acts as the identity and inverse of λm is $-\lambda m$.

WORKING RULES TO SHOW THAT (G, $*$) IS A GROUP

Step I. *Take arbitrary a, b of G and show that $a * b$ is a uniquely defined element of G.*

Step II. *Take arbitrary a, b, c of G and show that $a * (b * c)$ and $(a * b) * c$ are equal. This establishes the associativity of '$*$'.*

Step III. *Find an element 'e' in G such that $a * e = a = e * a$ for all $a \in G$. e works as the identity element.*

Step IV. *For $a \in G$, find a number $b \in G$ such that $a * b = e = b * a$. b is the inverse of a.*

Step V. *Take arbitrary a, b in G and show that $a * b = b * a$. This establishes, the commutativity of '$*$'.*

Example 1. *Show that the set* **Z** *is an abelian group with binary operation '$*$' defined as $a * b = a + b + 1$ for $a, b \in$* **Z**.

Sol. For $a, b \in \mathbf{Z}$, the value of $a * b$ *i.e.,* $a + b + 1$ is a uniquely determined integer.

$\therefore$ '$*$' is binary operation on **Z**.

Associativity. Let $a, b, c \in \mathbf{Z}$.

$$\therefore \quad a * (b * c) = a * (b + c + 1) = a + (b + c + 1) + 1 = a + b + c + 2$$

and

$$(a * b) * c = (a + b + 1) * c = (a + b + 1) + c + 1 = a + b + c + 2$$

$$\therefore \quad a * (b * c) = (a * b) * c \quad \therefore \quad * \text{ is associative.}$$

Existence of identity. $-1 \in \mathbf{Z}$.

For $a \in \mathbf{Z}$, $\quad a * (-1) = a + (-1) + 1 = a$

and $\quad (-1) * a = (-1) + a + 1 = a.$

$$\therefore \quad a * (-1) = a = (-1) * a, \quad \forall a \in \mathbf{Z}$$

$\therefore -1$ is an identity element.

Existence of inverse. Let $a \in \mathbf{Z}$. $\quad \therefore -a - 2 \in \mathbf{Z}$.

$$a * (-a - 2) = a + (-a - 2) + 1 = -1$$

and

$$(-a - 2) * a = (-a - 2) + a + 1 = -1.$$

$$\therefore \quad a * (-a - 2) = -1 = (-a - 2) * a$$

$\therefore \quad -a - 2$ is an inverse of a.

$\therefore \quad (\mathbf{Z}, *)$ is a group.

Commutativity. For $a, b \in \mathbf{Z}$,

$$a * b = a + b + 1 = b + a + 1 = b * a.$$

$\therefore (\mathbf{Z}, *)$ is an abelian group.

Example 2. *Show that the set of all positive rational numbers forms an abelian group under the binary operation '$*$' defined by*

$$a * b = \frac{ab}{4}.$$

Sol. Let G be the set of all positive rational numbers.

For $a, b \in$ G, the value of $a * b$ *i.e.*, $\frac{ab}{4}$ is a uniquely determined element of G.

$\therefore$ $*$ is a binary operation on G.

Associativity. Let $a, b, c \in$ G.

$$\therefore \qquad a * (b * c) = a * \left(\frac{bc}{4}\right) = \frac{a\left(\frac{bc}{4}\right)}{4} = \frac{abc}{16}$$

and
$$(a * b) * c = \left(\frac{ab}{4}\right) * c = \frac{\left(\frac{ab}{4}\right)c}{4} = \frac{abc}{16}.$$

$\therefore \qquad a * (b * c) = (a * b) * c \quad \therefore \quad *$ is associative.

Existence of identity. $\quad 4 \in$ G

For $a \in$ G, $\qquad a * 4 = \frac{a \times 4}{4} = a$ and $4 * a = \frac{4 \times a}{4} = a$

$\therefore \qquad a * 4 = a = 4 * a \quad \forall\, a \in$ G

$\therefore$ 4 is an identity element.

Existence of inverse. Let $a \in$ G. $\quad \therefore \quad \frac{16}{a} \in$ G

$$a * \frac{16}{a} = \frac{a \times \frac{16}{a}}{4} = 4 \quad \text{and} \quad \frac{16}{a} * a = \frac{\frac{16}{a} \times a}{4} = 4$$

$$\therefore \qquad a * \frac{16}{a} = 4 = \frac{16}{a} * a$$

$\therefore$ $\frac{16}{a}$ is an inverse of a.

$\therefore$ (G, $*$) is a group.

Commutativity. For $a, b \in$ G,

$$a * b = \frac{ab}{4} = \frac{ba}{4} = b * a.$$

$\therefore$ (G, $*$) is an abelian group.

Example 3. *Show that :* $\qquad A = \left\{\begin{bmatrix} 1 & 0 \\ 0 & 1 \end{bmatrix}, \begin{bmatrix} -1 & 0 \\ 0 & 1 \end{bmatrix}, \begin{bmatrix} 1 & 0 \\ 0 & -1 \end{bmatrix}, \begin{bmatrix} -1 & 0 \\ 0 & -1 \end{bmatrix}\right\}$

is a multiplicative group.

Sol. We have $\qquad A = \left\{\begin{bmatrix} 1 & 0 \\ 0 & 1 \end{bmatrix}, \begin{bmatrix} -1 & 0 \\ 0 & 1 \end{bmatrix}, \begin{bmatrix} 1 & 0 \\ 0 & -1 \end{bmatrix}, \begin{bmatrix} -1 & 0 \\ 0 & -1 \end{bmatrix}\right\}.$

It can be easily verified that the usual multiplication is a binary operation on A.

Let $\qquad$ P, Q, R $\in$ A.

$\therefore \qquad$ P.(Q.R) = (P.Q).R

Let the matrix $\begin{bmatrix} 1 & 0 \\ 0 & 1 \end{bmatrix}$ be denoted by I.

$\therefore \qquad P \cdot I = P = I \cdot P \;\forall\; P \in A$

$\therefore$ I is the identity element for A.

We have $\begin{bmatrix} -1 & 0 \\ 0 & 1 \end{bmatrix}\begin{bmatrix} -1 & 0 \\ 0 & 1 \end{bmatrix} = I$

Also, $\begin{bmatrix} 1 & 0 \\ 0 & -1 \end{bmatrix}\begin{bmatrix} 1 & 0 \\ 0 & -1 \end{bmatrix} = I, \begin{bmatrix} -1 & 0 \\ 0 & -1 \end{bmatrix}\begin{bmatrix} -1 & 0 \\ 0 & -1 \end{bmatrix} = I$

$\therefore$ Each element in A is inverse of itself.

$\therefore$ Every element in A has an inverse.

$\therefore$ A is a group under multiplication.

Example 4. *Let G be the set of all 2 × 2 matrices* $\begin{bmatrix} a & b \\ -b & a \end{bmatrix}$, *where a and b are real numbers, not both zero (i.e.,* $a^2 + b^2 \neq 0$*). Show that G is an infinite abelian group w.r.t. usual matrix multiplication.*

Sol. We have $$G = \left\{\begin{bmatrix} a & b \\ -b & a \end{bmatrix} : a, b \in \mathbf{R}, a^2 + b^2 \neq 0\right\}.$$

It can be easily verified that the usual multiplication is a binary operation on G.

Let A, B, C ∈ G.

$\therefore \qquad A(BC) = (AB)C.$

Let $$I = \begin{bmatrix} a & b \\ -b & a \end{bmatrix}, \text{ where } a = 1, b = 0.$$

$$\therefore \qquad I = \begin{bmatrix} 1 & 0 \\ 0 & 1 \end{bmatrix} \text{ and } a^2 + b^2 = 1 + 0 = 1 \neq 0$$

For A ∈ G, we have AI = A = IA

$\therefore$ I is an identity element for G.

Let $$A = \begin{bmatrix} a & b \\ -b & a \end{bmatrix} \in G$$

$|A| = a^2 + b^2 \neq 0 \quad \therefore \quad A^{-1}$ exists

$A_{11} = a, A_{12} = b, A_{21} = -b, A_{22} = a$

$$\therefore \qquad \text{adj. } (A) = \begin{bmatrix} a & b \\ -b & a \end{bmatrix}' = \begin{bmatrix} a & -b \\ b & a \end{bmatrix}$$

$$\therefore \qquad A^{-1} = \frac{1}{|A|} \text{ adj. } (A) = \frac{1}{a^2 + b^2}\begin{bmatrix} a & -b \\ b & a \end{bmatrix}$$

$$= \begin{bmatrix} \dfrac{a}{a^2 + b^2} & -\dfrac{b}{a^2 + b^2} \\ -\left(-\dfrac{b}{a^2 + b^2}\right) & \dfrac{a}{a^2 + b^2} \end{bmatrix}$$

Also $$\left(\frac{a}{a^2+b^2}\right)^2+\left(-\frac{b}{a^2+b^2}\right)^2=\frac{1}{a^2+b^2}\neq 0$$

$\therefore$ $A^{-1} \in G$

We also have $AA^{-1} = I = A^{-1}A$.

$\therefore$ A^{-1} is an inverse of A.

$\therefore$ G is an infinite group.

Let $$A=\begin{bmatrix} a & b \\ -b & a \end{bmatrix}, B=\begin{bmatrix} l & m \\ -m & l \end{bmatrix} \in G$$

$\therefore$ $$AB=\begin{bmatrix} a & b \\ -b & a \end{bmatrix}\begin{bmatrix} l & m \\ -m & l \end{bmatrix}=\begin{bmatrix} al-bm & am+bl \\ -bl-am & -bm+al \end{bmatrix}$$

and $$BA=\begin{bmatrix} l & m \\ -m & l \end{bmatrix}\begin{bmatrix} a & b \\ -b & a \end{bmatrix}=\begin{bmatrix} la-mb & lb+ma \\ -ma-lb & -mb+la \end{bmatrix}.$$

$\therefore$ AB = BA. $\therefore$ G is also abelian.

Example 5. *Let $G = \{(a, b) : a \neq 0 ; a, b \in \mathbf{R}\}$ and let $*$ be a binary operation on G defined by $(a, b) * (c, d) = (ac, bc + d)$ $\forall$ $(a, b), (c, d) \in G$. Show that $(G, *)$ is a non-abelian group.*

Sol. We have $G = \{(a, b) : a \neq 0 ; a, b \in \mathbf{R}\}$.

Let $(a, b), (c, d) \in G$.

$\therefore$ $$(a, b) * (c, d) = (ac, bc + d)$$
$$a, c \neq 0 \Rightarrow ac \neq 0$$

Also, the numbers ac and $bc + d$ are uniquely defined.

$\therefore$ $*$ is a binary operation on G.

Associativity. Let $(a, b), (c, d), (e, f) \in G$.

$$(a, b) * ((c, d) * (e, f)) = (a, b) * (ce, de + f)$$
$$= (a(ce), b(ce) + de + f) = (ace, bce + de + f)$$

Also, $$((a, b) * (c, d)) * (e, f) = (ac, bc + d) * (e, f)$$
$$= ((ac)e, (bc + d)e + f) = (ace, bce + de + f)$$

$\therefore$ is associative.

Existence of identity element. $(1, 0) \in G$.

For $(a, b) \in G$,

$$(a, b) * (1, 0) = (a(1), b(1) + 0) = (a, b)$$

and $$(1, 0) * (a, b) = (1(a), 0(a) + b) = (a, b)$$

$\therefore$ $$(a, b) * (1, 0) = (a, b) = (1, 0) * (a, b) \quad \forall\ (a, b) \in G$$

$\therefore$ (1, 0) is an identity element.

Existence of inverse. Let $(a, b) \in G$.

$\therefore$ $a \neq 0$ $\therefore$ $\left(\frac{1}{a}, -\frac{b}{a}\right) \in G$

$$(a, b) * \left(\frac{1}{a}, -\frac{b}{a}\right) = \left(a\left(\frac{1}{a}\right), b\left(\frac{1}{a}\right) + \left(-\frac{b}{a}\right)\right) = (1, 0)$$

and $$\left(\frac{1}{a}, -\frac{b}{a}\right) * (a, b) = \left(\frac{1}{a}(a), \left(-\frac{b}{a}\right)a + b\right) = (1, 0)$$

$\therefore \left(\frac{1}{a}, -\frac{b}{a}\right)$ is an inverse of (a, b).

$\therefore$ (G, $*$) is a group.

Non-commutativity. $(2, 3), (1, 4) \in$ G.

$$(2, 3) * (1, 4) = (2(1), 3(1) + 4) = (2, 7)$$

and $$(1, 4) * (2, 3) = (1(2), 4(2) + 3) = (2, 11)$$

$\therefore$ $$(2, 3) * (1, 4) \neq (1, 4) * (2, 3).$$

$\therefore$ We cannot have

$$(a, b) * (c, d) = (c, d) * (a, b) \quad \forall\ (a, b), (c, d) \in \text{G}.$$

$\therefore$ $*$ is non-commutative. $\therefore$ (G, $*$) is a non-abelian group.

6. ADDITION MODULO m

Let $\{0, 1, 2, \ldots\ldots, m - 1\}$ be the set of first m non-negative integers. The binary operation '$\oplus$' is defined on this set by $a \oplus b = r$, where r is the least non-negative remainder obtained by dividing $a + b$ by m. This binary operation $\oplus$ is called **'addition modulo m'**.

7. ADDITIVE GROUP OF INTEGERS MODULO m

Let $\quad$ G $= \{0, 1, 2, \ldots\ldots, m - 1\}$.

By definition for $a, b \in$ G, $a \oplus b = r$, where $0 \leq r < m$.

$\therefore \quad r \in$ G

$\therefore \quad a \oplus b \in \text{G}\ \forall\ a, b \in$ G

$\therefore \quad \oplus$ is a binary operation on G.

Associativity. Let $a, b, c \in$ G.

$$\begin{aligned} a \oplus (b \oplus c) &= a \oplus (b + c) && (\because\ b \oplus c \equiv b + c \text{ (modulo } m)) \\ &= \text{least non-negative remainder when } a + (b + c) \text{ is divided by } m \\ &= \text{least non-negative remainder when } (a + b) + c \text{ is divided by } m \\ &= (a + b) \oplus c \\ &= (a \oplus b) \oplus c && (\because\ a + b \equiv a \oplus b \text{ (modulo } m)) \end{aligned}$$

$\therefore \oplus$ is associative.

Existence of identity element. $0 \in$ G and we have

$$a \oplus 0 = a = 0 \oplus a\ \forall\ a \in \text{G}$$

$\therefore$ 0 is an identity element.

Existence of inverse.

$$0 \oplus 0 = 0 = 0 \oplus 0$$

$\therefore$ Inverse of 0 is 0.

Let $\quad a\ (> 0) \in$ G. $\quad \therefore m - a \in$ G

and $$a \oplus (m - a) = 0 = (m - a) \oplus a \qquad (\because\ a + (m - a) = m = m(1) + 0)$$

$\therefore \quad m - a$ is an inverse of a.

$\therefore \quad$ (G, $\oplus$) is a group of order m.

Commutativity.

$a \oplus b$ = least non-negative integer when $a + b$ is divided by m
= least non-negative integer when $b + a$ is divided by m
$= b \oplus a$

$\therefore$ (G, $\oplus$) is an abelian group of order m.

8. MULTIPLICATION MODULO m

Let $\{0, 1, 2, ..., m - 1\}$ be the set of first m non-negative integers. The binary operation '$\odot$' is defined on this set by $a \odot b = r$, where r is the least non-negative remainder obtained by dividing ab by m. This binary operation $\odot$ is called **multiplication modulo m**.

9. MULTIPLICATIVE GROUP OF INTEGERS MODULO p, WHERE p IS A PRIME NUMBER

Let $G = \{1, 2, 3, ..., p - 1\}$.

By definition for $a, b \in G$, $a \odot b = r$, where $0 \le r < p$.

Let $r = 0$. $\quad \therefore \ p / ab$

$\Rightarrow$ p/a or p/b $\quad$ ($\because$ p is prime)

Both are impossible because $1 \le a, b < p$

$\therefore$ $0 < r < p$ *i.e.*, $1 \le a \odot b < p$

$\therefore$ $\odot$ is a binary operation on G.

Associativity. Let $a, b, c \in G$.

$a \odot (b \odot c) = a \odot (bc)$ $\quad$ ($\because$ $b \odot c \equiv bc \pmod p$)
= least positive remainder when $a(bc)$ is divided by p
= least positive remainder when $(ab)c$ is divided by p
$= (ab) \odot c$
$= (a \odot b) \odot c$ $\quad$ ($\because$ $ab \equiv a \odot b \pmod p$)

$\therefore$ $\odot$ is associative.

Existence of identity element. $1 \in G$ and we have

$$a \odot 1 = a = 1 \odot a \ \forall \ a \in G$$

$\therefore$ 1 is an identity element.

Existence of inverse. Let $a \in G$. Consider the products :

$$a \odot 1, a \odot 2, a \odot 3, ..., a \odot (p - 1).$$

We claim that these are all distinct numbers.

If possible, let $a \odot m = a \odot n$, where $1 \le m < n < p$.

$\Rightarrow$ am and an leave the same least positive remainder when divided by p.

$\Rightarrow$ $p/(an - am)$ $\Rightarrow$ $p / a(n - m)$ $\Rightarrow$ p/a or $p/(n - m)$

Both are impossible because $1 \le a, n - m < p$.

$\therefore$ Our supposition is wrong.

$\therefore$ $a \odot 1, a \odot 2, a \odot 3, ..., a \odot (p - 1)$ are all distinct elements of G. These are $p - 1$ in number.

$\therefore$ One of these numbers must be equal to 1.

Let $a \odot k = 1$. $\therefore$ $k \odot a$ is also equal to 1.

$\therefore$ $a \odot k = 1 = k \odot a$

$\therefore$ k is an inverse of a.

$\therefore$ (G, $\odot$) is a group of order $p - 1$.

Commutativity.

$$a \odot b = \text{least non-negative integer when } ab \text{ is divided by } p$$
$$= \text{least non-negative integer when } ba \text{ is divided by } p$$
$$= b \odot a$$

$\therefore$ (G, $\odot$) is an abelian group of order $p - 1$.

Remark 1. Let G = {1, 2, 3, ..., $m - 1$}, where m is not a prime number.

In this case, there will exist $a, b \in$ G such that $ab = m$ and so $a \odot b = 0$, which is not in G.

$\therefore \odot$ is not a binary operation on G. For example, if $m = 10$, then 2, 5 $\in$ G and $2 \odot 5 = 0 \notin$ G.

Remark 2. We cannot include 0 in the group G = {1, 2, 3,..., $p - 1$}, because the inverse of '0' will not exist in G.

10. EULER'S ϕ-FUNCTION

If n is a positive integer, then the Euler ϕ-function, $\phi(n)$, is defined as $\phi(1) = 1$ and for $n > 1$, $\phi(n)$ as the number of positive integers less than n and relatively prime to n.

For example, if $n = 9$, then $\phi(9) = 6$ because there are 6 numbers 1, 2, 4, 5, 7, 8 which are less than 9 and relatively prime to 9. If p is a prime number, then we have $\phi(p) = p - 1$.

Example 6. *Let n > 1 be a positive integer. Let G be the set of all positive integers less than n and relatively prime to n. Show that G is an abelian group of order $\phi(n)$ under multiplication modulo n.*

Sol. Let G = $\{a : 0 < a < n, (n, a) = 1, a \in \mathbf{Z}\}$

$$\therefore \qquad o(G) = \phi(n)$$

By definition, for $a, b \in$ G, $a \odot b = r$, where $0 \le r < n$.

Let $r = 0$. $\quad \therefore n / ab$

We have $(a, n) = 1$, $(b, n) = 1$

$\therefore \quad ax + ny = 1,\ bz + nw = 1$ for some integers x, y, z, w.

$$\Rightarrow \quad ax(1) + ny = 1 \quad \Rightarrow \quad ax(bz + nw) + ny = 1$$

$$\Rightarrow \quad (xz)ab + (axw + y)n = 1$$

$$\Rightarrow \quad (ab, n) = 1 \quad \Rightarrow \quad n \text{ does not divide } ab$$

$\therefore$ Our supposition is wrong.

$$\therefore \quad 0 < r < n \text{ i.e., } 0 < a \odot b < n$$

$\therefore \odot$ is a binary operation on G.

Associativity. Let $\quad a, b, c \in$ G

$$a \odot (b \odot c) = a \odot (bc) \qquad (\because \quad b \odot c \equiv bc \pmod{n})$$
$$= \text{least positive remainder when } a(bc) \text{ is divided by } n$$
$$= \text{least positive remainder when } (ab)c \text{ is divided by } n$$
$$= (ab) \odot c$$
$$= (a \odot b) \odot c \qquad (\because \quad ab \equiv a \odot b \pmod{n})$$

$\therefore \odot$ is associative.

Existence of identity element. $1 \in$ G and we have

$$a \odot 1 = a = 1 \odot a \ \forall \ a \in G$$

$\therefore$ 1 is an identity element.

Existence of inverse. Let $a \in$ G. Consider the products $a \odot b$, where $b \in$ G. We claim that these are all distinct numbers.

If possible, let $a \odot j = a \odot k$, $\quad$ where $j, k \in$ G and $0 < j < k < n$.

$\Rightarrow \quad aj$ and ak leave the same least positive remainder when divided by n.

$\Rightarrow \quad n/(ak - aj) \quad \Rightarrow \quad n / a(k - j)$

$\Rightarrow \quad n/(k - j)$ $\quad (\because \quad (n, a) = 1$ and using **Gauss theorem**)

This is impossible because $0 < k - j < n$.

$\therefore$ Our supposition is wrong.

$\therefore$ The numbers $a \odot b$, where $b \in$ G are all distinct elements of G. These are $\phi(n)$ in number.

$\therefore$ One of these numbers must be equal to 1.

Let $a \odot c = 1. \quad \therefore \quad c \odot a$ is also equal to 1

$\therefore \quad a \odot c = 1 = c \odot a$

$\therefore$ c is an inverse of a.

$\therefore$ (G, $\odot$) is a group of order $\phi(n)$.

It is easy to check that $a \odot b = b \odot a \quad \forall\, a, b \in$ G.

$\therefore$ (G, $\odot$) is an abelian group of order $\phi(n)$.

Remark. If $n = 8$, then G = {1, 3, 5, 7} and the corresponding composition table is :

$\odot$	1	3	5	7
1	1	3	5	7
3	3	1	7	5
5	5	7	1	3
7	7	5	3	1

Example 7. *By using composition table, show that G = {0, 1, 2, 3, 4, 5, 6} is an abelian group under addition modulo 7.*

Sol. We have G = {0, 1, 2, 3, 4, 5, 6}.

The composition table is as follows :

$\oplus$	0	1	2	3	4	5	6
0	0	1	2	3	4	5	6
1	1	2	3	4	5	6	0
2	2	3	4	5	6	0	1
3	3	4	5	6	0	1	2
4	4	5	6	0	1	2	3
5	5	6	0	1	2	3	4
6	6	0	1	2	3	4	5

From the table, we see that $\oplus$ is a binary operation on G.

Associativity.

$$2 \oplus (4 \oplus 5) = 2 \oplus 2 = 4 \quad \text{and} \quad (2 \oplus 4) \oplus 5 = 6 \oplus 5 = 4$$

$$\therefore \quad 2 \oplus (4 \oplus 5) = (2 \oplus 4) \oplus 5$$

∴ Associativity holds for 2, 4, 5. Similarly, associativity can be verified for other triplets of numbers of G.

∴ $\oplus$ is associative.

Existence of identity element. $0 \in$ G and from the table, we have

$$a \oplus 0 = a = 0 \oplus a \quad \forall\, a \in \text{G}$$

∴ 0 is an identity element.

Existence of inverse. From the table, we have

$$0 \oplus 0 = 0 = 0 \oplus 0, \qquad 1 \oplus 6 = 0 = 6 \oplus 1$$
$$2 \oplus 5 = 0 = 5 \oplus 2, \qquad 3 \oplus 4 = 0 = 4 \oplus 3$$

∴ **The inverses of 0, 1, 2, 3, 4, 5, 6 are 0, 6, 5, 4, 3, 2, 1 respectively.**

∴ G is a group under addition modulo 7.

Commutativity. From the table, we have

$$a \oplus b = b \oplus a \quad \forall\, a, b \in \text{G}.$$

∴ G is an abelian group under addition modulo 7.

Example 8. *By using composition table, show that G = {1, 2, 3, 4, 5, 6} is an abelian group under multiplication modulo 7.*

Sol. We have G = {1, 2, 3, 4, 5, 6}.

6 is equal to 7 – 1 and 7 is a prime number.

The composition table is :

$\odot$	1	2	3	4	5	6
1	1	2	3	4	5	6
2	2	4	6	1	3	5
3	3	6	2	5	1	4
4	4	1	5	2	6	3
5	5	3	1	6	4	2
6	6	5	4	3	2	1

From the table, we see that $\odot$ is a binary operation on G.

Associativity. $3 \odot (4 \odot 2) = 3 \odot 1 = 3$ and $(3 \odot 4) \odot 2 = 5 \odot 2 = 3$

∴ $3 \odot (4 \odot 2) = (3 \odot 4) \odot 2$

∴ Associativity holds for 3, 4, 2. Similarly, associativity can be verified for other triplets of numbers of G.

∴ $\odot$ is associative.

Existence of identity element. $1 \in$ G and from the table, we have

$$a \odot 1 = a = 1 \odot a \quad \forall\, a \in \text{G}$$

∴ 1 is an identity element.

Existence of inverse. From the table, we have

$$1 \odot 1 = 1 = 1 \odot 1, \qquad 2 \odot 4 = 1 = 4 \odot 2,$$
$$3 \odot 5 = 1 = 5 \odot 3, \qquad 6 \odot 6 = 1 = 6 \odot 6$$

∴ The inverses of 1, 2, 3, 4, 5, 6 are 1, 4, 5, 2, 3, 6 respectively.

∴ G is a group under multiplication modulo 7.

Commutativity. From the table, we have

$$a \odot b = b \odot a \quad \forall\, a, b \in G.$$

$\therefore$ G is an abelian group under multiplication modulo 7.

EXERCISE 1

1. If G = {– 1, 1}, then show that (G, .) is an abelian group of order 2, where the operation represents the ordinary multiplication.
2. (*i*) Show that **N** does not form group with respect to usual addition.
   (*ii*) Show that **N** does not form group with respect to usual multiplication.
3. Show that the set of all vectors V form an infinite abelian group w.r.t. vector addition.
4. Let $G = \{z \in \mathbf{C} : |z| = 1\}$. Show that G is an infinite abelian group w.r.t. usual multiplication.
5. Show that the set $\{1, \omega, \omega^2\}$ of all cube roots of unity is a commutative group w.r.t. usual multiplication.
6. Show that the set of all positive real numbers forms an abelian group under the binary operation '$*$' defined by $a * b = \dfrac{ab}{5}$.
7. Show that the set of all rational numbers of the form $2^m 3^n$, where $m, n \in \mathbf{Z}$, forms an abelian group w.r.t. usual multiplication.
8. Show that the set $G = \{a + b\sqrt{2} : a, b \in \mathbf{Q}\}$ is an infinite abelian group w.r.t. usual addition.
9. Show that the set of matrices $A_\alpha = \begin{bmatrix} \cos\alpha & -\sin\alpha \\ \sin\alpha & \cos\alpha \end{bmatrix}$, $\alpha \in \mathbf{R}$ forms an abelian group under matrix multiplication.
10. If $G = \left\{\begin{bmatrix} a & 0 \\ 0 & 0 \end{bmatrix} : a \neq 0, a \in \mathbf{R}\right\}$, show that G is an abelian group under matrix multiplication.
11. Let $G = \{a : a \in \mathbf{R}, a \neq -1\}$. Show that G is a group under the operation '$*$' defined by
$$a * b = a + b + ab.$$
12. Show that $G = \{2^n : n \in \mathbf{Z}\}$ is an abelian group with respect to multiplication.
13. Show that $G = \{3n : n \in \mathbf{Z}\}$ is an abelian group with respect to addition.
14. Let $G = \mathbf{Z} \times \mathbf{Z}$ and $*$ be a binary operation on G defined by $(a, b) * (c, d) = (a + c, b + d)$ for (a, b), $(c, d) \in G$. Show that $(G, *)$ is an infinite abelian group.
15. Let $G = \mathbf{R} \times \mathbf{R} - \{(0, 0)\}$ and $*$ be a binary operation on G defined by $(a, b) * (c, d) = (ac - bd, ad + bc)$. Show that $(G, *)$ is an infinite abelian group.
16. Prove that the set of all 2 × 2 matrices forms an abelian group w.r.t. matrix addition

ANSWERS

2. (*i*) Identity element does not exist in **N**
   (*ii*) Inverse of elements do not exist in **N**.

HINTS

9. Try $e = A_0$ and $(A_\alpha)^{-1} = A_{-\alpha}$.
10. Try $e = \begin{bmatrix} 1 & 0 \\ 0 & 0 \end{bmatrix}$ and inverse of $\begin{bmatrix} a & 0 \\ 0 & 0 \end{bmatrix}$ as $\begin{bmatrix} 1/a & 0 \\ 0 & 0 \end{bmatrix}$.
15. $(a, b) * (c, d) = (a, b)$ implies $ac - bd - a = 0$
$$bc + ad - b = 0.$$

$$\therefore \qquad \frac{c}{b^2+a^2} = \frac{d}{-ab+ab} = \frac{1}{a^2+b^2}.$$

$$\therefore \qquad (c, d) = (1, 0)$$

Try (1, 0) as an identity element.

$(a, b) * (c, d) = (1, 0)$ implies $\quad ac - bd - 1 = 0$

$\qquad bc + ad - 0 = 0.$

$$\therefore \qquad \frac{c}{0+a} = \frac{d}{-b+0} = \frac{1}{a^2+b^2}$$

$$\therefore \qquad (c, d) = \left(\frac{a}{a^2+b^2}, -\frac{b}{a^2+b^2}\right)$$

Try $\left(\frac{a}{a^2+b^2}, -\frac{b}{a^2+b^2}\right)$ as an inverse of (a, b).

---

11. POWER OF AN ELEMENT

Let (G, .) be any group. For any $a \in$ G, we define :

$$a^0 = e, a^1 = a, \quad a^2 = a \,.\, a, \quad a^3 = a \,.\, a^2, \, ..., \, a^n = a \,.\, a^{n-1}$$

and $\quad a^{-1} =$ inverse of a, $\quad a^{-2} = (a^{-1})^2, \quad a^{-3} = (a^{-1})^3, \, ..., \, a^{-n} = (a^{-1})^n.$

The above definition implies $a^m \,.\, a^n = a^{m+n}$ and $(a^m)^n = a^{mn}$ for $a \in$ G and $m, n \in$ **Z**.

Remark. For the group (**R**, +), a^n means na for $a \in$ **R** and $n \in$ **Z**.

12. GENERAL PROPERTIES OF A GROUP

Theorem 1. (Uniqueness of identity element and inverse). *If (G, .) is a group then*

(i) The identity element of G is unique

(ii) Every element has unique inverse.

Proof. (*i*) Let e and f be two identities of the group G.

$$\therefore \qquad f \,.\, e = f = e \,.\, f \qquad (\because \ e \text{ is identity})$$

and $$e \,.\, f = e = f \,.\, e \qquad (\because \ f \text{ is identity})$$

$\therefore$ We have $e = f$.

$\therefore$ Identity element is unique.

(*ii*) Let b and c be two inverses of $a \in$ G.

$$\therefore \qquad a \,.\, b = e = b \,.\, a$$

and $$a \,.\, c = e = c \,.\, a$$

$$\therefore \qquad b = b \,.\, e = b \,.\, (a \,.\, c) = (b \,.\, a) \,.\, c = e \,.\, c = c$$

$$\therefore \qquad b = c$$

$\therefore$ Inverse of a is unique.

Theorem 2. (Inverse of inverse and inverse of product). *If (G, .) is a group then*

(i) $(a^{-1})^{-1} = a \quad \forall\, a \in G \qquad$ *(ii)* $(a \,.\, b)^{-1} = b^{-1} \,.\, a^{-1} \quad \forall\, a, b \in G.$

Proof. (*i*) For $a \in$ G, we have

$$a \,.\, a^{-1} = e = a^{-1} \,.\, a \qquad ...(1)$$

(1) also imply that the inverse of a^{-1} is a.

$\therefore \qquad (a^{-1})^{-1} = a.$

(*ii*) We have $(a \,.\, b) \,.\, (b^{-1} \,.\, a^{-1}) = a \,.\, [b \,.\, (b^{-1} \,.\, a^{-1})]$

$$= a \,.\, [(b \,.\, b^{-1}) \,.\, a^{-1}] = a \,.\, (e \,.\, a^{-1}) = a \,.\, a^{-1} = e$$

Also $\qquad (b^{-1} \,.\, a^{-1}) \,.\, (a \,.\, b) = b^{-1} \,.\, [a^{-1} \,.\, (a \,.\, b)]$

$$= b^{-1} \,.\, [(a^{-1} \,.\, a) \,.\, b] = b^{-1} \,.\, (e \,.\, b) = b^{-1}\, b = e$$

$\therefore \qquad (a \,.\, b) \,.\, (b^{-1} \,.\, a^{-1}) = e = (b^{-1} \,.\, a^{-1}) \,.\, (a \,.\, b)$

$\therefore \qquad (a \,.\, b)^{-1} = b^{-1} \,.\, a^{-1}.$

Theorem 3. (Left cancellation law). *Let (G, .) be a group. If $a \,.\, b = a \,.\, c$, then prove that $b = c$.*

Proof. We have $\qquad a \,.\, b = a \,.\, c. \qquad \therefore \qquad a^{-1} \,.\, (a \,.\, b) = a^{-1} \,.\, (a \,.\, c)$

$\Rightarrow \qquad (a^{-1} \,.\, a) \,.\, b = (a^{-1} \,.\, a) \,.\, c \qquad \Rightarrow \qquad e \,.\, b = e \,.\, c \qquad \Rightarrow \quad b = c.$

IMPORTANT RESULTS

1. *In a group, identity element is unique.*
2. *In a group, every element has unique inverse.*
3. *If (G, .) is a group, then $(a^{-1})^{-1} = a \quad \forall\, a \in G$.*
4. *If (G, .) is a group, then $(a \,.\, b)^{-1} = b^{-1} \,.\, a^{-1} \quad \forall\, a, b \in G$.*
5. *If (G, .) is a group and $a \,.\, b = a \,.\, c$, then $b = c$.*
6. *If (G, .) is a group and $b \,.\, a = c \,.\, a$, then $b = c$.*

Example 1. *If every element of a group is its own inverse, then show that the group is abelian.*

Sol. Let (G, .) be a group such that $x^{-1} = x \quad \forall\, x \in G$.

Let $a, b \in G$.

$\therefore \qquad a \,.\, b = (a \,.\, b)^{-1} = b^{-1} \,.\, a^{-1} = b \,.\, a$

$\therefore \qquad a \,.\, b = b \,.\, a$

$\therefore$ (G, .) is abelian.

Example 2. *Let (G, .) be a group. If $a, b \in G$, then solve the equation $a \,.\, x \,.\, a = b$.*

Sol. We have $\qquad a \,.\, x \,.\, a = b. \qquad \therefore \qquad a^{-1} \,.\, (a \,.\, x \,.\, a) \,.\, a^{-1} = a^{-1} \,.\, b \,.\, a^{-1}$

$\Rightarrow \qquad (a^{-1} \,.\, a) \,.\, x \,.\, (a \,.\, a^{-1}) = a^{-1} \,.\, b \,.\, a^{-1} \qquad \Rightarrow \qquad e \,.\, x \,.\, e = a^{-1} \,.\, b \,.\, a^{-1}$

$\Rightarrow \qquad x = a^{-1} \,.\, b \,.\, a^{-1}.$

Example 3. *If (G, .) is a finite group, show that there exists a positive integer N such that*

$$a^N = e \quad \forall\, a \in G.$$

Sol. Let $G = \{a_1, a_2, \ldots\ldots, a_n\}$

For $a_i \in G$, consider $a_i, a_i^2, a_i^3, \ldots\ldots,$

Since all these elements are in G, which is finite, we must have $a_i^l = a_i^m$ for some l, m.

Let $l < m. \qquad \therefore \qquad a_i^l = a_i^{m-l}\, a_i^l$

$\Rightarrow \qquad e = a_i^{m-l}$ i.e., $a_i^{\lambda_i} = e, \qquad$ where $\lambda_i = m - l$.

This is true for all i, $1 \le i \le n$.

Let $\qquad N = \text{L.C.M.}\ \{\lambda_1, \lambda_2, \ldots\ldots, \lambda_n\}$

$\therefore \qquad a_i^N = e \quad \forall\, i : 1 \le i \le n.$

$\therefore$ The result holds.

Example 4. *If (G, .) is a group in which* $(a \,.\, b)^i = a^i \,.\, b^i$ *for three consecutive integers i,* $\forall\, a, b \in G$*, show that G is abelian.*

Sol. Let
$$(a \,.\, b)^m = a^m \,.\, b^m \qquad ...(1)$$
$$(a \,.\, b)^{m+1} = a^{m+1} \,.\, b^{m+1} \qquad ...(2)$$
$$(a \,.\, b)^{m+2} = a^{m+2} \,.\, b^{m+2} \qquad ...(3)$$

(2) $\Rightarrow \quad (a \,.\, b)^m \,.\, (a \,.\, b) = a^{m+1} \,.\, b^{m+1}$

$\Rightarrow \quad (a^m \,.\, b^m) \,.\, (a \,.\, b) = a^{m+1} \,.\, b^{m+1}$ [Using (1)]

$\Rightarrow \quad a^m \,.\, (b^m \,.\, a) \,.\, b = a^m \,.\, (a \,.\, b^m) \,.\, b$

$\Rightarrow \quad b^m \,.\, a = a \,.\, b^m$ ...(4)

(3) $\Rightarrow \quad (a \,.\, b)^{m+1} \,.\, (a \,.\, b) = a^{m+2} \,.\, b^{m+2}$

$\Rightarrow \quad (a^{m+1} \,.\, b^{m+1}) \,.\, (a \,.\, b) = a^{m+2} \,.\, b^{m+2}$ [Using (2)]

$\Rightarrow \quad a^{m+1} \,.\, (b^{m+1} \,.\, a) \,.\, b = a^{m+1} \,.\, (a \,.\, b^{m+1}) \,.\, b$

$\Rightarrow \quad b^{m+1} \,.\, a = a \,.\, b^{m+1} \quad \Rightarrow \quad b^{m+1} \,.\, a = a \,.\, (b^m \,.\, b)$

$\Rightarrow \quad b^{m+1} \,.\, a = (a \,.\, b^m) \,.\, b \quad \Rightarrow \quad (b^m \,.\, b) \,.\, a = (b^m \,.\, a) \,.\, b$

$\Rightarrow \quad b^m \,.\, (b \,.\, a) = b^m \,.\, (a \,.\, b) \quad \Rightarrow \quad b \,.\, a = a \,.\, b \quad \forall\, a, b \in G.$

$\therefore$ G is abelian.

Example 5. *Prove that if (G, .) is an abelian group, then for all a, b* $\in$ *G and n* $\in$ **Z**, *we have* $(a \,.\, b)^n = a^n \,.\, b^n$.

Sol. Let $a, b \in G$.

Case I. n $\in$ N. We shall prove that
$$(a \,.\, b)^n = a^n \,.\, b^n \qquad ...(1)$$
by using **P.M.I.**

Let $n = 1$.

$\therefore \quad$ L.H.S. of (1) $= (a \,.\, b)^1 = a \,.\, b$

and $\quad$ R.H.S. of (1) $= a^1 \,.\, b^1 = a \,.\, b$

$\therefore$ (1) is true for $n = 1$.

Let (1) be true for $n = k$.

$\therefore \quad (a \,.\, b)^k = a^k \,.\, b^k$ ...(2)

Let $\quad n = k + 1$.

$\therefore \quad (a \,.\, b)^{k+1} = (a \,.\, b)^k \,(a \,.\, b) = (a^k \,.\, b^k)\,(a \,.\, b)$ (Using (2))

$= (a^k \,.\, b^k)\,(b \,.\, a) = a^k \,.\, (b^k \,.\, (b \,.\, a))$

$= a^k \,.\, ((b^k \,.\, b) \,.\, a) = a^k \,.\, (b^{k+1} \,.\, a)$

$= a^k \,.\, (a \,.\, b^{k+1}) = (a^k \,.\, a) \,.\, b^{k+1} = a^{k+1} \,.\, b^{k+1}.$

$\therefore$ (1) is true for $n = k + 1$.

$\therefore$ By P.M.I., $\quad (a \,.\, b)^n = a^n \,.\, b^n \quad \forall\, n \in$ **N**.

Case II. n = 0 $\quad (a \,.\, b)^0 = e = e \,.\, e = a^0 \,.\, b^0$

$\therefore \quad (a \,.\, b)^0 = a^0 \,.\, b^0.$

Case III. n is – ve integer. Let $n = -p$.

$\therefore \quad p \in$ **N**.

Now $(a \,.\, b)^n = (a \,.\, b)^{-p} = [(a \,.\, b)^p]^{-1} = (a^p \,.\, b^p)^{-1}$ (By case I)

$= (b^p \,.\, a^p)^{-1}$ ($\because$ G is abelian)

$= (a^p)^{-1} \,.\, (b^p)^{-1} = a^{-p} \,.\, b^{-p} = a^n \,.\, b^n$.

$\therefore$ $(a \,.\, b)^n = a^n \,.\, b^n$ when n is – ve integer

Combining, we get

$$(a \,.\, b)^n = a^n \,.\, b^n \ \forall n \in \mathbf{Z}.$$

EXERCISE 2

1. Let (G, .) be a group. If $b \,.\, a = c \,.\, a$, then show that $b = c$.
2. If in a group G, $a^2 = a$, $a \in$ G, show that $a = e$.
3. Show that a group up to order 2 is always abelian.
4. Let (G, .) be a group. If $a, b \in$ G, prove that G is abelian if and only if $(a \,.\, b)^2 = a^2 \,.\, b^2$.
5. If (G, .) is a group, then show that $(a \,.\, b \,.\, a^{-1})^n = a \,.\, b^n \,.\, a^{-1}$ $\forall\, a, b \in$ G and $n \in \mathbf{Z}$.

Hints

4. $(a \,.\, b)^2 = a^2 \,.\, b^2 \Rightarrow (a \,.\, b) \,.\, (a \,.\, b) = a \,.\, (a \,.\, b) \,.\, b$
$\Rightarrow a \,.\, (b \,.\, a) \,.\, b = a \,.\, (a \,.\, b) \,.\, b \Rightarrow b \,.\, a = a \,.\, b.$
Conversely, $(a \,.\, b)^2 = (a \,.\, b) \,.\, (a \,.\, b) = a \,.\, (b \,.\, a) \,.\, b = a \,.\, (a \,.\, b) \,.\, b.$
5. First prove for $n \in \mathbf{N}$ by using the principle of mathematical induction.

13. ORDER OF AN ELEMENT

Let (G, .) be a group with identity element e. For $a \in$ G, if there exists a smallest natural number m such that $a^m = e$, then m is called the **order** of a and write $o(a) = m$.

If no such natural number exists then we say that a is of *infinite order*.

14. SOME IMPORTANT THEOREMS

Theorem 1. *For a finite group, the order of every element is finite and cannot exceed the order of the group.*

Proof. Let (G, .) be a finite group. Let a be an arbitrary element of G.

Consider the elements : $a, a^2, a^3, a^4, \ldots\ldots$.

These are all elements of G. Since G is finite, all these elements cannot be distinct.

$\therefore$ $a^k = a^l$ for some natural numbers k and l.

Let $k > l$. $\therefore$ $a^{k-l} \,.\, a^l = a^l \Rightarrow a^{k-l} = e$ ($\because$ $k = (k - l) + l$)

$\therefore$ By definition of $o(a)$, we have $o(a) \leq k - l$

$\therefore$ $o(a)$ is finite.

Let $o(a) = n$. If possible, let us suppose that $n > o(\text{G})$.

Since $o(a) = n$, the elements :

$a, a^2, a^3, \ldots\ldots, a^{n-1}, a^n$ $(= e)$ are all distinct and these are all elements of G. This is impossible, because these are n distinct elements and G has less than n elements.

$\therefore$ Our supposition is wrong. $\therefore$ $o(a) \leq o(\text{G})$.

Theorem 2. *In a group, the order of an element and its inverse are same.*

Proof. Let (G, .) be a group.

Let $a \in G$.

Case I. a is of infinite order

$\therefore \quad a^n \neq e \quad \forall\, n \in \mathbf{N}$

$\Rightarrow \quad (a^n)^{-1} \neq e^{-1} \quad \forall\, n \in \mathbf{N} \Rightarrow a^{-n} \neq e \quad \forall\, n \in \mathbf{N}$

$\Rightarrow \quad (a^{-1})^n \neq e \quad \forall\, n \in \mathbf{N}$

$\Rightarrow$ a^{-1} is of inifinite order.

Case II. a is of finite order

$\therefore$ a^{-1} is also finite order. Let $o(a) = m$ and $o(a^{-1}) = n$

$o(a) = m \Rightarrow a^m = e \Rightarrow (a^m)^{-1} = e^{-1} \Rightarrow a^{-m} = e \Rightarrow (a^{-1})^m = e \Rightarrow n \leq m.$

$o(a^{-1}) = n \Rightarrow (a^{-1})^n = e \Rightarrow a^{-n} = e \Rightarrow (a^{-n})^{-1} = e^{-1} \Rightarrow a^n = e \Rightarrow m \leq n.$

Combining, we get $m = n$. *i.e.*, $o(a) = o(a^{-1})$.

IMPORTANT RESULTS

1. *For a finite group, the order of every element is finite and cannot exceed the order of the group.*
2. *In a group, the order of an element and its inverse are same.*

Example 1. *In the group* ($\mathbf{Q}^o$, .) *find the order of elements of the group.*

Sol. In the group ($\mathbf{Q}^o$, .), the identity element is 1.

$(1)^1 = 1 \quad \therefore \quad o(1) = 1$

Also $-1 \in \mathbf{Q}^o$ and $(-1)^2 = 1$. $\quad \therefore \quad o(-1) = 2$

Let x be any element of $\mathbf{Q}^o$ different from 1 and – 1.

$\therefore \quad x^n \neq 1 \quad \forall\, n \in \mathbf{N}$

$\therefore$ The order of x is infinite.

$\therefore$ $o(1) = -$ **1**, $o(-1) =$ **2** and all other elements are of **infinite** order.

Example 2. *Let G = {0, 1, 2, 3, 4, 5}. Find the order of elements of the group G under the binary operation 'addition modulo 6'.*

Sol. We have G = {0, 1, 2, 3, 4, 5}.

Let $\oplus$ denote the binary operation 'addition modulo 6'.

The identity element is 0. $\therefore$ $o(0) =$ **1.**

Now $(1)^1 = 1, \quad (1)^2 = 1 \oplus 1 = 2, \quad (1)^3 = 1 \oplus (1)^2 = 1 \oplus 2 = 3,$

$(1)^4 = 1 \oplus (1)^3 = 1 \oplus 3 = 4, \quad (1)^5 = 1 \oplus (1)^4 = 1 \oplus 4 = 5,$

$(1)^6 = 1 \oplus (1)^5 = 1 \oplus 5 = 0.$

$\therefore$ $o(1) =$ **6.**

$(2)^1 = 2, \quad (2)^2 = 2 \oplus 2 = 4, \quad (2)^3 = 2 \oplus (2)^2 = 2 \oplus 4 = 0$

$\therefore$ $o(2) =$ **3.**

$(3)^1 = 3, \quad (3)^2 = 3 \oplus 3 = 0$

$\therefore$ $o(3) =$ **2.**

$(4)^1 = 4, \quad (4)^2 = 4 \oplus 4 = 2, \quad (4)^3 = 4 \oplus (4)^2 = 4 \oplus 2 = 0$

$\therefore \quad o(4) = \mathbf{3}.$

$(5)^1 = 5, \quad (5)^2 = 5 \oplus 5 = 4, \quad (5)^3 = 5 \oplus (5)^2 = 5 \oplus 4 = 3,$

$(5)^4 = 5 \oplus (5)^3 = 5 \oplus 3 = 2, \quad (5)^5 = 5 \oplus (5)^4 = 5 \oplus 2 = 1,$

$(5)^6 = 5 \oplus (5)^5 = 5 \oplus 1 = 0$

$\therefore \quad o(5) = \mathbf{6}.$

Example 3. *Let G = {1, 2, 3, 4, 5, 6}. Find the order of elements of the group G under the binary, operation 'multiplication modulo 7'.*

Sol. We have G = {1, 2, 3, 4, 5, 6}.

Let $\odot$ denote the binary operation 'multiplication modulo 7'.

The identity element is 1. $\therefore o(1) = \mathbf{1}$.

Now $(2)^1 = 2, \quad (2)^2 = 2 \odot 2 = 4, \quad (2)^3 = 2 \odot (2)^2 = 2 \odot 4 = 1.$

$\therefore \quad o(2) = \mathbf{3}.$

$(3)^1 = 3, \quad (3)^2 = 3 \odot 3 = 2, \quad (3)^3 = 3 \odot (3)^2 = 3 \odot 2 = 6,$

$(3)^4 = 3 \odot (3)^3 = 3 \odot 6 = 4, \quad (3)^5 = 3 \odot (3)^4 = 3 \odot 4 = 5,$

$(3)^6 = 3 \odot (3)^5 = 3 \odot 5 = 1.$

$\therefore \quad o(3) = \mathbf{6}.$

$(4)^1 = 4, \quad (4)^2 = 4 \odot 4 = 2, \quad (4)^3 = 4 \odot (4)^2 = 4 \odot 2 = 1.$

$\therefore \quad o(4) = \mathbf{3}.$

$(5)^1 = 5, \quad (5)^2 = 5 \odot 5 = 4, \quad (5)^3 = 5 \odot (5)^2 = 5 \odot 4 = 6,$

$(5)^4 = 5 \odot (5)^3 = 5 \odot 6 = 2, \quad (5)^5 = 5 \odot (5)^4 = 5 \odot 2 = 3,$

$(5)^6 = 5 \odot (5)^5 = 5 \odot 3 = 1.$

$\therefore \quad o(5) = \mathbf{6}.$

$(6)^1 = 6, \quad (6)^2 = 6 \odot 6 = 1.$

$o(6) = \mathbf{2}.$

Remark. Here note that the number 7 is prime.

Example 4. *If (G, .) is a group, then show that $o(a) = o(x^{-1} . a . x)$ $\forall$ a, x $\in$ G.*

Sol. We first show that $(x^{-1} . a . x)^m = x^{-1} . a^m . x \quad \forall\, m \in \mathbf{N}$.

$$\text{Now} \quad (x^{-1} . a . x)^m = \underbrace{(x^{-1} . a . x) . (x^{-1} . a . x) \ldots\ldots (x^{-1} . a . x)}_{m \text{ times}}$$

$$= x^{-1} . a . (x . x^{-1}) . a . (x . x^{-1}) \ldots\ldots a . x$$

$$= x^{-1} a . e . a\, e \ldots\ldots a\, x$$

$$= x^{-1} . a . a . \ldots\ldots . a . x = x^{-1} . a^m . x.$$

$$\therefore \quad (x^{-1} . a . x)^m = x^{-1} . a^m . x \quad \forall\, m \in \mathbf{N} \qquad \ldots(1)$$

Case I. a is of infinite order

$\therefore \quad a^n \neq e \quad \forall\, n \in \mathbf{N}$

$\Rightarrow \quad a^n \neq x . e . x^{-1} \quad \forall\, n \in \mathbf{N}$

$\Rightarrow \quad x^{-1} . a^n . x \neq x^{-1} . (x . e . x^{-1}) . x \quad \forall\, n \in \mathbf{N}$

$\Rightarrow \quad (x^{-1} . a . x)^n \neq e \quad \forall\, n \in \mathbf{N}.$ (By using (1))

$\Rightarrow \quad x^{-1} . a . x$ is of infinite order.

Case II. a is of finite order

$\therefore$ $x^{-1} . a . x$ is also of finite order.

Let $o(a) = m$ and $o(x^{-1} . a . x) = n$.

$$o(a) = m \Rightarrow a^m = e \Rightarrow a^m = x . e . x^{-1} \Rightarrow x^{-1} . a^m . x = e$$
$$\Rightarrow (x^{-1} . a . x)^m = e \Rightarrow n \le m.$$
$$o(x^{-1} . a . x) = n \Rightarrow (x^{-1} . a . x)^n = e \Rightarrow x^{-1} . a^n . x = e \Rightarrow a^n = x . e . x^{-1}$$
$$\Rightarrow a^n = e \Rightarrow m \le n.$$

Combining, we get $m = n$ i.e., $o(a) = o(x^{-1} . a . x)$.

Example 5. *Let (G, .) be a group and $a \in G$ be of order m. Show that $a^n = e$ if and only if m/n.*

Sol. $o(a) = m$.

$\therefore$ m is the smallest natural number such that $a^m = e$.

Let m/n. $\therefore$ $n = km$ for some $k \in \mathbf{N}$

$\therefore$ $a^n = a^{km} = (a^m)^k = e^k = e$

$\therefore$ $a^n = e$

Conversely, let $a^n = e$. $\therefore$ $n \ge m$

By division algorithm, $\exists$ integers q, r, where $0 \le r < m$ such that $n = mq + r$.

$\therefore$ $a^{mq+r} = e$

$\Rightarrow$ $a^{mq} . a^r = e$ $\Rightarrow$ $(a^m)^q . a^r = e$

$\Rightarrow$ $e^q . a^r = e$ $\Rightarrow$ $e . a^r = e$

$\Rightarrow$ $a^r = e$ $\Rightarrow$ $r = 0$ $(\because$ $o(a) = m$ and $0 \le r < m)$

$\therefore$ $n = mq$ or m/n.

EXERCISE 3

1. In a group, show that the identity element is the only element of order one.
2. If (G, .) is a group, show that $o(a^k) \le o(a)$ $\forall a \in G, k \in \mathbf{Z}$.
3. Let G = {± 1, ± i}. Find the order of each element in the group G under usual multiplication as the binary operation.
4. In the group (**Z**, +), show that order of identity is one and all other elements are of infinite order.
5. (*i*) Let G = {0, 1, 2, 3, 4}. Find the order of elements of the group G under the binary operation 'addition modulo 5'.

   (*ii*) Let G = {0, 1, 2, 3, 4, 5, 6, 7}. Find the order of elements of the group G under the binary operation 'addition modulo 8'.
6. (*i*) Let G = {1, 2, 3, 4}. Find the order of elements of the group G under the binary operation 'multiplication modulo 5'.

   (*ii*) Let G = {1, 2, 3, 4, 5, 6, 7, 8, 9, 10}. Find the order of elements of the group G under the binary operation 'multiplication modulo 11'.
7. Let (G, .) be a group and $a \in$ G is of order m. If m and n are relatively prime, then show that order of a^n is also m.

ANSWERS

3. $o(1) = 1, o(-1) = 2, o(i) = 4, o(-i) = 4$

5. (*i*) $o(0) = 1, o(1) = 5, o(2) = 5, o(3) = 5, o(4) = 5$

(*ii*) $o(0) = 1, o(1) = 8, o(2) = 4, o(3) = 8, o(4) = 2, o(5) = 8, o(6) = 4, o(7) = 8$

6. (*i*) $o(1) = 1, o(2) = 4, o(3) = 4, o(4) = 2$

(*ii*) $o(1) = 1, o(2) = 10, o(3) = 5, o(4) = 5, o(5) = 5, o(6) = 10, o(7) = 10, o(8) = 10, o(9) = 5, o(10) = 2.$

HINTS

2. $o(a) = m$ implies $a^m = e$ and $(a^k)^m = a^{mk} = (a^m)^k = e^k = e.$

7. $a^m = e \Rightarrow (a^m)^n = e^m \Rightarrow a^{mn} = e \Rightarrow (a^n)^m = e \Rightarrow o(a^n) \le m.$

Let $o(a^n) = k. \quad \therefore \quad k \le m$

$(m, n) = 1$ implies $m\lambda + n\mu = 1$ for some $\lambda, \mu \in \mathbf{Z}$.

$\therefore \quad a = a^1 = a^{m\lambda + n\mu} = a^{m\lambda} . a^{n\mu} = (a^m)^\lambda . a^{n\mu} = e^\lambda . a^{n\mu} = e . a^{n\mu} = a^{n\mu}$

$\therefore \quad a^k = (a^{n\mu})^k = a^{n\mu k} = [(a^n)^k]^\mu = e^\mu = e$

$\therefore \quad m \le k. \qquad (\because o(a) = m)$

Combining, we get $m = k$.

---

15. HOMOMORPHISM

A mapping ϕ from a group G into a group G′ is called a **homomorphism** of G into G′ if $\phi(ab) = \phi(a)\,\phi(b) \quad \forall\, a, b \in G.$

If binary operation in G is $*$ then ab stands for $a * b$ and if the binary operation in G′ is $*'$, then $\phi(a) *' \phi(b)$ is written briefly $\phi(a)\,\phi(b)$.

Examples :

(*i*) Let G be the group of all integers under usual addition.

Define $\phi : G \to G$ by $\phi(x) = 5x \quad \forall\, x \in G$.

Let $x_1, x_2 \in G$.

$$\therefore \quad \phi(x_1 + x_2) = 5(x_1 + x_2) = 5x_1 + 5x_2 = \phi(x_1) + \phi(x_2).$$

$\therefore \phi : G \to G$ is a homomorphism.

(*ii*) Let G be the group of all integers under usual addition and G′ be group of all non-zero rational numbers under usual multiplication.

Define $\phi : G \to G'$ by $\phi(x) = 2^x \quad \forall\, x \in G$.

Let $x, y \in G$.

$$\phi(x + y) = 2^{x+y} = 2^x . 2^y = \phi(x)\,\phi(y)$$

$$\therefore \quad \phi(x + y) = \phi(x)\,\phi(y).$$

$\therefore \quad \phi : G \to G'$ is a homomorphism.

16. HOMOMORPHIC IMAGE OF IDENTITY AND INVERSE

Theorem. *If ϕ is a homomorphism of group G into group G′, then*

(i) $\phi(e) = e'$, where e and e′ are identity elements of the groups G and G′ respectively.

(ii) $\phi(a^{-1}) = [\phi(a)]^{-1} \; \forall\, a \in G$.

Proof. (*i*) For $a \in G$, we have

$$\phi(a) = \phi(ea) = \phi(e)\,\phi(a)$$

Also $\phi(a) = e'\phi(a)$.

$\therefore$ $\phi(e)\,\phi(a) = e'\,\phi(a)$

$\therefore$ By cancellation law in G′, we have $\phi(e) = e'$.

(ii) $\phi(a)\,\phi(a^{-1}) = \phi(a\,a^{-1}) = \phi(e) = e'$

Also $\phi(a^{-1})\,\phi(a) = \phi(a^{-1}\,a) = \phi(e) = e'$

$\therefore$ $\phi(a)\,\phi(a^{-1}) = e' = \phi(a^{-1})\,\phi(a)$

$\therefore$ By definition $[\phi(a)]^{-1} = \phi\,(a^{-1})$ *i.e.*, $\phi(a^{-1}) = [\phi(a)]^{-1}$.

17. KERNEL OF A HOMOMORPHISM

If ϕ is a homomorphism of group G into group G′, then the set

$$\{a \in G : \phi(a) = \text{identity element of } G'\}$$

is called the **kernel of the homomorphism** ϕ and write **ker** ϕ.

Since $\phi(e) = e'$, so $e \in \ker \phi$.

$\therefore$ ker ϕ is always a non-empty set.

Example 1. *If f is a homomorphism of group G into group G′ and g, a homomorphism of group G′ into group G″, show gof is a homomorphism of group G into group G″.*

Sol. By definition *gof* is a mapping from G into G″.

For $a, b \in G$, we have

$(gof)\,(ab) = g\,(f(ab))$

$= g(f(a)\,f(b))$ ($\because$ $f : G \to G'$ is a homomorphism)

$= g(f\,(a))\,g(f(b))$ ($\because$ $g : G' \to G''$ is a homomorphism)

$= (gof)(a)\,(gof)(b)$.

$\therefore$ *gof* is a homomorphism of G into G″.

Example 2. *Let G be the group of non-zero real numbers under usual multiplication. Let G′ be the group of elements containing 1 and – 1 under usual multiplication. Define* $\phi : G \to G'$ *by* $\phi(x) = \begin{cases} 1 & \text{if } x > 0 \\ -1 & \text{if } x < 0 \end{cases}$. *Show that* $\phi : G \to G'$ *is a homomorphism.*

Sol. Let $x, y \in G$. There are four possibilities

Case I. x > 0, y > 0

$\therefore$ $xy > 0$ and $\phi(xy) = 1$ and $\phi(x)\,\phi(y) = (1)\,(1) = 1$.

$\therefore$ $\phi(xy) = \phi(x)\,\phi(y)$

Case II. x > 0, y < 0

$\therefore$ $xy < 0$ and $\phi(xy) = -1$ and $\phi(x)\,\phi(y) = (1)\,(-1) = -1$.

$\therefore$ $\phi(xy) = \phi(x)\,\phi(y)$

Case III. x < 0, y > 0

$\therefore$ $xy < 0$ and $\phi\,(xy) = -1$ and $\phi(x)\,\phi(y) = (-1)\,(1) = -1$.

$\therefore$ $\phi(xy) = \phi(x)\,\phi(y)$

Case IV. x < 0, y < 0

$\therefore$ $xy > 0$ and $\phi\,(xy) = 1$ and $\phi(x)\,\phi(y) = (-1)\,(-1) = 1$.

$\therefore$ $\phi(xy) = \phi(x)\,\phi(y)\ \forall\ x, y \in G$

$\therefore$ $\phi : G \to G'$ is a homomorphism.

Example 3. *Let G be the group of all real 2 × 2 matrices* $\begin{pmatrix} a & b \\ c & d \end{pmatrix}$ *such that* $ad - bc \neq 0$ *under matrix multiplication. Let G′ be the group of all non-zero real numbers under multiplication. Define* $\phi : G \to G'$ *by* $\phi\begin{pmatrix} a & b \\ c & d \end{pmatrix} = ad - bc$.

Show that ϕ *is a homomorphism of G onto G′.*

Sol. Let $\begin{pmatrix} a & b \\ c & d \end{pmatrix}$ and $\begin{pmatrix} a' & b' \\ c' & d' \end{pmatrix}$ be in G.

$$\therefore \quad \phi\left[\begin{pmatrix} a & b \\ c & d \end{pmatrix}\begin{pmatrix} a' & b' \\ c' & d' \end{pmatrix}\right] = \phi\left[\begin{pmatrix} aa' + bc' & ab' + bd' \\ ca' + dc' & cb' + dd' \end{pmatrix}\right]$$

$$= (aa' + bc')(cb' + dd') - (ca' + dc')(ab' + bd')$$

$$= aa'cb' + aa'dd' + bc'cb' + bc'dd' - ca'ab' - ca'bd' - dc'ab' - dc'bd'$$

$$= aa'dd' + bc'cb' - ca'bd' - dc'ab'$$

$$= a'd'(ad - bc) - b'c'(ad - bc) = (ad - bc)(a'd' - b'c')$$

$$= \phi\begin{pmatrix} a & b \\ c & d \end{pmatrix} \phi\begin{pmatrix} a' & b' \\ c' & d' \end{pmatrix}$$

$\therefore$ ϕ is a homomorphism from G into G′.

ϕ is onto. Let $m \in$ G′.

$\therefore m \neq 0$ and $m \in \mathbf{R}$ $\quad \therefore \begin{pmatrix} m & 0 \\ 0 & 1 \end{pmatrix} \in$ G $\quad (\because \; m(1) - 0(0) = m \neq 0)$

and $$\phi\begin{pmatrix} m & 0 \\ 0 & 1 \end{pmatrix} = m(1) - 0(0) = m.$$

$\therefore$ Every element in G′ has a pre-image in G.

$\therefore$ ϕ is onto.

$\therefore$ $\phi : G \to G'$ is an onto homomorphism.

18. ISOMORPHISM

A mapping ϕ from a group G into a group G′ is called an **isomorphism** of G into G′ if

(*i*) ϕ is a homomorphism *i.e.*, $\phi(ab) = \phi(a)\,\phi(b) \quad \forall\, a, b \in G$

(*ii*) ϕ is one-one.

Remark. If $\phi : G \to G'$ is a mapping and we write $\phi(ab) = \phi(a)\,\phi(b)$, then it should be clearly understood that

(*i*) ab is that element of the group G which is obtained by applying the binary operation of G on the ordered pair $(a, b) \in G \times G$.

(*ii*) $\phi(a)\,\phi(b)$ is that element of the group G′ which is obtained by applying the binary operation of G′ on the ordered pair $(\phi(a), \phi(b)) \in G' \times G'$.

19. ORDER OF ISOMORPHIC IMAGE OF AN ELEMENT

Theorem. *If* ϕ *is an isomorphism of group G into group G′, then*

$$o(a) = o(\phi(a)) \quad \forall\, a \in G.$$

Proof. $\phi : G \to G'$ is an isomorphism of G into G′. Let e and e' be the identity elements of the groups G and G′ respectively. Let $a \in G$.

Case I . o(a) is finite. Let $o(a) = n$.

$\Rightarrow \quad a^n = e \quad \Rightarrow \quad \phi(a^n) = \phi(e)$

$\Rightarrow \quad \phi(aa \text{ ..... } n \text{ times}) = e' \quad \Rightarrow \quad \phi(a)\,\phi(a) \text{ ...... } n \text{ times} = e'$

$\Rightarrow \quad (\phi(a))^n = e' \quad \Rightarrow \quad o(\phi(a)) \le n.$

Let $\quad o(\phi(a)) = m. \quad \therefore \quad m \le n$

$\Rightarrow \quad (\phi(a))^m = e' \quad \Rightarrow \quad \phi(a)\,\phi(a) \text{ ...... } m \text{ times} = e'$

$\Rightarrow \quad \phi\,(aa \text{ ...... } m \text{ times}) = e' \quad \Rightarrow \quad \phi(a^m) = \phi(e)$

$\Rightarrow \quad a^m = e \qquad (\because \quad \phi \text{ is one-one})$

$\Rightarrow \quad n \le m \qquad (\because \quad o(a) = n)$

$\therefore \; m = n$ i.e., $\quad o(\phi(a)) = o(a).$

Case II. o(a) is infinite. If possible, let $o(\phi(a))$ be finite. Let $o(\phi(a)) = m$.

$\Rightarrow \quad (\phi(a))^m = e' \quad \Rightarrow \quad \phi(a)\,\phi(a) \text{ ...... } m \text{ times} = e'$

$\Rightarrow \quad \phi(aa \text{ ...... } m \text{ times}) = e'$

$\Rightarrow \quad \phi(a^m) = \phi(e) \quad \Rightarrow \quad a^m = e. \qquad (\because \quad \phi \text{ is one-one})$

$\therefore \quad o(a) \le m$, which is impossible. $\quad \therefore \quad o(\phi(a))$ cannot be finite.

$\therefore \quad o(\phi(a))$ is also infinite. $\quad \therefore \quad a$ and $\phi(a)$ have same order.

20. KERNEL OF AN ISOMORPHISM

Theorem. *Let ϕ be a homomorphism from a group G into group G′. The homomorphism ϕ is an isomorphism of G into G′ if and only if $\ker \phi = \{e\}$.*

Proof. $\phi : G \to G'$ is a homomorphism of group G into group G′.

Let ϕ be an isomorphism of G into G′. We have $\phi(e) = e'$.

$\therefore \quad e \in \ker \phi$.

Let a be any other element in $\ker \phi$.

$\therefore \quad \phi(a) = e'. \quad \therefore \quad \phi(a) = \phi(e)$

$\therefore \quad a = e \qquad (\because \quad \phi \text{ is one-one})$

$\therefore \quad \ker \phi = \{e\}$.

Conversely, let $\ker \phi = \{e\}$. We shall show that ϕ is one-one.

Let for $a, b \in G$, we have $\phi(a) = \phi(b)$.

$\Rightarrow \quad \phi(a)\,(\phi(b))^{-1} = \phi(b)\,(\phi(b))^{-1} \quad \Rightarrow \quad \phi(a)\,\phi(b^{-1}) = e'$

$\Rightarrow \quad \phi\,(ab^{-1}) = e' \quad \Rightarrow \quad ab^{-1} \in \ker \phi$

$\Rightarrow \quad ab^{-1} = e \quad \Rightarrow \quad (ab^{-1})\,b = eb$

$\Rightarrow \quad a(b^{-1}\,b) = b \quad \Rightarrow \quad ae = b \quad \Rightarrow \quad a = b.$

$\therefore \phi$ is one-one.

$\therefore \phi$ is an isomorphism of G into G′.

21. ISOMORPHIC GROUPS

Two groups G and G′ are called **isomorphic groups** if there exists an isomorphism of G onto G′. If groups G and G′ are isomorphic groups then we write $G \approx G'$.

Remark. If groups G and G′ are isomorphic then there may exist one or more than one isomorphism from G onto G′.

IMPORTANT RESULTS

1. *If ϕ is a homomorphism of group G into group G′, then $\phi(e) = e'$.*
2. *If ϕ is a homomorphism of group G into group G′, then*
$$\phi(a^{-1}) = [\phi(a)]^{-1} \quad \forall\, a \in G.$$
3. *If ϕ is an isomorphism of a group G into group G′, then*
$$o(a) = o(\phi(a)) \quad \forall\, a \in G.$$
4. *Let $\phi : G \to G'$ be a homomorphism. The homomorphism ϕ is an isomorphism of G into G′ if and only if ker $\phi = \{e\}$.*

Example 4. *Let G be any group and a be any fixed element in G. Define a map $\phi : G \to G$ by the formula $\phi(x) = axa^{-1}$, $\forall\, x \in G$. Prove that ϕ is an isomorphism of G onto itself.*

Sol. For $x \in G$, $\phi(x) = axa^{-1}$ is a unique element of G.

$\therefore$ ϕ is well defined.

ϕ is a homomorphism. Let $x_1, x_2 \in G$.

$$\therefore \quad \phi(x_1x_2) = a(x_1x_2)\, a^{-1} = ax_1\, a^{-1}ax_2\, a^{-1}$$
$$= (a\, x_1\, a^{-1})\, (ax_2\, a^{-1}) = \phi(x_1)\, \phi(x_2)$$
$$\therefore \quad \phi(x_1x_2) = \phi(x_1)\, \phi(x_2)\ \forall\, x_1, x_2 \in G.$$

$\therefore$ ϕ is a homomorphism.

ϕ is one-one. Let $x_1, x_2 \in G$ and $\phi(x_1) = \phi(x_2)$.

$$\Rightarrow \quad ax_1a^{-1} = ax_2\, a^{-1} \quad \Rightarrow \quad a^{-1}(ax_1\, a^{-1})\, a = a^{-1}\,(ax_2\, a^{-1})a$$
$$\Rightarrow \quad (a^{-1}\, a)\, x_1\, (a^{-1}\, a) = (a^{-1}\, a)\, x_2\, (a^{-1}\, a)$$
$$\Rightarrow \quad ex_1\, e = e\, x_2\, e \quad \Rightarrow \quad x_1 = x_2.$$
$$\therefore \quad \phi(x_1) = \phi(x_2) \quad \Rightarrow \quad x_1 = x_2 \ \therefore \phi \text{ is one-one.}$$

ϕ is onto. Let $x \in G$. $\therefore$ $a^{-1}xa \in G$ and

$$\phi(a^{-1}x\, a) = a(a^{-1}\, x\, a)\, a^{-1}$$
$$= (aa^{-1})\, x\, (aa^{-1}) = ex\, e = x.$$

$\therefore$ For $x \in G$, $\exists\, a^{-1}\, xa \in G : \phi\, (a^{-1}\, xa) = x$

$\therefore$ ϕ is onto.

$\therefore$ ϕ is an isomorphism of G onto itself.

Example 5. *Let G be the additive group of real numbers and G′, the multiplicative group of positive real numbers. If $\phi : G \to G'$ is a mapping defined by $\phi(x) = e^x\ \forall\, x \in G$, then show that ϕ is an isomorphism of G onto G′.*

Sol. For $x \in G$, $\phi(x) = e^x$ is a unique positive real number.

$\therefore$ ϕ is well defined.

ϕ is a homomorphism. Let $x_1, x_2 \in G$.

$$\therefore \quad \phi\,(x_1 + x_2) = e^{x_1 + x_2} = e^{x_1}\, e^{x_2} = \phi(x_1)\, \phi(x_2)$$
$$\therefore \quad \phi(x_1 + x_2) = \phi(x_1)\, \phi(x_2) \quad \forall\, x_1, x_2 \in G$$

$\therefore$ ϕ is a homomorphism

ϕ is one-one. Let $x_1, x_2 \in$ G and $\phi(x_1) = \phi(x_2)$.

$\Rightarrow \quad e^{x_1} = e^{x_2} \quad \Rightarrow \log e^{x_1} = \log e^{x_2}$

$\Rightarrow \quad x_1 \log e = x_2 \log e \quad \Rightarrow x_1 = x_2.$

$\therefore \quad \phi(x_1) = \phi(x_2) \quad \Rightarrow x_1 = x_2 \quad \therefore \quad \phi$ is one-one.

ϕ is onto. Let $y \in$ G′

Let $\log y = x$. $\therefore$ $x \in \mathbf{R}$ and $y = e^x$.

$\therefore$ $\phi(x) = y$ and $x \in$ G. $\therefore$ ϕ is onto.

$\therefore$ $\phi :$ G $\to$ G′ is an isomorphism.

Example 6. *Show that the mapping $x \to x^{-1}$ of a group G onto itself is an isomorphism if and only if G is abelian, x being any arbitrary element of G.*

Sol. We have $\phi :$ G $\to$ G defined by $\phi(x) = x^{-1}$ for $x \in$ G.

ϕ is well defined because inverse of an element of a group is unique.

Let G be abelian.

For $x, y \in$ G, $\quad \phi(xy) = (xy)^{-1} = y^{-1}x^{-1} = x^{-1}y^{-1} = \phi(x)\,\phi(y)$

$\therefore \quad \phi(xy) = \phi(x)\,\phi(y) \quad \forall\, x, y \in$ G

$\therefore$ ϕ is a homomorphism.

Let $\phi(x) = \phi(y)$ for $x, y \in$ G.

$\Rightarrow \quad x^{-1} = y^{-1} \quad \Rightarrow (x^{-1})^{-1} = (y^{-1})^{-1} \quad \Rightarrow x = y.$

$\therefore$ ϕ is one-one

Let $z \in$ G. $\therefore$ $z^{-1} \in$ G and $\phi(z^{-1}) = (z^{-1})^{-1} = z$.

$\therefore$ ϕ is onto.

$\therefore$ $\phi :$ G $\to$ G is an isomorphism.

Conversely, let $\phi :$ G $\to$ G be an isomorphism.

$\therefore$ For $x, y \in$ G, $\quad \phi(xy) = \phi(x)\,\phi(y)$

$\Rightarrow \quad (xy)^{-1} = x^{-1}y^{-1} \quad \Rightarrow [(xy)^{-1}]^{-1} = (x^{-1}\,y^{-1})^{-1}$

$\Rightarrow \quad xy = (y^{-1})^{-1}\,(x^{-1})^{-1}$

$\Rightarrow \quad xy = yx \quad \Rightarrow$ G is abelian.

WORKING RULES TO SHOW TWO GIVEN GROUPS TO BE ISOMORPHIC

Rule I. *Designate the groups by G and G′. Also observe the binary operations of the groups.*

Rule II. *Construct a mapping ϕ from G into G′ keeping in mind that it is to be one-one, onto and a homomorphism.*

Rule III. *An isomorphism is to preserve identities, inverses and orders. These facts may be used while constructing the mapping ϕ.*

Rule IV. *Show that the mapping ϕ is well defined.*

Rule V. *Show that ϕ is one-one, onto and a homomorphism.*

EXERCISE 4

1. Let (G, $*$) and (G′, $*'$) be groups. Define a mapping $\phi : G \to G'$ by $\phi(x) = e'$ $\forall x \in G$, where e' is the identity element of the group G′. Show that ϕ is a homomorphism from G into G′.

2. Let G be the group of all complex number under usual addition. Define $\phi : G \to G$ by $\phi(z) = \bar{z}$ $\forall z \in G$. Show that ϕ is an isomorphism from G onto G.

3. Show that the mapping $f : \mathbf{C} \to \mathbf{R}$ such that $f(x + iy) = x$ is a homomorphism of the additive group of complex numbers onto additive group of real numbers. Also find ker f.

4. Let $\mathbf{C}_0$ and $\mathbf{R}_0$ be the multiplicative groups of non-zero complex numbers and of non-zero real numbers respectively. Define $\phi : \mathbf{C}_0 \to \mathbf{R}_0$ by $\phi(z) = |z|$ for $z \in \mathbf{C}_0$. Show that ϕ is a homomorphism of $\mathbf{C}_0$ into R_0 and ϕ is not onto. Also find ker ϕ.

5. **Let G be the multiplicative group of all $n \times n$ non-singular matrices with elements as real numbers and let G′ be the multiplicative group of all non-zero real numbers. Show that the mapping $\phi : G \to G'$ defined by $\phi(A) = \det A$ for $A \in G$ is a homomorphism of G onto G′. Also find ker ϕ.**

6. **If G be the multiplicative group of positive real numbers and G′, the additive group of real numbers. Show that the groups G and G′ are isomorphic.**

7. Show that the set G of all matrices of the type $\begin{bmatrix} a & 0 \\ 0 & 1 \end{bmatrix}$, where $a \in \mathbf{R}$, $a \neq 0$ form a multiplication group and is isomorphic to the group G′ of non-zero real numbers under multiplication.

8. If ϕ is an isomorphism from group G onto group G′, then show that ϕ^{-1} is an isomorphism from G′ onto G.

ANSWERS

3. ker $f = \{iy : y \in \mathbf{R}\}$

4. Pre-images of negative real numbers are not defined.

$$\ker \phi = \{z : |z| = 1, z \in \mathbf{C}_0\}.$$

5. ker ϕ is the set of all $n \times n$ non-singular matrices with elements as real numbers and whose determinant is 1.

HINTS

1. $\phi(x * y) = e' = e' *' e' = \phi(x) *' \phi(y)$.

5. If $k \in G'$, then $\phi(A) = k$, where $A = \begin{bmatrix} k & 0 & 0 & \ldots & 0 \\ 0 & 1 & 0 & \ldots & 0 \\ 0 & 0 & 1 & \ldots & 0 \\ \ldots & \ldots & \ldots & \ldots & \ldots \\ 0 & 0 & 0 & \ldots & 1 \end{bmatrix}$

6. Try $\phi : G \to G'$ defined by $\phi(x) = \log x$ $\forall x \in G$.

7. Define $\phi : G \to G'$ by $\phi\left(\begin{bmatrix} x & 0 \\ 0 & 1 \end{bmatrix}\right) = x$, $x\ (\neq 0) \in \mathbf{R}$.

8. For $x', y' \in G'$, $\exists\, x, y \in G : \phi(x) = x', \phi(y) = y'$

$\therefore$ $\phi^{-1}(x') = x, \phi^{-1}(y') = y$.

Now $\phi(x)\,\phi(y) = x'y'$.

$\Rightarrow$ $\phi(xy) = x'y' \Rightarrow \phi^{-1}(x'y') = xy = \phi^{-1}(x')\,\phi^{-1}(y')$.

11

Subgroups

1. COMPLEX

Let (G, .) be a group. A non-empty subset H of G is called a **complex** of the group (G, .).

2. SUBGROUP

Let (G, .) be a group A non-empty subset H of G is called a **subgroup** of the group (G, .), if under the binary operation '.' of G, the set H itself forms a group.

∴ By definition, every subgroup of G is a complex of G but a complex of G may not be a subgroup of G.

Since every set is a subset of itself, so if (G, .) is a group, then G is a subgroup of (G, .).

If 'e' is identity element of group G, then the set $\{e\}$ is also a subgroup of G.

The subgroups $\{e\}$ and G are called **trivial subgroups** of the group G.

∴ Every group has at least one subgroup.

Remark 1. In practice we generally write $a \,.\, b$ as ab. In other words, ab denotes the element of the group G when the binary operation '.' of G is applied on the ordered pair (a, b). If binary operation on G is $*$, then ab represents $a * b$.

Remark 2. If a non-empty set H of group G is a group under some other binary operation on H and not with the binary operation of G, then H cannot be said a subgroup of G.

Illustrations. (*i*) **Z**, **Q**, **R** are all subgroups of the group (**C**, +).

(*ii*) **Q**°, **R**° are subgroups of the group (**C**°, .).

Here **Q**°, **R**°, **C**° represents the set of non-zero rational numbers, non-zero real numbers and non-zero complex numbers respectively.

(*iii*) Let G = $\{1, -1, i, -i\}$ and H = $\{1, -1\}$.

Since G and H are both groups under usual multiplication, we can say that H is a subgroup of (G, .).

(*iv*) The set of even integers is a subgroup of the additive group of integers.

(*v*) The set of positive real numbers is a subgroup of multiplicative group of non-zero real numbers.

Theorem. *Let H be a subgroup of a group G. Then :*

(*i*) *The identity element of the group H is the same as that of the group G.*

(*ii*) *The inverse of any element of the group H is the same as the inverse of the element regarding as an element of G.*

(*iii*) *The order of any element of H is the same as the order of the element regarding as an element of the group G.*

Proof. (*i*) Let e and e' be the identity elements of the groups H and G respectively.

$\therefore$ For $a \in$ H, $ae = a = ea$ ...(1)

Also $a \in$ H implies $a \in$ G.

$\therefore$ $ae' = a = e'a$ ...(2)

(1) and (2) $\Rightarrow$ $ae = ae'$

By cancellation law $e = e'$.

(*ii*) Let e be the identity element of the subgroup H and the group G.

Let $a \in$ H. Let b be the inverse of a in H.

$\therefore$ $ab = e = ba$ ...(1)

Let c be the inverse of a in G.

$\therefore$ $ac = e = ca$ ...(2)

(1) and (2) $\Rightarrow$ $ab = ac$ $\Rightarrow$ $b = c$.

(*iii*) Let $a \in$ H. Let m and n be the order of a as an element of H and G respectively.

$\therefore$ $a^m = e$ and $a^n = e$

$o(a) = m$ $\Rightarrow$ $m \le n$

$o(a) = n$ $\Rightarrow$ $n \le m$

Combining, we get $m = n$.

3. TESTS FOR A COMPLEX TO BE A SUBGROUP

A complex of a group may or may not be a subgroup of the group. The direct method for testing a complex to be a subgroup is to verify all the properties of a group for the given complex w.r.t. the induced composition of the group. In the theorems to follow, we shall establish some results for testing a complex to be a subgroup.

Theorem 1. *Let G be a group with binary operation denoted multiplicatively. Let H be a non-empty subset (i.e., a complex) of G. H is a subgroup of G if and only if :*

(*i*) $ab \in H \quad \forall\, a, b \in H$ (*ii*) $a^{-1} \in H \quad \forall\, a \in H$.

Proof. Let H be a subgroup of G. Let $a, b \in$ H. Since the binary operation of G is also a binary operation on H, we have $ab \in$ H.

Let $a \in$ H. Since H is a group in itself, we have $a^{-1} \in$ H.

Conversely, let $ab \in$ H, $a^{-1} \in$ H whenever $a, b \in$ H.

Since $ab \in$ H $\forall\, a, b \in$ H, the binary operation of G is also a binary operation on H. Alternatively, we say that H is closed w.r.t. binary operation of G.

Associativity. Let $a, b, c \in$ H.

$\therefore$ $a, b, c \in$ G. $\therefore$ $a(bc) = (ab)c$.

Existence of identity element. Let $a \in$ H.

$\therefore$ By second condition, $a^{-1} \in$ H. $\therefore$ By first condition, $aa^{-1} \in$ H *i.e.*, $e \in$ H

Let $b \in$ H. $\therefore$ $b \in$ G $\therefore$ $be = b = eb$

$\therefore$ e is an identity element of H.

Existence of inverse. Let $a \in$ H. $\therefore$ By second condition, $a^{-1} \in$ H.

$\therefore$ H is a group w.r.t. binary operation of G.

$\therefore$ H is a subgroup of G.

Theorem 2. *Let G be a group with binary operation denoted multiplicatively. Let H be a non-empty subset (i.e., a complex) of G. H is a subgroup of G if and only if $ab^{-1} \in H \quad \forall\, a, b \in H$.*

Proof. Let H be a subgroup of G. Let $a, b \in$ H. Since H is a group in itself, we have $b^{-1} \in$ H. Also, H is closed w.r.t. operation of G, we have $ab^{-1} \in$ H.

Conversely, let $ab^{-1} \in$ H whenever $a, b \in$ H.

Associativity. Let $a, b, c \in$ H.

$\therefore\quad a, b, c \in$ G. $\quad\therefore\quad a(bc) = a(bc)$.

Existence of identity element. Let $a \in$ H.

$\therefore$ By the given condition, $aa^{-1} \in$ H *i.e.,* $e \in$ H.

Let $b \in$ H. $\therefore b \in$ G

$\therefore\quad be = b = eb$

$\therefore\quad e$ is the identity element of H.

Existence of inverse. Let $a \in$ H.

$\therefore$ By the given condition, $ea^{-1} \in$ H *i.e.,* $a^{-1} \in$ H.

Closure property. Let $a, b \in$ H.

$\therefore$ Inverse of b *i.e.,* $b^{-1} \in$ H.

$\therefore$ By the given condition, $a(b^{-1})^{-1} \in$ H *i.e.,* $ab \in$ H.

$\therefore$ H is closed w.r.t. operation of G. $\quad\therefore$ H is a group w.r.t. operation of G.

$\therefore$ H is a subgroup of G.

Remark. If the binary operation of G is denoted additively then the statement of the above theorem would take the following form :

Let G be a group with binary operation denoted additively. Let H be a non-empty subset (*i.e., a* complex) of G. H is a subgroup of G if and only if

$$a - b \in H \quad \forall\, a, b \in H.$$

Theorem 3. *Let G be a group with binary operation denoted multiplicatively. Let H be a non-empty finite subset (i.e., a finite complex) of G. H is a subgroup of G if and only if $ab \in H$ $\forall\, a, b \in H$.*

Proof. Let H be a subgroup of G. Let $a, b \in$ H. Since the binary operation of G is also a binary operation on H, we have $ab \in$ H.

Conversely, let $ab \in$ H whenever $a, b \in$ H.

$\therefore$ The binary operation of G is also a binary operation on H.

Associativity. Let $a, b, c \in$ H.

$\therefore\quad a, b, c \in$ G. $\quad\therefore\quad a(bc) = (ab)c$.

Existence of identity element. Let $a \in$ H.

$\therefore\quad a^2 = aa \in$ H, $a^3 = aa^2 \in$ H, $a^4 = aa^3 \in$ H, ...

$\therefore\quad a, a^2, a^3, a^4,$...... all lie in H.

Since H is finite, there must be repetitions in the above elements.

Let $a^l = a^m$ for some integers l, m, where $l > m$.

$$\Rightarrow \qquad a^l\, a^{-m} = a^m\, a^{-m} \;\Rightarrow\; a^{l-m} = a^0 \;\Rightarrow\; a^{l-m} = e$$

$$\therefore \qquad e \in H$$

Let $b \in$ H. $\therefore b \in$ G

$\therefore$ $be = b = eb$

$\therefore$ e is the identity element of H.

Existence of inverse. We have $a^{l-m} = e$, where $l > m$.

$\therefore$ $a^{l-m} = e$, where $l - m \geq 1$

$\Rightarrow$ $a\, a^{l-m-1} = e$, where $l - m - 1 \geq 0$

Also $a^{l-m-1}\, a = a^{l-m-1+1} = a^{l-m} = e$

$\therefore$ $aa^{l-m-1} = e = a^{l-m-1}\, a$

$\therefore$ a^{l-m-1} is the inverse of a.

$\therefore$ H is a group w.r.t. binary operation of G.

$\therefore$ H is a subgroup of G.

Remark 1. The set H = {1, – 1} is closed w.r.t. multiplication and is a finite subset of **R**.

$\therefore$ H is a subgroup of the group (**R**, .).

Remark 2. In the above theorem, the subset H must be a finite non-empty subset of G, otherwise the result may not hold good. For example, (**Z**, +) is a group and **N** is a non-empty subset of **Z** and is also closed under addition. Here **N** is not a subgroup of (**Z**, +) because **N** is not finite.

Theorem 4. *If ϕ is a homomorphism of a group (G, $*$) into a group (G′, $*'$). Prove that ϕ(G) is a subgroup of the group (G′, $*'$).*

Proof. We have

$$\phi(\text{G}) = \{\phi(g) : g \in \text{G}\}$$

Let e and e' be the identity elements of the groups G and G′ respectively

$\therefore$ $\phi(e) = e' \Rightarrow e' \in \phi(\text{G})$

Let $\phi(a), \phi(b) \in \phi(\text{G})$.

$\therefore$ $\phi(a) *' (\phi(b))^{-1} = \phi(a) *' \phi(b^{-1}) = \phi(a * b^{-1}) \in \phi(\text{G})$.

$(\because a, b \in \text{G} \Rightarrow a * b^{-1} \in \text{G})$

$\therefore$ ϕ(G) is a subgroup of the group (G′, $*'$).

Example 1. *Let G be a group with binary operation denoted multiplicatively. For $a \in$ G, let H = $\{a^n : n \in \mathbf{Z}\}$. Show that H is a subgroup of G.*

Sol. We have H = $\{a^n : n \in \mathbf{Z}\}$.

H $\neq \phi$, because $a^1 = a \in$ H.

Let $\alpha, \beta \in$ H.

$\therefore$ $\alpha = a^m$, $\beta = a^n$ for some $m, n \in \mathbf{Z}$.

$\therefore$ $\alpha\beta^{-1} = a^m (a^n)^{-1} = a^m a^{-n} = a^{m-n} \in$ H because $m - n \in \mathbf{Z}$.

$\therefore$ $\alpha\beta^{-1} \in$ H $\forall\, \alpha, \beta \in$ H. $\therefore$ H is a subgroup of G.

Example 2. *Let G be a group with binary operation denoted multiplicatively. The set $\{h \in$ G $: hx = xh \ \ \forall\, x \in$ G$\}$ is called the* **centre** *of the group G. Show that the centre of G is a subgroup of G.*

Sol. Let H = $\{h \in \text{G} : hx = xh \ \forall\, x \in \text{G}\}$

Let $h, h' \in$ H.

$\therefore$ $hx = xh \quad \forall\, x \in$ G ...(1)

$h'x = xh' \quad \forall\, x \in$ G ...(2)

Let $y \in G$. $\therefore$ $(hh')y = h(h'y) = h(yh')$ (Using (2))

$= (hy)h' = (yh)h' = y(hh')$. (Using (1))

$\therefore$ $(hh')y = y(hh') \quad \forall\, y \in G$

$\therefore$ $hh' \in H$

Let $y \in G$. $\therefore$ $y^{-1} \in G$

$\therefore$ (1) $\Rightarrow$ $hy^{-1} = y^{-1}h \Rightarrow (hy^{-1})^{-1} = (y^{-1}h)^{-1}$

$\Rightarrow$ $(y^{-1})^{-1} h^{-1} = h^{-1} (y^{-1})^{-1} \Rightarrow yh^{-1} = h^{-1}y \Rightarrow h^{-1}y = yh^{-1}$

$\therefore$ $h^{-1} \in H$

$\therefore$ $hh', h^{-1} \in H$ whenever $h, h' \in H$.

$\therefore$ H is a subgroup of G.

Example 3. *Let G be a group with binary operation denoted multiplicatively. Let H be a subgroup of G. For $x \in G$, define $xHx^{-1} = \{xhx^{-1} : h \in H\}$. Show that xHx^{-1} is a subgroup of G.*

Sol. We have $xHx^{-1} = \{xhx^{-1} : h \in H\}$.

Let $xhx^{-1}, xh'x^{-1} \in xHx^{-1}$.

Now $(xhx^{-1})(xh'x^{-1}) = xh(x^{-1}x)h'x^{-1}$

$= xheh'x^{-1} = x(hh')\,x^{-1} \in xHx^{-1}$ $(\because\ hh' \in H)$

Also $(xhx^{-1})^{-1} = (x^{-1})^{-1}\, h^{-1}\, x^{-1} = xh^{-1}\, x^{-1} \in xHx^{-1}$ $(\because\ h^{-1} \in H)$

$\therefore$ $(xhx^{-1})(xh'x^{-1}),\ (xhx^{-1})^{-1} \in xHx^{-1}$

$\therefore$ xHx^{-1} is a subgroup of G.

4. PRODUCT AND INVERSE OF COMPLEXES

Let H and K be any two complexes of a group G with binary operation denoted multiplicatively.

The product HK of complexes H and K is defined as

$$HK = \{hk : h \in H, k \in k\}.$$

Since H and K are non-empty, HK is also non-empty.

Also, $hk \in G$ for every $h \in H$, $k \in K$, because G is a group.

$\therefore$ $HK \subseteq G$.

$\therefore$ HK is a complex of group G.

For the complex H, we define the inverse (H^{-1}) of H as

$$H^{-1} = \{h^{-1} : h \in H\}.$$

Since H is non-empty, H^{-1} is also non-empty.

Also $h^{-1} \in G \quad \forall\, h \in H$, because G is a group.

$\therefore$ H^{-1} is also a complex of group G.

Theorem 1. *Let H and K be two subgroups of a group G with binary operation denoted multiplicatively. The complex HK of the group G is a subgroup of G if and only if HK = KH.*

Proof. H and K are subgroups of G.

Let the complex HK of G be a subgroup of G.

Let $x \in HK$. $\therefore$ $x^{-1} \in HK$

Let $x^{-1} = hk$, where $h \in H$, $k \in K$.

Now $x = (x^{-1})^{-1} = (hk)^{-1} = k^{-1}\, h^{-1} \in KH.$

$\therefore$ $HK \subseteq KH$

Now, let $kh \in$ KH, where $k \in$ K and $h \in$ H.

$\Rightarrow \quad k^{-1} \in K, h^{-1} \in H \quad \Rightarrow \quad h^{-1}k^{-1} \in HK \quad \Rightarrow \quad (kh)^{-1} \in HK$

$\Rightarrow \quad [(kh)^{-1}]^{-1} \in HK \quad \Rightarrow \quad kh \in HK.$

$\therefore \quad KH \subseteq HK$

Combining, we get $\quad HK = KH.$

Conversely, let $\quad HK = KH.$

Let $hk, h'k' \in HK$.

$\therefore \quad (hk)(h'k') = h(kh')k' = h(h''\,k'')\,k' = (hh'')(k''k') \in HK.$

$(kh' \in KH \Rightarrow kh' \in HK \Rightarrow kh' = h''k''$ for $h'' \in H, k'' \in K)$

$\therefore$ The binary operation of G is also a binary operation on HK.

Associativity. Let $a, b, c \in$ HK.

$\therefore \quad a, b, c \in G. \quad \therefore \quad a(bc) = (ab)c.$

Existence of identity element. $e \in$ H and $e \in$ K.

$\therefore \quad ee \in HK.$ *i.e.*, $e \in HK$

Let $a \in$ HK. $\quad \therefore \quad a \in G$

$\therefore \quad ae = a = ea$

$\therefore$ e is the identity element of HK.

Existence of inverse. Let $hk \in$ HK, where $h \in$ H and $k \in$ K.

$\therefore \quad (hk)^{-1} = k^{-1}h^{-1} \in KH$

$\Rightarrow \quad (hk)^{-1} \in HK. \qquad (\because \quad HK = KH)$

$\therefore$ Inverse of elements exists in HK.

$\therefore$ HK is a subgroup of G.

Remark 1. If G is an abelian group and H, K are subgroups of G, then the complex HK of G is always a subgroup of G. This happens because HK and KH are equal sets when G is abelian.

Remark 2. If the binary operation of group G is denoted by '$*$' then the statement of the above theorem takes the following form :

'*Let H and K be two subgroups of a group (G, $*$). The complex H $*$ K of the group (G, $*$) is a subgroup of (G, $*$) if and only if H $*$ K = K $*$ H.*

Theorem 2. *Let H be a subgroup of a group G with binary operation denoted multiplicatively. Prove that H^{-1} = H. Also show that the converse is not true.*

Proof. Let $\quad h \in H$

$\Rightarrow \quad h^{-1} \in H \qquad (\because \quad$ H is a subgroup)

$\Rightarrow \quad (h^{-1})^{-1} \in H^{-1}$ *i.e.*, $h \in H^{-1}$

$\therefore \quad H \subseteq H^{-1}$

Let $\quad h^{-1} \in H^{-1} \quad \therefore \quad h \in H$

$\Rightarrow \quad h^{-1} \in H \qquad (\because \quad$ H is a subgroup)

$\therefore \quad H^{-1} \subseteq H$

Combining, we get $\quad H^{-1} = H.$

Let G = {± 1, ± i}. With usual multiplication as binary operation, G is a group. Let H = {– 1}.

$\therefore$ H^{-1} = H because $(-1)^{-1} = -1$.

Here H is not a subgroup of G, because identity element is not in H.

$\therefore$ The converse part of the theorem is not true.

5. ORDER OF PRODUCT OF TWO FINITE SUBGROUPS

Theorem. *If H and K are finite subgroups of a group G, then*

$$o(HK) = \frac{o(H)\, o(K)}{o(H \cap K)}.$$

Proof. Let the binary operation of the group G be denoted multiplicatively.

$\therefore$ $\quad HK = \{hk : h \in H, k \in K\}$

Case I. $H \cap K = \{e\}$. Let $h_1k_1, h_2k_2 \in HK$ and $h_1 \neq h_2$.

Let $\quad h_1k_1 = h_2k_2$

$\Rightarrow \quad h_2^{-1}(h_1k_1)k_1^{-1} = h_2^{-1}(h_2k_2)k_1^{-1} \quad \Rightarrow \quad h_2^{-1}h_1 = k_2k_1^{-1}$ ...(1)

Now $h_2^{-1}h_1 \in H$ and $k_2k_1^{-1} \in K$

$\therefore$ (1) $\Rightarrow h_2^{-1}h_1 \in H \cap K \Rightarrow h_2^{-1}h_1 = e \Rightarrow h_1 = h_2$, which is absurd.

$\therefore \quad h_1k_1 \neq h_2k_2$

$\therefore \quad h_1 \neq h_2 \quad \Rightarrow \quad h_1k_1 \neq h_2k_2$

Similarly, $\quad k_1 \neq k_2 \quad \Rightarrow \quad h_1k_1 \neq h_2k_2$

In other words, $\quad h_1k_1 = h_2k_2 \Rightarrow h_1 = h_2, k_1 = k_2$.

$\therefore$ The elements $hk : h \in H, k \in K$ are all distinct and there is no collapsing.

$\therefore \quad o(HK) = o(H)\, o(K)$

$\therefore \quad o(HK) = \dfrac{o(H)\, o(K)}{o(H \cap K)}.$ $\quad (\because \; o(H \cap K) = 1)$

Case II. $H \cap K \neq \{e\}$. Let $h \in H, k \in K$. $\therefore$ $hk \in HK$.

Let $p \in H \cap K$. $\therefore$ $p \in H$ and $p \in K$ and

$hk = h(pp^{-1})k = (hp)(p^{-1}k)$, where $hp \in H$ and $p^{-1}k \in K$.

$\therefore$ hk is duplicated at least $o(H \cap K)$ times in the list of HK.

Now, let $h'k'$ be any duplication of hk in the list of HK.

$\Rightarrow \quad hk = h'k' \quad \Rightarrow \quad h^{-1}(hk)\,k'^{-1} = h^{-1}(h'k')\,k'^{-1}$

$\Rightarrow \quad kk'^{-1} = h^{-1}h' = q$ (say).

$\therefore$ $q \in H$ and $q \in K$ and thus $q \in H \cap K$

Also $\quad h^{-1}h' = q \quad \Rightarrow \quad h' = hq$ and

$\quad kk'^{-1} = q \quad \Rightarrow \quad k = qk' \quad \Rightarrow \quad k' = q^{-1}k$

$\therefore \quad h'k' = (hq)(q^{-1}k)$, where $q \in H \cap K$.

$\therefore$ The duplication $h'k'$ is one among those already accounted for.

$\therefore$ hk is duplicated exactly $o(H \cap K)$ times in the list of HK.

$\therefore \quad o(HK) \times o(H \cap K) = o(H)\, o(K)$

$\Rightarrow \quad o(HK) = \dfrac{o(H)\, o(K)}{o(H \cap K)}.$

$\therefore$ In both cases, we have

$$\mathbf{o(HK) = \frac{o(H)\, o(K)}{o(H \cap K)}}.$$

6. CONSTRUCTION OF NEW SUBGROUPS

With the help of given subgroups of a group, we can construct more subgroups.

Theorem. *If H_1 and H_2 are two subgroups of a group G, then $H_1 \cap H_2$ is also a subgroup of G.*

Proof. $e \in H_1, e \in H_2 \Rightarrow e \in H_1 \cap H_2$

$\therefore$ $H_1 \cap H_2$ is a non-empty subset of G.

Let $a, b \in H_1 \cap H_2$.

$\therefore$ $a, b \in H_1$, and $a, b \in H_2$

$\Rightarrow ab \in H_1, ab \in H_2 \Rightarrow ab \in H_1 \cap H_2$

$\therefore$ $H_1 \cap H_2$ is closed w.r.t. the binary operation of G.

Associativity. Let $a, b, c \in H_1 \cap H_2$.

$\therefore$ $a, b, c \in G$ $\therefore$ $a(bc) = (ab)c$.

Existence of identity element. We have seen that $e \in H_1 \cap H_2$.

Let $a \in H_1 \cap H_2$. $\therefore$ $a \in G$

$\therefore$ $ae = a = ea$

$\therefore$ e is identity element of $H_1 \cap H_2$.

Existence of inverse. Let $a \in H_1 \cap H_2$.

$\Rightarrow a \in H_1, a \in H_2 \Rightarrow a^{-1} \in H_1, a^{-1} \in H_2 \Rightarrow a^{-1} \in H_1 \cap H_2$.

$\therefore$ Every element of $H_1 \cap H_2$ has its inverse in $H_1 \cap H_2$.

$\therefore$ $H_1 \cap H_2$ is a group w.r.t. the binary operation on G.

$\therefore$ $H_1 \cap H_2$ is a subgroup of G.

Remark 1. It can also be proved that the intersection of any number of subgroups of a group is also a subgroup of the group under consideration.

Remark 2. The union of two subgroups of a group may not be a subgroup of the group under consideration.

For example, let G = **Z**. $\therefore$ (G, +) is a group.

Let $H_1 = \{2n : n \in \mathbf{Z}\}$ and $H_2 = \{5n : n \in \mathbf{Z}\}$.

$\therefore$ H_1 and H_2 are both subgroups of G.

Now $H_1 \cup H_2 = \{0, \pm 2, \pm 4, \pm 5, \pm 6, \pm 8, \pm 10, ...\}$

Here $2 \in H_1 \cup H_2, 5 \in H_1 \cup H_2$

but $2 + 5 = 7 \notin H_1 \cup H_2$.

$\therefore$ $H_1 \cup H_2$ is not closed under usual addition.

$\therefore$ $H_1 \cup H_2$ is not a subgroup of G.

IMPORTANT RESULTS

1. *If H is a subgroup of a group G, then*

   (i) the identity element of H and G are same.

   (ii) the inverse of an element of H w.r.t. H and G are same.

   (iii) the order of an element of H w.r.t. H and G are same.

2. *Let G be a group with binary operation denoted multiplicatively. A complex H of G is a subgroup of G if*

   either $ab \in H, a^{-1} \in H \quad \forall a, b \in H$

or $ab^{-1} \in H \quad \forall\, a, b \in H$

or $ab \in H \quad \forall\, a, b \in H$ *provided H is finite.*

3. *Let H and K be subgroups of a group G with binary operation denoted multiplicatively. The complex HK of G is a subgroup of G if and only if HK = KH.*
4. *Let H be a subgroup of a group G with binary operation denoted multiplicatively. Then $H^{-1} = H$. The converse of this is not true.*
5. *Intersection of two subgroups of a group is also a subgroup of the group.*

EXERCISE 1

1. Let (G, +) be a group. Let H be a non-empty subset of G. Prove that :
   (*i*) if H is a subgroup of G, then $a + b, -b \in H \quad \forall\, a, b \in H$.
   (*ii*) if $a + b, -b \in H \quad \forall\, a, b \in H$, then H is a subgroup of G.
2. Let (G, $*$) be a group. Let H be a non-empty subset of G. Prove that :
   (*i*) if H is a subgroup of G, then for $a, b \in H$, we have $a * c \in H$, where c is the inverse of b.
   (*ii*) if for $a, b \in H$, we have $a * c \in H$, where c is the inverse of b, then H is a subgroup of G.
3. The non-empty subset $\{1, 3, 3^2, 3^3, ...\}$ is not a subgroup of the group $\{......, 3^{-3}, 3^{-2}, 3^{-1}, 1, 3, 3^2, 3^3, ...\}$ under usual multiplication. Comment.
4. Show that the set of all elements 'a' of an abelian group G for which $a^2 = e$, is a subgroup of G.
5. Let G be a group. For $g \in G$, let $H(g) = \{a \in G : ag = ga\}$. Show that H($g$) is a subgroup of G.
6. Let G be the additive group of all real numbers and H be the multiplicative group of all positive real numbers. Is H a subgroup of G ?
7. Let G be the additive group of integers. Prove that the set of all multiples of integers by a fixed integer k is a subgroup of G.

ANSWERS

3. The inverses of the elements in the given subset does not exist in given subset.
6. No.

HINTS

4. Let $H = \{a \in G : a^2 = e\}$.
   Let $\quad a, b \in H$.
   $\therefore \quad (ab)^2 = (ab)(ab) = (ab)(ba) = a(b^2)\, a = aea = a^2 = e$ and $(a^{-1})^2 = a^{-2} = (a^2)^{-1} = e^{-1} = e$.
6. H is a complex of G. For H to be a subgroup of G, it must be a group with respect to the binary operation of G.

7. COSETS

Let G be a group with binary operation denoted multiplicatively. Let H be a subgroup of G.

For $a \in G$, we define $Ha = \{ha : h \in H\}$.

Ha is called a **right coset** of H in G generated by a.

Similarly, $a\mathrm{H} = \{ah : h \in H\}$ is called a **left coset** of H in G, generated by a.

H$a \neq \phi$ because $a = ea \in Ha$. Also $ha \in Ha$ implies $ha \in G$.

$\therefore$ Ha is a complex of G. Similarly, aH is a complex of G.

If G is an abelian group, then we have $Ha = aH$, because $ha = ah \ \forall h \in H$.

H itself is also a right (and left) coset of G because $He = H = eH$.

Remark 1. If the binary operation of G is denoted additively, then right coset of H in G generated by a is given by $H + a = \{h + a : h \in H\}$.

Also $a + H = \{a + h : h \in H\}$.

Example 1. *Let f be a homomorphism of a group G into a group G′ with kernel K. Let a be a given element of G such that f(a) = a′ ∈ G′. Then prove that the set of all those elements of G which have the image a′ in G′ is the coset Ka of K in G.*

Sol. We have $f : G \to G'$ and $f(a) = a'$.

Let $B = \{b \in G : f(b) = a'\}$. We shall show that $B = Ka$.

Let $ka \in Ka$

$\therefore \quad f(ka) = f(k)\, f(a) = e'\, f(a) = f(a) = a'$

$\therefore \quad ka \in B \quad \therefore \quad Ka \subseteq B$

Let $b \in B \quad \therefore \quad f(b) = a'$

$\therefore \quad f(a) = f(b) \qquad (\because f(a) = a')$

$\Rightarrow \quad f(a)(f(a))^{-1} = f(b)(f(a))^{-1}$

$\Rightarrow \quad e' = f(b)\, f(a^{-1}) \Rightarrow e' = f(ba^{-1}) \Rightarrow ba^{-1} \in K$

Let $ba^{-1} = k$ for some $k \in K$

$\Rightarrow \quad (ba^{-1})a = ka \Rightarrow b(a^{-1}a) = ka$

$\Rightarrow \quad be = ka \Rightarrow b = ka$

$\Rightarrow \quad b \in Ka \qquad (\because k \in K)$

$\therefore \quad B \subseteq Ka$

$\therefore \quad B = Ka.$

$\therefore$ The result holds.

Theorem 1. *Let H be a subgroup of a group G. Any two right (or left) cosets of H in G are either equal or disjoint.*

Proof. Let the binary operation of G be denoted multiplicatively.

Let Ha and Hb be any two right cosets of H in G, where $a, b \in G$.

If $Ha \cap Hb = \phi$, we have nothing to prove. Let $Ha \cap Hb \neq \phi$ and $c \in Ha \cap Hb$.

$\therefore \quad c = h_1a$ and $c = h_2b$ for some $h_1, h_2 \in H$.

$\Rightarrow \quad h_1a = h_2b \quad \Rightarrow \quad h_1^{-1}(h_1a)\, b^{-1} = h_1^{-1}(h_2\, b)\, b^{-1}$

$\Rightarrow \quad (h_1^{-1} h_1)ab^{-1} = (h_1^{-1} h_2)(bb^{-1})$

$\Rightarrow \quad ab^{-1} = h_1^{-1} h_2 \quad \Rightarrow \quad ab^{-1} \in H.$

Let $ha \in Ha$.

Now $ha = (ha)e = (ha)(b^{-1}b) = (hab^{-1})\, b = h'\, b \in Hb \qquad (h' = hab^{-1} \in H)$

$\therefore \quad Ha \subseteq Hb$

Let $hb \in Hb$.

Now $hb = (hb)e = (hb)(a^{-1}a) = (hba^{-1})a = (h(ab^{-1})^{-1})a = h''a \in Ha$
$(h'' = h(ab^{-1})^{-1} \in H)$

$\therefore \quad Hb \subseteq Ha$

Combining, we get $Ha = Hb$.

$\therefore$ If two right cosets of H in G are not disjoint then they must be equal.

Similarly, we can prove that any two left cosets of H in G are either equal or disjoint.

Theorem 2. *Let H be a subgroup of a group G. There is one-one correspondence between any two right cosets of H in G.*

Proof. Let the binary operation of G be denoted multiplicatively. Let Ha and Hb be any two cosets of H in G, where $a, b \in$ G.

Define $\phi : Ha \to Hb$ by $\phi(ha) = hb \quad \forall\, h \in H$.

ϕ is well defined. Let $h_1a, h_2a \in Ha$.

$$h_1a = h_2a \quad \Rightarrow \quad h_1 = h_2 \quad \Rightarrow \quad h_1b = h_2b \quad \Rightarrow \quad \phi(h_1a) = \phi(h_2a)$$

$\therefore$ ϕ is well defined.

ϕ is one-one. Let $h_1a, h_2a \in Ha$.

$$\phi(h_1a) = \phi(h_2a) \quad \Rightarrow \quad h_1b = h_2b \quad \Rightarrow \quad h_1 = h_2 \quad \Rightarrow \quad h_1a = h_2a$$

$\therefore$ ϕ is one-one

ϕ is onto. Let $hb \in Hb$.

$\Rightarrow \qquad h \in H \quad \Rightarrow \quad ha \in Ha$

Also $\qquad \phi(ha) = hb$

$\therefore$ ϕ is onto.

$\therefore$ There is one-one correspondence between cosets Ha and Hb of H in G.

Corollary. Let H be a subgroup of a group G. There is one-one correspondence between any two left cosets of H in G.

Proof. Here define $\phi : aH \to bH$ by $\phi(ah) = bh \quad \forall\, ah \in aH$ and show that ϕ is well defined, one-one and onto.

Remark. Since He is equal to H, the subgroup H is also a right coset of H in G.

$\therefore$ There is one-one correspondence between the elements of H and any arbitrary but fixed right coset of H in G.

$\therefore$ Every right coset of H in G has same number of elements as in H.

$\therefore$ If H is finite, then $o(H) = o(Ha) \quad \forall\, a \in G$

In case H is infinite, then no right coset of H in G can be finite.

$\therefore$ If H is infinite, then all right cosets of H in G are infinite.

Theorem 3. *Let H be a subgroup of a group G. The union of all right cosets of H in G is equal to G.*

Proof. Let the binary operation of the group G be denoted multiplicatively. We shall prove that $\bigcup_{a \in G} Ha = G$.

Let $b \in G$. $\therefore \quad hb \in G \quad \forall\, h \in H$

$\Rightarrow \qquad Hb \subseteq G$

$\therefore \quad \bigcup_{a \in G} Ha \subseteq G$

Also, $\qquad c \in G \quad \Rightarrow \quad c = ec \in Hc \quad \Rightarrow \quad c \in \bigcup_{a \in G} Ha$

Combining, we get $\bigcup_{a \in G} Ha = G$.

Corollary. Let H be a subgroup of a group G. The union of all left cosets of H in G is equal to G.

Theorem 4. *Let H be a subgroup of a group G. The right cosets of H in G partitions the group G.*

Proof. Let the binary operation of the group G be denoted multiplicatively. We know that :

(*i*) No right coset of H in G is empty.

(*ii*) Any two right cosets of H in G are either equal or disjoint.

(*iii*) Union of all right cosets of H in G is equal to G.

$\therefore$ The set of all right cosets of H in G forms a partition of the group G.

Corollary. Let H be a subgroup of a group G. The left cosets of H in G partitions the group G.

8. INDEX OF A SUBGROUP IN A GROUP

Let H be a subgroup of a group G. The number of distinct right cosets of H in G is called the **index** of the subgroup H in the group G and it is denoted by [G : H].

9. RELATION OF CONGRUENCE MODULO A SUBGROUP

Definition. Let H be a subgroup of a group G with binary operation denoted multiplicatively. We define a relation R on G as follows :

For $a, b \in$ G, we say aRb if and only if $ab^{-1} \in$ H and write aRb as $a \equiv b$ (mod H) and read it as a is congruent to b modulo H.

Theorem. *Let H be a subgroup of a group G with binary operation denoted multiplicatively. The relation of 'congruence modulo H' is an equivalence relation on G and the right cosets of H in G partitions the group G.*

Proof. For $a, b \in$ G, let the relation 'congruence modulo H' be denoted by R.

$\therefore$ For $a, b \in$ G, we have a R b if and only if $ab^{-1} \in$ H and we write

$$a \text{ R } b \text{ as } a \equiv b \pmod{\text{H}}.$$

R is reflexive. Let $a \in$ G.

$\therefore$ $aa^{-1} = e \in$ H

$\therefore$ $a \equiv a$ (mod H) *i.e.*, a R a.

$\therefore$ R is reflexive.

R is symmetric. Let $a, b \in$ G and a R b.

$\therefore$ $a \equiv b$ (mod H) *i.e.*, $ab^{-1} \in$ H

$\Rightarrow$ $(ab^{-1})^{-1} \in$ H $\Rightarrow$ $ba^{-1} \in$ H $\Rightarrow$ $b \equiv a$ (mod H) *i.e.*, b R a.

$\therefore$ R is symmetric.

R is transitive. Let $a, b, c \in$ G and a R b, b R c.

$\therefore$ $a \equiv b$ (mod H) and $b \equiv c$ (mod H)

$\Rightarrow$ $ab^{-1} \in$ H and $bc^{-1} \in$ H $\Rightarrow$ $(ab^{-1})(bc^{-1}) \in$ H $\Rightarrow$ $ac^{-1} \in$ H

$\Rightarrow$ $a \equiv c$ (mod H) *i.e.*, aRc

$\therefore$ R is transitive.

$\therefore$ R is an equivalence relation.

$\therefore$ The relation of congruence modulo H partitions G into mutually disjoint equivalence classes.

Let $a \in$ G. The equivalence class of a is the set $\{x \in$ G : a R $x\}$.

We denote this set as $[a]$.

For $a \in$ G, the set Ha is a right coset of H in G generated by a.

We claim that $[a]$ = Ha.

Let $b \in [a]$.

$\therefore$ a R b *i.e.*, $a \equiv b$ (mod H)

$\Rightarrow$ $ab^{-1} \in$ H $\Rightarrow$ $(ab^{-1})^{-1} \in$ H $\Rightarrow$ $ba^{-1} \in$ H

$\Rightarrow$ $ba^{-1} = h$ for some $h \in$ H $\Rightarrow$ $b = ha$ $\Rightarrow$ $b \in$ Ha

$\therefore$ $[a] \subseteq$ Ha.

Let $c \in$ Ha.

$\therefore$ $c = h'a$ for some $h' \in$ H

$\Rightarrow \quad ca^{-1} = h' \quad \Rightarrow \quad ca^{-1} \in H \quad \Rightarrow \quad c \equiv a \pmod{H}$

$\Rightarrow \quad c\,R\,a \quad \Rightarrow \quad c \in [a]$

$\therefore \quad Ha \subseteq [a]$

Combining, we get $[a] = Ha$.

$\therefore$ The set of right cosets of H in G partitions the group G.

Remark. We have also proved the above result without using the relation "congruence modulo a subgroup."

Example 2. *Let $G = \{\pm 1, \pm i\}$ be a group w.r.t. usual multiplication. The set $H = \{1, -1\}$ is a subgroup of G. Find all right cosets of H in G.*

Sol. We have $G = \{\pm 1, \pm i\}$ and $H = \{1, -1\}$.

$\therefore \quad H1 = \{1(1), -1(1)\} = \{1, -1\} = \mathbf{H}$

$H(-1) = \{1(-1), -1(-1)\} = \{-1, 1\} = \mathbf{H}$

$Hi = \{1(i), -1(i)\} = \{\mathbf{i, -i}\}$

and $\quad H(-i) = \{1(-i), -1(-i)\} = \{\mathbf{-i, i}\}$.

Example 3. *Let G be the group of all integers under usual addition and $H = \{5n : n \in \mathbf{Z}\}$. Find the right cosets of H in G generated by 0, 1, 2, 3, 4. Also verify that these cosets are mutually disjoint and their union is G.*

Sol. We have $\quad G = \{......, -2, -1, 0, 1, 2, ...\}$

and $\quad H = \{......, -10, -5, 0, 5, 10, ...\}$

$\therefore$ Right cosets of H in G generated by 0, 1, 2, 3, 4 are H + 0, H + 1, H + 2, H + 3, H + 4 respectively.

$H + 0 = \{...... , -10 + 0, -5 + 0, 0 + 0, 5 + 0, 10 + 0, ...\}$

$= \{......, -10, -5, 0, 5, 10, ...\}$

$H + 1 = \{......, -10 + 1, -5 + 1, 0 + 1, 5 + 1, 10 + 1, ...\}$

$= \{......, -9, -4, 1, 6, 11, ...\}$

$H + 2 = \{......, -10 + 2, -5 + 2, 0 + 2, 5 + 2, 10 + 2, ...\}$

$= \{......, -8, -3, 2, 7, 12, ...\}$

$H + 3 = \{......, -10 + 3, -5 + 3, 0 + 3, 5 + 3, 10 + 3, ...\}$

$= \{......, -7, -2, 3, 8, 13, ...\}$

$H + 4 = \{......, -10 + 4, -5 + 4, 0 + 4, 5 + 4, 10 + 4, ...\}$

$= \{......, -6, -1, 4, 9, 14, ...\}$.

There is no element common in any of the two right cosets.

$\therefore$ These right cosets are mutually disjoint.

Also every element of G is in exactly one of the right cosets.

$\therefore$ Union of all right cosets is G.

Example 4. *Let H be a subgroup of a group G with binary operation denoted multiplicatively. Show that $Hh = H = hH \quad \forall\, h \in H$.*

Sol. Let $\quad h \in H$.

$\therefore \quad Hh = \{h'h : h' \in H\}$ and $hH = \{hh' : h' \in H\}$.

Let $h'h \in Hh$ for some $h' \in H$

$\therefore \quad h'h \in H \quad (\because$ H is a subgroup)

$\therefore \quad Hh \subseteq H$

Let $\quad h'' \in H$.

$\therefore \quad h'' = h''e = h''(h^{-1}h) = (h''h^{-1})h \in Hh \quad (\because \; h''h^{-1} \in H)$

$\therefore \quad H \subseteq Hh$

Combining, we get $\quad Hh = H.$

Similarly, we can show that $hH = H$.

$$\therefore \qquad Hh = H = hH.$$

IMPORTANT RESULTS

1. *If H is any subgroup of a group G, then any two right (or left) cosets of H in G are either equal or disjoint.*
2. *If H is any subgroup of a group G, then there is one-one correspondence between any two right or left cosets of H in G.*
3. *If H is any subgroup of a group G, then the union of all right or left cosets of H in G is equal to G.*
4. *If H is any subgroup of a group G, then the right or left cosets of H in G partitions the group G.*
5. *If H is a subgroup of a group G, then the relation of 'congruence modulo H' is an equivalence relation on G and the right cosets of H in G partitions the group G.*

EXERCISE 2

1. Let $G = \{1, \omega, \omega^2\}$ be a group w.r.t. usual multiplication and $H = \{1\}$. Find all left and right cosets of H in G.
2. Let $G = \{1, -1, i, -i\}$ be a group w.r.t. usual multiplication and $H = \{1, -1\}$. Find all the left cosets of H in G.
3. Let G be the group of all integers under usual addition and $H = \{4n : n \in \mathbf{Z}\}$. Find the right cosets of H in G generated by 0, 1, 2, 3. Also verify that these cosets are mutually disjoint and their union is G.
4. Let G be the group of all integers under usual addition. Let $H = \{3n : n \in \mathbf{Z}\}$. Show that :
   (*i*) right coset of H in G generated by $k \in G$ is same as the left coset of H in G generated by k.
   (*ii*) right cosets of H in G generated by 1, 4, 7, 10, are all same.
   (*iii*) right cosets of H in G generated by 0, 1, 2 form a partition of G.
5. Let H be a subgroup of a group G with binary operation denoted multiplicatively. Prove that :
   (*i*) $Ha = Hb$ iff $ab^{-1} \in H$ $\qquad$ (*ii*) $aH = bH$ iff $a^{-1}b \in H$.

ANSWERS

1. $(1)H = \{1\}$, $\omega H = \{\omega\}$, $\omega^2 H = \{\omega^2\}$
   $H(1) = \{1\}$, $H\omega = \{\omega\}$, $H\omega^2 = \{\omega^2\}$.
2. $(1)H = \{1, -1\}$, $(-1)H = \{-1, 1\}$, $iH = \{i, -i\}$, $(-i)H = \{-i, i\}$.
3. $H + 0 = \{......, -8, -4, 0, 4, 8, ...\}$
   $H + 1 = \{......, -7, -3, 1, 5, 9, ...\}$
   $H + 2 = \{......, -6, -2, 2, 6, 10, ...\}$
   $H + 3 = \{......, -5, -1, 3, 7, 11, ...\}$.

10. LAGRANGE'S THEOREM

If H is a subgroup of a group G, then the set G is equal to the union of all mutually disjoint right cosets of H in G. This result is true irrespective of the fact whether G is finite or infinite. In case, the group G is finite, then we shall have a very interesting result given by the **Lagrange's theorem**.

Statement. *The order of each subgroup of a finite group divides the order of the group.*

Proof. Let G be a finite group with binary operation denoted multiplicatively. Let H be a subgroup of the group G. Since G is finite, the subgroup H must also be finite.

Let $o(G) = n$ and $o(H) = m$.

We shall show that m divides n.

Let $H = \{h_1, h_2, \ldots\ldots, h_m\}$

$H = G \Rightarrow m = n \Rightarrow m/n.$

If $H \neq G$ then $\exists$ at least one $a_1 \in G - H$.

Now $Ha_1 = \{h_1a_1, h_2a_1, \ldots\ldots, h_ma_1\}$

Ha_1 has m distinct elements because $h_i\, a_1 = h_j\, a_1 \Rightarrow h_i = h_j$, which is impossible.

We know that any two right cosets of a subgroup in the group are either equal or disjoint.

Since $a_1 \notin H$, the cosets H and Ha_1 are disjoint. The subgroup H is also a right coset of H in G because $H = He$. There are two possibilities :

If $H \cup Ha_1 = G$, then $m + m = n \Rightarrow 2m = n \Rightarrow m/n.$

If $H \cup Ha_1 \neq G$ then $\exists$ at least one $a_2 \in G - (H \cup Ha_1)$.

Now $Ha_2 = \{h_1a_2, h_2a_2, \ldots\ldots, h_ma_2\}$

Ha_2 has m distinct elements because $h_i\, a_2 = h_j\, a_2 \Rightarrow h_i = h_j$, which is impossible.

$Ha_2 \cap H = \phi$ because $a_2 \notin H$.

$Ha_2 \cap Ha_1 = \phi$ because $h_i\, a_2 = h_j\, a_1$

$\Rightarrow a_2\, a_1^{-1} = h_i^{-1} h_j \Rightarrow a_2\, a_1^{-1} \in H$

$\Rightarrow (a_2\, a_1^{-1})\, a_1 \in Ha_1 \Rightarrow a_2 \in Ha_1$, which is impossible.

There are two possibilities :

If $H \cup Ha_1 \cup Ha_2 = G$, then $m + m + m = n \Rightarrow 3m = n \Rightarrow m/n.$

If $H \cup Ha_1 \cup Ha_2 \neq G$, then $\exists$ at least one $a_3 \in G - \{H \cup Ha_1 \cup Ha_2\}$.

Again, we can show that either

$$H \cup Ha_1 \cup Ha_2 \cup Ha_3 = G$$

or $$\exists \text{ at least one } a_4 \in G - \{H \cup Ha_1 \cup Ha_2 \cup Ha_3\}.$$

Ultimately this process will end because G is finite.

Let $H \cup Ha_1 \cup Ha_2 \cup Ha_3 \cup \ldots\ldots \cup Ha_{k-1} = G$

$\therefore \quad m + m + m + m + \ldots\ldots\ k \text{ times} = n$

$\Rightarrow \quad km = n \Rightarrow m/n$

$\therefore$ Order of subgroup H divides the order of group G.

Remark. We have $H \cup Ha_1 \cup Ha_2 \cup \ldots\ldots \cup Ha_{k-1} = G$.

$\therefore$ Distinct right cosets of H in G are H, $Ha_1, Ha_2, \ldots\ldots, Ha_{k-1}$ and these are k in number.

$\therefore$ Index of H in G = k

$$\therefore \quad [G : H] = \frac{o(G)}{o(H)}. \qquad \left(\because\ k = \frac{n}{m} = \frac{o(G)}{o(H)}\right)$$

$\therefore$ **If H is a subgroup of a finite group G, then** $[G : H] = \dfrac{o(G)}{o(H)}$.

11. ORDER OF ELEMENTS IN A FINITE GROUP

Theorem 1. *Let G be a finite group with binary operation denoted multiplicatively. If $a \in G$, then $o(a)/o(G)$ and $a^{o(G)} = e$.*

Proof. We have $a \in G$.

o(a) is finite. Consider the positive integral powers $a, a^2, a^3, \ldots\ldots$ of a. Since G is a group, these elements are in G. G is finite, so $a, a^2, a^3, \ldots\ldots$ cannot be all distinct.

Let $a^k = a^l$, where $k > l$

$\Rightarrow \quad a^{k-l}\, a^l = ea^l \Rightarrow a^{k-l} = e \Rightarrow o(a) \le k - l$

$\therefore$ $o(a)$ is finite. Let $o(a) = m$.

Define $H = \{\ldots\ldots, a^{-2}, a^{-1}, e, a, a^2, \ldots\ldots\}$.

H is a subgroup of G. H is a non-empty subset of G.

Also $a^{k_1}, a^{k_2} \in H$ implies $a^{k_1}(a^{k_2})^{-1} = a^{k_1-k_2} \in H$.

$\therefore$ H is a subgroup of G.

o(H) = m. Let $a^t \in H$.

By division algorithm, there exists integers q and r : $t = mq + r$ and $0 \le r < m$

$$\therefore \quad a^t = a^{mq+r} = a^{mq}\, a^r = (a^m)^q\, a^r = e^q\, a^r = ea^r = a^r \qquad (\because\ o(a) = m)$$

$\therefore$ Every element of H is of the type a^r, where $0 \le r < m$

$$\therefore \quad H = \{e, a, a^2, \ldots\ldots, a^{m-1}\} \qquad (\because\ a^0 = e)$$

We claim that the elements of H are all distinct.

If possible, let $a^{\lambda_1} = a^{\lambda_2}$, where $\lambda_1 > \lambda_2$ and $a^{\lambda_1}, a^{\lambda_2} \in H$.

$$\therefore \quad a^{\lambda_1-\lambda_2}\, a^{\lambda_2} = ea^{\lambda_2} \text{ or } a^{\lambda_1-\lambda_2} = e$$

This is impossible because $o(a) = m$ and $\lambda_1 - \lambda_2 < m$.

$\therefore$ The elements of H are all distinct. $\therefore$ $o(H) = m$

$\therefore$ By **Lagrange's theorem,** $o(H)/o(G)$.

$\Rightarrow \quad m/o(G) \Rightarrow o(a)/o(G)$.

Let $\quad o(G) = m\mu$

$$\therefore \quad a^{o(G)} = a^{m\mu} = (a^m)^\mu = e^\mu = e$$

$$\therefore \quad a^{o(G)} = e.$$

$\therefore$ For any $a \in G$, we have $o(a)/o(G)$ and $a^{o(G)} = e$.

Example 1. *By using Lagrange's theorem, show that* $5^{\phi(9)} \equiv 1 \pmod 9$.

Sol. Let G be the set of all positive integers less than 9 and relatively prime to 9.

$$\therefore \quad G = \{1, 2, 4, 5, 7, 8\}$$

$$o(G) = \phi(9) = 6.$$

G is a group with 'multiplication modulo 9' as the binary operation.

Now $5 \in G$.

$\therefore$ As a consequence of **Lagrange's theorem,**

$5 \odot 5 \odot \ldots\ldots \phi(9)$ times $= 1$, where $\odot$ denotes multiplication modulo 9.

$\Rightarrow$ $(5 \times 5 \times \ldots\ldots \phi(9)$ times$) - 1$ is a multiple of 9.

$\Rightarrow$ $5^{\phi(9)} - 1$ is a multiple of 9 $\Rightarrow$ $5^{\phi(9)} \equiv 1 \pmod 9$.

Verification. $\quad 5^{\phi(9)} = 5^6 = 15625 = (1736 \times 9) + 1$

$\therefore \quad 5^{\phi(9)} \equiv 1 \pmod 9$.

12. EULER'S THEOREM

We know that the Euler ϕ-function, $\phi(n)$, is defined for all positive integers n and $\phi(1) = 1$ and for $n > 1$, $\phi(n)$ is the number of positive integers less than n and relatively prime to n. For example, positive numbers less than 12 and relatively prime to 12 are 1, 5, 7, 11 and these are 4 in number.

$\therefore \qquad \phi(12) = 4.$

Statement. *If n is a positive integer and a is an integer relatively prime to n, then $a^{\phi(n)} \equiv 1 (mod\ n)$.*

Proof. Let $n = 1$. $\phi(1) = 1$. $\therefore$ $a^{\phi(1)} - 1 = a - 1 = 1. (a - 1)$

$\therefore$ $a^{\phi(1)} \equiv 1 \pmod 1$

$\therefore$ The result is true for $n = 1$.

Let $n > 1$. Let G be the set of all positive integers less than n and relatively prime to n.

We know that the set G is an abelian group of order $\phi(n)$ with 'multiplication modulo n' as the binary operation.

By division algorithm, let $a = nq + r$, where $0 \le r < n$.

$\therefore$ $r \in$ G because a is relatively prime to n.

$\therefore$ As a consequence of **Lagrange's theorem**

$r \odot r \odot$ ...... $\phi(n)$ times = 1, where $\odot$ denotes multiplication modulo n

$\Rightarrow$ $(r \times r \times$ ...... $\phi(n)$ times$) - 1$ is a multiple of n

$\Rightarrow$ $r^{\phi(n)} - 1$ is a multiple of n

$\Rightarrow$ $(a - nq)^{\phi(n)} - 1$ is a multiple of n

$\Rightarrow$ $a^{\phi(n)} - 1$ is a multiple of n.

$\therefore$ We have $\qquad$ $\mathbf{a^{\phi(n)} \equiv 1 \ (mod\ n).}$

Example 2. *By using Lagrange's theorem, show that $4^p \equiv 4$ mod p, where p is the prime number 11.*

Sol. $p = 11$ is a prime number. Let G = {1, 2, 3, ..., 10}

$\therefore$ $o(G) = 10$

G is a group with 'multiplication modulo 11' as binary operation.

Now $4 \in$ G.

$\therefore$ As a consequence of **Lagrange's theorem,**

$4 \odot 4 \odot$ ...... 10 times = 1, where $\odot$ denotes multiplication modulo 11.

$\Rightarrow$ $(4 \times 4 \times$ ...... 10 times$) - 1$ is a multiple of 11.

$\Rightarrow$ $4^{10} - 1$ is a multiple of 11

$\Rightarrow \qquad 11/(4^{10} - 1) \qquad \Rightarrow \qquad 11/4(4^{10} - 1) \qquad \Rightarrow \qquad 11/(4^{11} - 4)$

$\Rightarrow \qquad \mathbf{4^{11} \equiv 4 \ (mod\ 11).}$

Verification. $4^{11} = 4194304 = (381300 \times 11) + 4$

$\therefore \qquad 4^{11} \equiv 4 \pmod{11}.$

13. FERMAT'S THEOREM

For a positive integer n, we have $\phi(n) = 1$ if $n = 1$ and for $n > 1$, $\phi(n)$ is the number of positive integers less than n and relatively prime to n. In case n is a prime number p, then 1, 2, 3, ..., $p - 1$ are all positive numbers less than p and relatively prime to p.

$\therefore \qquad \phi(p) = p - 1.$

Statement. *If p is any prime number and a is an integer, then*

$$a^p \equiv a \ (mod\ p).$$

Proof. The integer a may or may not be relatively prime to p.

Case I. (a, p) = 1.

Let G be the set of all positive integers 1, 2, 3, ..., $p - 1$.

We know that G is an abelian group of order $p - 1$ with 'multiplication modulo p' as the binary operation.

By division algorithm, let $a = pq + r$, where $0 \le r < p$.

$\therefore$ $r \in$ G because $(a, p) = 1$.

$\therefore$ As a consequence of **Lagrange's theorem,**

$r \odot r \odot$ ...... $(p - 1)$ times = 1, where $\odot$ denotes multiplication modulo p

$\Rightarrow$ $(r \times r \times$ ...... $(p - 1)$ times $- 1)$ is a multiple of p

$\Rightarrow$ $r^{p-1} - 1$ is a multiple of p $\Rightarrow$ $(a - pq)^{p-1} - 1$ is a multiple of p

$\Rightarrow$ $a^{p-1} - 1$ is a multiple of p $\Rightarrow$ $a(a^{p-1} - 1)$ is a multiple of p

$\Rightarrow$ $a^p - a$ is a multiple of p $\Rightarrow$ $a^p \equiv a$ (mod p).

Case II. (a, p) ≠ 1

Since p is prime, we have $(a, p) = p$.

$\Rightarrow$ p/a $\Rightarrow$ p/a^p

$\therefore$ $p/(a^p - a)$ or $a^p \equiv a$ (mod p).

$\therefore$ We have $\mathbf{a^p \equiv a (mod\ p)}$.

IMPORTANT RESULTS

1. **Lagrange's theorem.** *The order of each subgroup of a finite group divides the order of the group.*
2. *If G is a finite group with binary operation denoted multiplicatively, then*
$$a^{o(G)} = e \quad \forall\, a \in G.$$
3. **Euler's theorem.** *If n is a positive integer and a is an integer relatively prime to n, then* $a^{\phi(n)} \equiv 1 \ (mod\ n)$.
4. **Fermat's theorem.** *If p is any prime number and a is an integer, then* $a^p \equiv a \ (mod\ p)$.

EXERCISE 3

1. If G is a group of prime order, show that G cannot have any proper subgroup *i.e.*, a subgroup other then $\{e\}$ and itself.
2. By using Lagrange's theorem, show that, $7^{\phi(10)} \equiv 1$ (mod 10).
3. By using Lagrange's theorem, show that, $6^p \equiv 6$ (mod p), where $p = 11$.
4. Show that $5^{10} - 3^{10}$ is a multiple of 11.

12

Normal Subgroups

1. INTRODUCTION

Let G be a group with binary operation denoted multiplicatively. Let H be any subgroup of G. If G is abelian then for any $g \in$ G and $h \in$ H, we have

$$ghg^{-1} = gg^{-1}h = eh = h \in \text{H}.$$

$$\therefore \quad ghg^{-1} \in \text{H} \quad \forall\, h \in \text{H and } g \in \text{G}.$$

This property may not be true when G is not abelian. In case, G is not abelian and we have $ghg^{-1} \in$ H $\forall\, h \in$ H and $g \in$ G, then the subgroup H of G is a peculiar type of subgroup, called *normal subgroup* of the group G.

2. NORMAL SUBGROUP

Let G be a group with binary operation denoted multiplicatively. A subgroup N of the group G is called a **normal subgroup** of G if

$$gng^{-1} \in \text{N} \quad \forall\, n \in \text{N and } g \in \text{G}.$$

Examples 1. Every subgroup of an abelian group G is a normal subgroup of the group G.

2. If G is any group, then the subgroups $\{e\}$ and G of G are normal subgroups of G. A group not having any normal subgroup except for $\{e\}$ and itself is called a **simple group.**

3. TESTS FOR A SUBGROUP TO BE A NORMAL SUBGROUP

We have seen that a subgroup of a group may or may not be a normal subgroup of the group. In the next few theorems, we shall discuss some tests to see whether a subgroup is normal or not.

Theorem 1. *A subgroup N of a group G, with binary operation denoted multiplicatively, is normal if and only if* $g\text{N}g^{-1} = \text{N} \quad \forall\, g \in \text{G}$.

Proof. Let N be a normal subgroup of the group G.

Let $n \in$ N and $g \in$ G.

$$\therefore \quad gng^{-1} \in \text{N} \quad \Rightarrow \quad g\text{N}g^{-1} \subseteq \text{N} \quad \forall\, g \in \text{G}.$$

Let $n \in$ N and $g \in$ G.

$$\therefore \quad g^{-1} \in \text{G}$$

$$\Rightarrow \quad g^{-1}n(g^{-1})^{-1} \in \text{N} \qquad (\because \text{ N is normal})$$

$$\Rightarrow \quad g^{-1}ng \in \text{N} \quad \Rightarrow \quad g(g^{-1}ng)g^{-1} \in g\text{N}g^{-1}$$

$$\Rightarrow \quad (gg^{-1})n(gg^{-1}) \in g\text{N}g^{-1} \quad \Rightarrow \quad ene \in g\text{N}g^{-1}$$

$$\Rightarrow \quad n \in g\text{N}g^{-1} \quad \Rightarrow \quad \text{N} \subseteq g\text{N}g^{-1} \quad \forall\, g \in \text{G}.$$

Combining, we get $gNg^{-1} = N \quad \forall\, g \in G$.

Conversely let $gNg^{-1} = N \quad \forall\, g \in G$.

Let $g \in G$ and $n \in N$.

$\therefore \quad gng^{-1} \in gNg^{-1}$

$\Rightarrow \quad gng^{-1} \in N \qquad (\because\ gNg^{-1} = N)$

$\therefore$ The subgroup N is normal.

Theorem 2. *A subgroup N of a group G is a normal subgroup of G if and only if the product of any two right cosets of N in G is again a right coset of N in G.*

Proof. Let the binary operation of the group G be denoted multiplicatively. Let N be a normal subgroup of the group G.

Let Ng and Ng' be any two right cosets of N in G.

Let $ng \in Ng$ and $n'g' \in Ng'$.

$$\therefore \quad ngn'g' = ngn'g^{-1}gg' = n(gn'g^{-1})\, gg' = (nn'')\, gg' \in Ngg'. \qquad (\text{Putting } n'' = gn'g^{-1} \in N)$$

$\therefore \quad NgNg' \subseteq Ngg'$

Now, let $ngg' \in Ngg'$.

$\therefore \quad ngg' = ngeg' = (ng)(eg') \in NgNg' \qquad (\because\ e \in N)$

$\therefore \quad Ngg' \subseteq NgNg'$

Combining, we get $NgNg' = Ngg'$

$\therefore$ The product of any two right cosets of N in G is again a right coset of N in G

Conversely, let product of any two right cosets of N in G is again a right coset of N in G.

Let $n \in N$ and $g \in G$.

$\therefore$ Ng and Ng^{-1} are two right cosets of N in G.

$\therefore$ $NgNg^{-1}$ is also a right coset of N in G.

Let $NgNg^{-1} = Nx$, where x is some element of G.

Now $e = egeg^{-1} \in NgNg^{-1}$ and $e = ee \in Ne$.

Also, any two right cosets of N in G are either equal or disjoint, we have $Nx = Ne$ *i.e.*, N

$\therefore \quad NgNg^{-1} = N$

Now $\quad gng^{-1} = egng^{-1} \in NgNg^{-1}$

$\therefore \quad gng^{-1} \in N \quad \forall\, n \in N, g \in G.$

$\therefore$ The subgroup N is normal.

Theorem 3. *If N_1 and N_2 are two normal subgroups of a group G, then the complex N_1N_2 is also a normal subgroup of G.*

Proof. Let the binary operation of group G be denoted multiplicatively.

Let $n_1 \in N_1$ and $n_2 \in N_2$.

Now $\quad n_1n_2 = n_1n_2(n_1^{-1}n_1) = (n_1n_2n_1^{-1})\, n_1 \in N_2N_1 \qquad (\because\ n_1n_2n_1^{-1} \in N_2)$

$\therefore \quad N_1N_2 \subseteq N_2N_1$

Also $\quad n_2n_1 = n_2n_1\,(n_2^{-1}\, n_2) = (n_2n_1n_2^{-1})\, n_2 \in N_1N_2 \qquad (\because\ n_2n_1n_2^{-1} \in N_1)$

$\therefore \quad N_2N_1 \subseteq N_1N_2.$

Combining, we get $N_1N_2 = N_2N_1$.

$\Rightarrow$ N_1N_2 is a subgroup of the group G because we know that the product of two subgroups of a group is also a subgroup of the group if they commute with each other.

Now let $g \in G$, $n_1 \in N_1$, $n_2 \in N_2$.

$\therefore$ $g(n_1n_2)g^{-1} = gn_1en_2g^{-1} = (gn_1g^{-1})(gn_2g^{-1}) \in N_1N_2$. $\quad (\because \; gn_1g^{-1} \in N_1, \; gn_2g^{-1} \in N_2)$

$\therefore$ N_1N_2 is a normal subgroup of G.

Theorem 4. *If N_1 and N_2 are two normal subgroup of a group G, then $N_1 \cap N_2$ is also a normal subgroup of G.*

The proof of this theorem is left as an exercise for the reader.

IMPORTANT RESULTS

1. *A subgroup N of a group G, with binary, operation denoted multiplicatively, is normal if and only if $gNg^{-1} = N \quad \forall g \in G$.*
2. *A subgroup N of a group G is a normal subgroup of G if and only if the product of any two right cosets of N in G is again a right coset of N in G.*
3. *The product of two normal subgroups of a group is also a normal subgroup of the group.*
4. *The intersection of two normal subgroups of a group is also a normal subgroup of the group.*

Example 1. *Show that a normal subgroup of a group commute with every complex of the group.*

Sol. Let G be a group with binary operation denoted multiplicatively. Let N be a normal subgroup of G. Let K be any complex of G. We shall show that NK = KN.

Let $n \in N$, $k \in K$.

Now $\quad nk = (kk^{-1})nk = k(k^{-1}nk) = k(k^{-1}n(k^{-1})^{-1}) \in KN \quad (\because \; k^{-1}n(k^{-1})^{-1} \in N)$

$\therefore$ $\quad NK \subseteq KN$

Also $\quad kn = (kn)(k^{-1}k) = (knk^{-1})k \in NK \quad (\because \; knk^{-1} \in N)$

$\therefore$ $\quad KN \subseteq NK$

Combining, we get $\quad NK = KN$.

Example 2. *If N is a normal subgroup of a group G with binary operation denoted multiplicatively and H is any subgroup of the group G, then show that NH is a subgroup of G.*

Sol. N is a normal subgroup of G.

$\therefore$ N commutes with every complex of G.

In particular, H is a complex of G.

$\therefore$ N commutes with H.

$\therefore$ $\quad NH = HN$

$\Rightarrow$ NH is a subgroup of the group G, because we know that the product of two subgroups of a group is also a subgroup of the group provided they commute with each other.

Example 3. *Let M and N be two normal subgroups of a group G such that $M \cap N = \{e\}$. Show that every element of M commute with every element of N.*

Sol. Let the binary operation of the group G be denoted multiplicatively.

Let $m \in M$ and $n \in N$.

Now $(mn)(nm)^{-1} = (mn)(m^{-1}n^{-1}) = (mnm^{-1})n^{-1} \in N$, because $mnm^{-1} \in N$ and $n^{-1} \in N$.

Also $(mn)(nm)^{-1} = (mn)(m^{-1}n^{-1}) = m(nm^{-1}n^{-1}) \in M$, because $m \in M$ and $nm^{-1}n^{-1} \in M$.

$\therefore \quad (mn)(nm)^{-1} \in M \cap N$

$\Rightarrow \quad (mn)(nm)^{-1} = e \qquad (\because \; M \cap N = \{e\})$

$\Rightarrow \quad mn = e(nm) \;\Rightarrow\; mn = nm.$

$\therefore$ The result holds.

Example 4. *Let H be the only subgroup of finite order m in the group G. Show that H is a normal subgroup of G.*

Sol. Let the binary operation of the group G be denoted multiplicatively. Let H be the only subgroup of G of order m.

Let g be any element of G.

$\therefore$ $gHg^{-1} = \{ghg^{-1} : h \in H\}$ is a complex of G because

$geg^{-1} = gg^{-1} = e \in gHg^{-1}$ and $gHg^{-1} \subseteq G$.

Let $gh_1g^{-1}, gh_2g^{-1} \in gHg^{-1}$.

$$\begin{aligned} \therefore \quad (gh_1g^{-1})(gh_2g^{-1})^{-1} &= (gh_1g^{-1})((g^{-1})^{-1}h_2^{-1}g^{-1}) \\ &= (gh_1g^{-1})\,(gh_2^{-1}g^{-1}) = gh_1(g^{-1}g)h_2^{-1}\,g^{-1} \\ &= g(h_1h_2^{-1})g^{-1} \in gHg^{-1} \qquad (\because \; h_1h_2^{-1} \in H) \end{aligned}$$

$\therefore$ gHg^{-1} is a subgroup of G.

Let $gh_i\,g^{-1}, gh_j\,g^{-1} \in gHg^{-1}$ and $gh_i\,g^{-1} = gh_j\,g^{-1}$.

$\Rightarrow \quad h_i = h_j.$

$\therefore$ gHg^{-1} has exactly m elements.

$\therefore$ $gHg^{-1} = H$, because there is only one subgroup of G of order m.

$\therefore$ $gHg^{-1} = H \quad \forall g \in G$

Let $h \in H$. $\therefore$ $ghg^{-1} \in H \quad \forall h \in H$ and $g \in G$

$\therefore$ H is a normal subgroup of G.

EXERCISE 1

1. Show that every group of prime order is a simple group.
2. Show that every subgroup of an abelian group is normal.
3. If N_1, N_2 and N_3 be any three normal subgroups of a group G, then show that $N_1 \cap N_2 \cap N_3$ is also a normal subgroup of the group G.
4. If H is a subgroup of a group G and N is a normal subgroup of G, then show that $H \cap N$ is a normal subgroup of the group H.
5. N is a normal subgroup of a group G. If H is any subgroup of G such that $N \subseteq H \subseteq G$, then show that N is a normal subgroup of the group H.

HINTS

4. $H \cap N$ being the intersection of two subgroups of G is also a subgroup of G.

$\therefore$ $H \cap N$ is a subgroup of H, because $H \cap N \subseteq H$.

Let $n \in H \cap N$, $h \in H$.

$n \in H \cap N \;\Rightarrow\; n \in N$ and $h \in H \;\Rightarrow\; h \in G \;\Rightarrow\; hnh^{-1} \in N$

Also $n \in H \cap N \;\Rightarrow\; n \in H$ and $h \in H \;\Rightarrow\; h^{-1} \in H \;\Rightarrow\; hnh^{-1} \in H$

$\therefore$ $hnh^{-1} \in H \cap N$.

---

4. QUOTIENT GROUP

Let G be a group with binary operation denoted multiplicatively. Let N be a normal subgroup of the group G. Let the set of all right cosets of N in G be denoted by G/N.

$\therefore$ G/N = $\{Ng : g \in G\}$ and its elements are complexes of the group G.

Let $Ng_1, Ng_2 \in G/N$.

Since N is a normal subgroup, of G, we have $(Ng_1)(Ng_2) = Ng_1g_2 \in G/N$.

Let $Ng_1 = Ng_1'$ and $Ng_2 = Ng_2'$. $\therefore$ $g_1'g_1^{-1}, g_2'g_2^{-1} \in N$

Now $g_1'g_2'(g_1g_2)^{-1} = g_1'g_2'g_2^{-1}g_1^{-1} = g_1'g_2'g_2^{-1}(g_1'^{-1}g_1')g_1^{-1}$

$= [g_1'(g_2'g_2^{-1})g_1'^{-1}](g_1'g_1^{-1}) \in N$

$\therefore$ $Ng_1g_2 = Ng_1'g_2'$ *i.e.*, $(Ng_1)(Ng_2) = (Ng_1')(Ng_2')$.

$\therefore$ Multiplication of complexes is a binary operation on G/N.

We shall show that G/N is a group under the binary operation 'multiplication of complexes'.

Associativity. Let $Ng_1, Ng_2, Ng_3 \in G/N$.

$\therefore$ $Ng_1(Ng_2\,Ng_3) = Ng_1Ng_2g_3 = Ng_1(g_2\,g_3)$

$= N(g_1g_2)g_3 = Ng_1g_2\,Ng_3 = (Ng_1Ng_2)Ng_3$

$\therefore$ $Ng_1(Ng_2\,Ng_3) = (Ng_1Ng_2)\,Ng_3$

$\therefore$ Multiplication is associative.

Existence of identity element. Let 'e' be the identity element of G.

$\therefore$ $Ne \in G/N$. Let $Ng \in G/N$.

$\therefore$ $NgNe = Nge = Ng$ and $Ne\,Ng = Neg = Ng$

$\therefore$ $NgNe = Ng = Ne\,Ng \quad \forall\, Ng \in G/N$

Also $Ne = N$.

$\therefore$ N is the identity element of G/N.

Existence of inverse. Let $Ng \in G/N$.

$\Rightarrow$ $g \in G \Rightarrow g^{-1} \in G \Rightarrow Ng^{-1} \in G/N$

Now $NgNg^{-1} = Ngg^{-1} = Ne = N$ and $Ng^{-1}Ng = Ng^{-1}g = Ne = N$.

$\therefore$ $NgNg^{-1} = N = Ng^{-1}Ng$

$\therefore$ Ng^{-1} is the inverse of Ng.

$\therefore$ G/N is a group under usual multiplication of complexes as the binary operation. This group is called the **quotient group** of G by N.

Example 1. *If N is a normal subgroup of a finite group G, then show that*

$$o(G/N) = \frac{o(G)}{o(N)}.$$

Sol. Since G is finite, the right cosets of N in G are also finite in number.

$\therefore$ The set G/N = $\{Ng : g \in G\}$ is finite.

$\therefore$ o(G/N) = No. of distinct right cosets of N in G

$$= \frac{\text{No. of elements in G}}{\text{No. of elements in N}} = \frac{o(G)}{o(N)}. \quad \text{(By \textbf{Lagrange's Theorem})}$$

$$\therefore \quad \mathbf{o(G/N)} = \frac{\mathbf{o(G)}}{\mathbf{o(N)}}.$$

Example 2. *Let N be a normal subgroup of a group G. If $g \in G$, be an element of finite order, then show that o(Ng) divides o(g).*

Sol. Let $o(g) = m. \quad \therefore \quad g^m = e.$

Now $(Ng)^m = (Ng)(Ng) \text{ ...... } m \text{ times}$

$= N(gg \text{ ...... } m \text{ times}) = Ng^m = Ne = N$

Let $o(Ng) = k.$

$\therefore$ k is the least positive integer such that $(Ng)^k = N$.

By division algorithm, $\exists$ integers q and r such that

$$m = qk + r, \quad \text{where } 0 \le r < k.$$

If possible, let $r \neq 0$.

Now $(Ng)^{qk+r} = (Ng)^m = N \quad \Rightarrow \quad ((Ng)^k)^q (Ng)^r = N$

$\Rightarrow \quad (N)^q (Ng)^r = N \quad \Rightarrow \quad (Ng)^r = N$

This violates the definition of $o(Ng)$.

$\therefore \quad r = 0$

$\Rightarrow \quad m = qk \quad \Rightarrow \quad k/m \quad \Rightarrow \quad o(Ng)/o(g).$

Theorem 1. *If ϕ is a homomorphism of a group G into group G′, then kernal of ϕ is a normal subgroup of G.*

Proof. Let the binary operations of the groups G and G′ be denoted multiplicatively.

$\therefore \quad \phi(ab) = \phi(a)\,\phi(b)\ \forall\ a, b \in G$

Let e and e' be the identity elements of the groups G and G′ respectively.

Now $\ker \phi = \{g \in G : \phi(g) = e'\}$

$\ker \phi$ is non-empty because $e \in \ker \phi$.

Let $a, b \in \ker \phi$.

$\therefore \quad \phi(a) = e', \phi(b) = e'.$

Now $\phi(ab^{-1}) = \phi(a)\,\phi(b^{-1}) = e'\,(\phi(b))^{-1} = e'(e')^{-1} = e'e' = e'$

$\therefore \quad ab^{-1} \in \ker \phi$

$\therefore$ $\ker \phi$ is a subgroup of G.

Let $g \in G$ and $a \in \ker \phi$.

$\therefore \quad \phi(gag^{-1}) = \phi(g)\,\phi(a)\,\phi(g^{-1}) = \phi(g)e'\,(\phi(g))^{-1} = \phi(g)\,(\phi(g))^{-1} = e'$

$\therefore \quad gag' \in \ker \phi \quad \forall\ g \in G, a \in \ker \phi.$

$\therefore$ $\ker \phi$ is a normal subgroup of G.

Theorem 2. *Let N be a normal subgroup of a group G with binary operation denoted multiplicatively. Define $\phi : G \to G/N$ by $\phi(g) = Ng \quad \forall\ g \in G$. Show that*

(i) ϕ is a homomorphism of G onto G/N

(ii) $\ker \phi = N$.

Proof. (*i*) We have $\phi : G \to G/N$ by $\phi(g) = Ng \quad \forall\ g \in G$.

Since N is a normal subgroup of G, the set $G/N = \{Ng : g \in G\}$ forms a group with multiplication of complexes as binary operation.

Let $g_1, g_2 \in G$.

$$\phi(g_1 g_2) = Ng_1g_2 = Ng_1Ng_2 = \phi(g_1)\,\phi(g_2)$$

$\therefore\quad \phi : G \to G/N$ is a homomorphism.

Let $Ng \in G/N$ for some $g \in G$.

$\therefore\qquad \phi(g) = Ng \quad \therefore\ \phi$ is onto.

(*ii*) Let $g \in \ker\phi$. $\quad\therefore\ \phi(g) =$ identity element of G/N.

$\therefore\qquad \phi(g) = N$

$\Rightarrow\qquad Ng = N \ \Rightarrow\ g \in N \ \Rightarrow\ \ker\phi \subseteq N.$

Let $n \in N$. $\quad\therefore\ n \in G$

$\therefore\qquad \phi(n) = Nn = N \qquad (\because\ n \in N)$

$\Rightarrow\ \phi(n) =$ identity element of G/N.

$\Rightarrow\qquad n \in \ker\phi \ \Rightarrow\ N \subseteq \ker\phi$

$\therefore\qquad \ker\phi = N.$

Remark. The mapping $\phi : G \to G/N$ given by $\phi(g) = Ng$ for $g \in G$ is called the **natural mapping** of G onto G/N.

Example 3. *Show that a homomorphism from a simple group is either trivial or one-one.*

Sol. Let G be a simple group and its binary operation denoted multiplicatively.

$\therefore$ G has no normal subgroup other than $\{e\}$ and G.

Let G′ be another group and $\phi : G \to G'$ be a homomorphism.

$\therefore$ $\ker\phi$ is a normal subgroup of G.

$\therefore$ $\ker\phi$ is either $\{e\}$ or G.

Case I. ker $\phi = \{e\}$

Let $g_1, g_2 \in G$ and $\phi(g_1) = \phi(g_2)$.

$\Rightarrow\qquad \phi(g_1)(\phi(g_2))^{-1} = e' \qquad \Rightarrow\ \phi(g_1)\,\phi(g_2^{-1}) = e'$

$\Rightarrow\qquad \phi(g_1g_2^{-1}) = e' \qquad \Rightarrow\ g_1g_2^{-1} \in \ker\phi$

$\Rightarrow\qquad g_1g_2^{-1} = e \qquad \Rightarrow\ g_1 = g_2.$

$\therefore$ ϕ is one-one.

Case II. ker ϕ = G

$\Rightarrow\qquad \phi(g) = e' \quad \forall\, g \in G$

$\therefore$ ϕ is trivial homomorphism.

5. THE FUNDAMENTAL THEOREM OF HOMOMORPHISM

Statement. *Let ϕ be a homomorphism of a group G onto a group G′. Then G/ker $\phi \approx$ G′.*

Proof. Let the binary operations of the groups G and G′ be denoted multiplicatively. $\phi : G \to G'$ is a homomorphism of G onto G′.

Let $\ker\phi = K$. $\therefore$ K is a normal subgroup of G.

$\therefore$ The set $G/K = \{Kg : g \in G\}$ is a group with multiplication of complexes as the binary operation.

Define $\psi : G/K \to G'$ by $\psi(Kg) = \phi(g) \quad \forall\, g \in G$.

ψ is well defined. Let $Kg_1, Kg_2 \in G/K$ and $Kg_1 = Kg_2$.

$\Rightarrow \quad g_1 g_2^{-1} \in K \quad \Rightarrow \quad \phi(g_1 g_2^{-1}) = e' \quad \Rightarrow \quad \phi(g_1)\,\phi(g_2^{-1}) = e'$

$\Rightarrow \quad \phi(g_1)(\phi(g_2))^{-1} = e' \quad \Rightarrow \quad \phi(g_1) = \phi(g_2)$

$\Rightarrow \quad \psi(Kg_1) = \psi(Kg_2).$

$\therefore$ ψ is well defined.

ψ is a homomorphism. Let $Kg_1, Kg_2 \in G/K$.

Now $\quad \psi(Kg_1Kg_2) = \psi(Kg_1g_2) = \phi(g_1g_2) = \phi(g_1)\,\phi(g_2) = \psi(Kg_1)\,\psi(Kg_2).$

$\therefore$ ψ is a homomorphism.

ψ is one-one. Let $Kg_1, Kg_2 \in K/G$ and $\psi(Kg_1) = \psi(Kg_2)$.

$\Rightarrow \quad \phi(g_1) = \phi(g_2) \quad \Rightarrow \quad \phi(g_1)\,(\phi(g_2))^{-1} = e'$

$\Rightarrow \quad \phi(g_1)\,\phi(g_2^{-1}) = e' \quad \Rightarrow \quad \phi(g_1 g_2^{-1}) = e'$

$\Rightarrow \quad g_1 g_2^{-1} \in K$ (*i.e.*, ker ϕ) $\quad \Rightarrow \quad Kg_1 = Kg_2.$

$\therefore$ ψ is one-one.

ψ is onto. Let $g' \in G'$. Since $\phi : G \to G'$ is onto, there exists at least one $g \in G : \phi(g) = g'$.

$\therefore$ $Kg \in G/K$ and $\psi(Kg) = \phi(g) = g'$

$\therefore$ ψ is onto.

$\therefore$ $\psi : G/K \to G'$ is a one-one homomorphism of G/K onto G′.

$\therefore$ The groups G/K and G′ are isomorphic *i.e.*, **G/K ≈ G′.**

IMPORTANT RESULTS

1. *If ϕ is a homomorphism of a group G into group G′ then kernal of ϕ is a normal subgroup of G.*
2. *Let N be a normal subgroup of a group G with binary operation denoted multiplicatively. If $\phi : G \to G/N$ be defined by $\phi(g) = Ng$, for $g \in G$ then ϕ is a homomorphism of G onto G/N and with kernal equal N.*
3. *If ϕ is a homomorphism of a group G onto a group G′, then $G/\ker\phi \approx G'$.*

EXERCISE 2

1. Let H be a subgroup of an abelian group G. Show that the quotient group of G by H is defined and is also abelian.

13

Cyclic Groups

1. DEFINITION

A group G is called a **cyclic group** if there exists an element a in G such that every element of G is of the form a^n, where n is an integer. The element a is called a **generator** of G and we write G = (a).

If the cyclic group G is generated by a ($\in$ G) then the elements of G are of the form :

$$......, a^{-3}, a^{-2}, a^{-1}, a^0(= e), a^1, a^2, a^3, ...$$

Some of these elements may be equal of each other.

A cyclic group may also have more than one generator *i.e.*, a generator of a cyclic group is not unique.

Also, a cyclic group may have any number of elements.

Illustrations : 1. Let G = $\{1, \omega, \omega^2\}$. We know that G is a group with usual multiplication as the binary operation.

We have $\omega^0 = 1$.

$\therefore$ $G = \{\omega^0, \omega, \omega^2\}$.

$\therefore$ G is a cyclic group and ω is a generator of G.

$\therefore$ $G = (\omega)$.

Also, $(\omega^2)^3 = (\omega^3)^2 = 1^2 = 1, (\omega^2)^2 = \omega^4 = \omega^3 . \omega = 1 . \omega = \omega$

$\therefore$ $G = \{(\omega^2)^3, (\omega^2)^2, \omega^2\}$

$\therefore$ 'ω^2' is also a generator of G.

2. Let G = $\{1, -1, i, -i\}$. G is a group with usual multiplication as the binary operation.

We have

$$i^0 = 1, i^2 = -1, i^3 = -i$$

$\therefore$ $G = \{i^0, i^2, i, i^3\}$

$\therefore$ G is a cyclic group and i is a generator of G.

$\therefore$ $G = (i)$.

Also $(-i)^0 = 1, (-i)^2 = -1, (-i)^3 = i$

$\therefore$ $G = \{(-i)^0, (-i)^2, (-i)^3, -i\}$

$\therefore$ '$-i$' is also a generator of G.

3. Let G = $\{1, 2, 3, 4, 5, 6\}$. Since, 7 is a prime number, the set G with binary operation, 'multiplication modulo 7' is a group. Let $\odot$ denote this binary operation.

We have $3^0 = 1, 3^1 = 3, \quad 3^2 = 3 \odot 3 = 2, \quad 3^3 = 3^2 \odot 3 = 2 \odot 3 = 6,$

$3^4 = 3^3 \odot 3 = 6 \odot 3 = 4, \quad 3^5 = 3^4 \odot 3 = 4 \odot 3 = 5,$

$\therefore$ $G = \{3^0, 3^2, 3^1, 3^4, 3^5, 3^3\}$

$\therefore$ G is a cyclic group and '3' is a generator of G. $\therefore$ G = (3).

Remark. If G is any group, then a^0 = identity element, for any $a \in G$.

2. PROPERTIES OF CYCLIC GROUPS

Theorem 1. *Every cyclic group is an abelian group.*

Proof. Let G be a cyclic group. Let a be a generator of G.

$\therefore$ $$G = \{a^i : i \in \mathbf{Z}\}$$

Let $a^r, a^s \in G$.

$\therefore$ $$a^r a^s = a^{r+s} = a^{s+r} = a^s a^r$$

$\therefore$ G is abelian.

Remark. A non-abelian group cannot be cyclic.

Theorem 2. *If a is a generator of a cyclic group G then a^{-1} (i.e., inverse of a) is also a generator of G.*

Proof. We have $G = \{a^i : i \in \mathbf{Z}\}$.

Now $a^i = (a^{-1})^{-i}$.

$\therefore$ Every element of G is an integral power of a^{-1}.

$\therefore$ a^{-1} is also a generator of G.

Theorem 3. *If a finite group of order n contains an element of order n, then the group must be cyclic.*

Proof. Let G be a finite group of order n. Let a ($\in$ G) be such that $o(a) = n$.

Let $$H = \{a^i : i \in \mathbf{Z}\}.$$

H is non-empty subset of G.

Let $a^r, a^s \in H$.

$\therefore$ $$a^r(a^s)^{-1} = a^r a^{-s} = a^{r-s} \in H$$

$\therefore$ H is a subgroup of G.

Let $a^i \in H$.

$\therefore$ By division algorithm, there exists integers $q, r \in \mathbf{Z}$ such that

$$i = nq + r, \text{ where } 0 \le r < n.$$

$\therefore$ $$a^i = a^{nq+r} = a^{nq} a^r = (a^n)^q a^r = e^q a^r = ea^r = a^r$$

$\therefore$ Each element of H is either a^0 (= e) or a^1 or a^2 or ...... or a^{n-1}.

$\therefore$ $$H = \{a^0 (= e), a, a^2, ..., a^{n-1}\}.$$

We claim that the elements of H are all distinct.

If possible, let $a^r = a^s$ for $0 \le r < s < n$.

$\Rightarrow$ $$a^r = a^{s-r} a^r \Rightarrow e = a^{s-r}$$

$\therefore$ $$a^{s-r} = e \text{ and } s - r < n$$

This is impossible because $o(a) = n$. $\therefore$ All elements of H are distinct.

$\therefore$ $$o(H) = n.$$

$\therefore$ $$H = G \qquad (\because H \subseteq G)$$

$\therefore$ G is a cyclic group and is generated by a.

Theorem 4. *Every group of prime order is cyclic.*

Proof. Let G be a group of order p, where p is a prime number. Since $p \geq 2$, G has at least two elements. Let e be the identity element of G. Let a be another element of G which is different from 'e'.

Let $o(a) = n \quad \therefore \quad n \geq 2.$

Let $H = \{a^i : i \in \mathbf{Z}\}.$

H is a non-empty subset of G.

Let $a^r, a^s \in H$.

$\therefore \quad a^r(a^s)^{-1} = a^r a^{-s} = a^{r-s} \in H$

$\therefore$ H is a subgroup of G.

Let $a^i \in H$.

$\therefore$ By division algorithm, there exists integers $q, r \in \mathbf{Z}$ such that

$$i = nq + r, \quad \text{where } 0 \leq r < n.$$

$\therefore \quad a^i = a^{nq+r} = a^{nq} a^r = (a^n)^q a^r = e^q a^r = ea^r = a^r$

$\therefore$ Each element of H is either a^0 (= e) or a^1 or a^2 or ...... or a^{n-1}.

$\therefore \quad H = \{a^0 (= e), a, a^2, ..., a^{n-1}\}$

We claim that the elements of H are all distinct.

If possible, let $a^r = a^s$ for $0 \leq r < s < n$.

$\Rightarrow \quad a^r = a^{s-r} a^r \quad \Rightarrow \quad e = a^{s-r}$

$\therefore \quad a^{s-r} = e$ and $s - r < n$

This is impossible because $o(a) = n$. $\therefore$ All elements of H are distinct.

$\therefore \quad o(H) = n.$

By **Lagrange's theorem,** we have n/p.

$\therefore \quad n = p \quad (\because \; n \geq 2$ and p is prime)

$\therefore \quad H = G \quad (\because \; H \subseteq G)$

$\therefore$ G is a cyclic group and is generated by a.

Theorem 5. *Every group of composite order has proper subgroups.*

Proof. Let G be a group of composite order mn, where $m, n > 1$ and $m, n \in \mathbf{Z}$. The group G may or may not be cyclic.

Case I. G is cyclic. Let a be a generator of G.

$\therefore \quad o(a) = mn \quad (\because \; o(G) = mn)$

$\Rightarrow \quad a^{mn} = e \quad \Rightarrow \quad (a^m)^n = e$

$\therefore$ $o(a^m)$ is finite and is less than or equal to n *i.e.*, $o(a^m) \leq n$.

Let $o(a^m) = \lambda.$

$\Rightarrow \quad (a^m)^\lambda = e \quad \Rightarrow \quad a^{m\lambda} = e \quad \Rightarrow \quad mn \leq m\lambda \quad \Rightarrow \quad n \leq \lambda$

$\Rightarrow \quad \lambda \geq n \quad \Rightarrow \quad o(a^m) \geq n$

$\therefore \quad o(a^m) = n.$

Let $H = (a^m)$ be the cyclic subgroup of G generated by a^m.

$\therefore \quad o(H) = n \quad (\because \; o(a^m) = n)$

Since $n < mn$, H is a proper subgroup of G.

Case II. G is not cyclic. There cannot be any element in G which is of order mn, otherwise G would come out to be cyclic.

$\therefore$ Order of each element of G is less than mn.

$\therefore$ There exists at least one element, say b, other than the identity element such that $1 < o(b) < mn$.

$\therefore$ The cyclic subgroup (b) is a proper subgroup of G.

Theorem 6. *Let G be a cyclic group generated by its element a and o(a) = n. For m < n, the element a^m is a generator of G if and only if (m, n) = 1.*

Note. The proof of this theorem is beyond the scope of this book.

Remark. The above theorem helps us in finding the generators of a cyclic group. For example, if a is a generator of a cyclic group of order 10, then a^3, a^7, a^9 are also generators of the group because each of 3, 7, 9 is relatively prime with 10.

Theorem 7. *Every subgroup of a cyclic group is cyclic.*

Proof. Let G be a cyclic group. Let a be a generator of G. Let H be a subgroup of G. If H = $\{e\}$ or H = G then H is cyclic. Let H be a proper subgroup of G. The elements of H are integral powers of a.

$$a^r \in H \;\Rightarrow\; (a^r)^{-1} \in H \;\Rightarrow\; a^{-r} \in H$$

$\therefore$ H contains elements which are positive as well as negative integral powers of a.

Let k be the least positive integer such that $a^k \in$ H. We shall show that H is generated by a^k. Let a^t be any arbitrary element of H.

By division algorithm, there exists integers q and r such that

$$t = kq + r \qquad \text{where } 0 \le r < k.$$

$\Rightarrow$ $$r = t - kq$$

Now $a^k \in H$. $\therefore$ $(a^k)^q \in H$

$\Rightarrow$ $$a^{kq} \in H \Rightarrow (a^{kq})^{-1} \in H \Rightarrow a^{-kq} \in H$$

$\Rightarrow$ $$a^t\, a^{-kq} \in H \Rightarrow a^{t-kq} \in H \qquad (\because\ r = t - kq)$$

$\Rightarrow$ $$a^r \in H$$

Since k is the least positive integer such that $a^k \in$ H, we have $r = 0$.

$\therefore$ $$t = kq$$

$\therefore$ $$a^t = a^{kq} = (a^k)^q$$

$\therefore$ Every element of H is some integral power of a^k.

$\therefore$ H is cyclic and is generated by a^k.

$\therefore$ Every subgroup of a cyclic group is cyclic.

Theorem 8. *Every infinite cyclic group has exactly two generators.*

Proof. Let G be an infinite cyclic group. Let a be a generator of G. We claim that no two distinct integral powers of a are equal.

If possible, let $a^i = a^j$, where $i > j$.

$\Rightarrow$ $$a^{i-j} a^j = a^j \Rightarrow a^{i-j} = e$$

$\Rightarrow$ $o(a)$ is finite. $(\because\ i - j \in \mathbf{N})$

This is impossible because a generator of an infinite cyclic group cannot be of finite order.

$\therefore$ $$a^i \ne a^j \quad \forall\, i \ne j,\ i, j \in \mathbf{Z}$$

$\therefore \quad G = \{......, a^{-3}, a^{-2}, a^{-1}, a^0 (= e), a, a^2, a^3, ...\}$

Let a^s be any element of G.

$\therefore \quad a^s = (a^{-1})^{-s}$ and $-s \in \mathbf{Z}$

$\therefore \quad a^{-1}$ is also a generator of G.

Also a and a^{-1} are distinct elements of G, otherwise we would have $a^2 = e$.

$\therefore \quad a$ and a^{-1} are two distinct generators of G. Now we shall show that there is no other generator of G.

If possible, let a^k $(k \neq \pm 1)$ be any other generator of G.

$\therefore \quad \exists\, \lambda \in \mathbf{Z} : (a^k)^\lambda = a \qquad (\because \; a \in G)$

$\Rightarrow \quad a^{k\lambda} = a \quad \Rightarrow \quad k\lambda = 1$. This is impossible because $k \neq \pm 1$.

$\therefore \quad a^k$ is not a generator of G.

$\therefore$ G has exactly two generators.

Theorem 9. *Every proper subgroup of an infinite cyclic group is infinite.*

Proof. Let G be a cyclic group. Let a be a generator of G. Let H be a proper subgroup of G. Since every subgroup of a cyclic group is cyclic, the subgroup H is also cyclic. Let k be the least positive integer such that a^k is in H.

$\therefore$ H is generated by a^k.

If possible, let H be finite and $o(\text{H}) = m$.

$\therefore \quad (a^k)^m = e$ *i.e.*, $a^{km} = e$

$\therefore \quad o(a)$ is finite and so G is finite.

This is impossible. $\therefore$ H is infinite.

$\therefore$ Every proper subgroup of an infinite cyclic group is infinite.

IMPORTANT RESULTS

1. *Every cyclic group is abelian.*
2. *If a is a generator of a cyclic group G then a^{-1} is also a generator of G.*
3. *If a finite group of order n contains an element of order n, then the group must be cyclic.*
4. *Every group of prime order is cyclic.*
5. *Every group of composite order has proper subgroups.*
6. *Let G be a cyclic group generated by an element a of G and $o(a) = n$. For $m < n$, the element a^m is a generator of G if and only if $(m, n) = 1$.*
7. *Every subgroup of a cyclic group is cyclic.*
8. *Every infinite cyclic group has exactly two generators.*
9. *Every proper subgroup of an infinite cyclic group is infinite.*

Example. *Show that every infinite cyclic group is isomorphic to the additive group of integers.*

Sol. Let G be an infinite cyclic group. Let a be a generator of G.

We claim that no two integral powers of a are equal.

If possible, let $a^i = a^j$, where $i > j$.

$\Rightarrow \quad a^{i-j}\, a^j = a^j \;\Rightarrow\; a^{i-j} = e$

$\Rightarrow$ $o(a)$ is finite $\quad (\because\; i - j \in \mathbf{N})$

This is impossible because a generator of an infinite cyclic group cannot be of finite order.

$\therefore \quad a^i \neq a^j \quad \forall\, i \neq j,\, i, j \in \mathbf{Z}$

$\therefore \quad G = \{......., a^{-3}, a^{-2}, a^{-1}, a^0\,(= e), a, a^2, a^3, ...\}$

Define $\phi : G \to \mathbf{Z}$ by

$$\phi(a^i) = i \quad \forall\, a^i \in G.$$

ϕ is well defined. Let $a^i, a^j \in G$.

$$a^i = a^j \;\Rightarrow\; i = j \;\Rightarrow\; \phi(a^i) = \phi(a^j)$$

$\therefore$ ϕ is well defined.

ϕ is a homomorphism. Let $a^i, a^j \in G$.

$$\phi(a^i a^j) = \phi(a^{i+j}) = i + j = \phi(a^i) + \phi(a^j)$$

$\therefore$ ϕ is a homomorphism.

ϕ is one-one. Let $a^i, a^j \in G$.

$$\phi(a^i) = \phi(a^j) \;\Rightarrow\; i = j \;\Rightarrow\; a^i = a^j$$

$\therefore$ ϕ is one-one.

ϕ is onto. Let $k \in \mathbf{Z}$.

$\therefore$ $a^k \in G$ and $\phi(a^k) = k$. $\therefore$ ϕ is onto.

$\therefore$ ϕ is an isomorphism from G onto **Z**. $\therefore$ $G \approx \mathbf{Z}$.

EXERCISE

1. Let G = {0, 1, 2, 3, 4, 5} be a group with binary operation 'addition modulo 6'. Show that G is a cyclic group and G = (5).
2. Show that every isomorphic image of a cyclic group is cyclic.
3. Explain the statement, "Every element of a group generates a subgroup."

HINT

2. Let G be a cyclic group generated by a. Let G′ be an isomorphic image of G under the isomorphism ϕ.

For $\phi(a^m) \in G'$, we have $\quad \phi(a^m) = \phi(a\, a \text{ ...... } m \text{ times})$

$$= \phi(a)\, \phi(a) \text{ ...... } m \text{ times} = (\phi\,(a))^m.$$

14

Permutation Groups

1. PERMUTATION

Let S be any non-empty set. A one-one mapping of S onto itself is called a **permutation** of S.

The set of all permutations of set S is denoted by A(S).

2. THEOREM

The set A(S) of all permutations of a non-empty set S is a group with composition of mappings as the binary operation.

Proof. Let $f, g \in$ A(S).

$\therefore$ f and g are one-one mappings of S onto itself.

The composition fog of f and g is a mapping of S to itself.

Let $s_1, s_2 \in$ S and $(fog)(s_1) = (fog)(s_2)$.

$\Rightarrow \quad f(g(s_1)) = f(g(s_2))$

$\Rightarrow \quad g(s_1) = g(s_2) \quad (\because \; f \text{ is one-one})$

$\Rightarrow \quad s_1 = s_2 \quad (\because \; g \text{ is one-one})$

$\therefore$ fog is one-one.

Let $s \in$ S. Since f is onto, $\exists\, s_1 \in \text{S} : f(s_1) = s$. Also, since g is onto, $\exists\, s_2 \in \text{S} : g(s_2) = s_1$.

$\therefore \quad f(g(s_2)) = s$ or $(fog)(s_2) = s$.

$\therefore$ fog is onto.

$\therefore$ fog is a permutation of S *i.e.*, $fog \in$ A(S).

$\therefore$ 'Composition of mappings' is a binary operation on A(S).

Associativity. Let $f, g, h \in$ A(s).

For $s \in$ S, $\quad [fo(goh)](s) = f((goh)(s)) = f(g(h(s))) = (fog)(h(s)) = [(fog)\, oh](s)$

$\therefore \quad fo(goh) = (fog)oh$

$\therefore$ 'o' is associative.

Existence of identity element. Let i be the identity mapping of S.

$\therefore$ $i \in$ A(S). Let $f \in$ A(S).

For $s \in$ S, $\quad (foi)(s) = f(i(s)) = f(s)$

and $\quad (iof)\,(s) = i(f(s)) = f(s)$

$\therefore \quad (foi)(s) = f(s) = (iof)(s) \;\forall\; s \in \text{S}$

$\Rightarrow \quad foi = f = iof \quad \forall\, f \in \text{A(S)}$

$\therefore$ 'i' is the identity element of A(S).

Existence of inverse. Let $f \in A(S)$.

For $s \in S, \exists$ exactly one $s' \in S : f(s') = s$.

Define $g : S \to S$ by $g(s) = s'$ provided $f(s') = s$.

$\therefore$ g is a mapping of S onto S.

Now, let $\quad g(s_1) = g(s_2), \quad$ where $g(s_1) = s_1'$ and $g(s_2) = s_2'$.

$\Rightarrow \quad f(s_1') = s_1$ and $f(s_2') = s_2$.

Now $g(s_1) = g(s_2)$ implies $s_1' = s_2'$

$\Rightarrow \quad f(s_1') = f(s_2') \quad \Rightarrow \quad s_1 = s_2$.

$\therefore$ g is one-one. $\quad \therefore \quad g \in A(S)$

Also $\quad (fog)(s_1) = f(g(s_1)) = f(s_1') = s_1 = i(s_1)$

and $\quad (gof)(s_1') = g(f(s_1')) = g(s_1) = s_1' = i\ (s_1')$

$\therefore \quad fog = i = gof$

$\therefore$ The inverse mapping g of f is the inverse of f.

$\therefore$ A(S) is a group with 'composition of mappings' as the binary operation.

3. PERMUTATION GROUP

If S is any non-empty set then the set A(S) of all permutations of S is a group with 'composition of mappings' as the binary operation. The subgroups of the group A(S) are called **permutation groups.** In the next theorem, we shall prove that every group is isomorphic to a permutation group.

4. CAYLEY'S THEOREM

Every group is isomorphic to a permutation group.

Proof. Let G be any group with binary operation denoted multiplicatively. Let A(G) be the set of all permutations of G. Therefore, A(G) is a group with composition of mappings as the binary operation. Let g be any element of G.

Define $f_g : G \to G$ by $f_g(x) = gx \quad \forall\ x \in G$.

Now $\quad x = y \quad \Rightarrow \quad gx = gy \quad \Rightarrow \quad f_g(x) = f_g(y)$

$\therefore$ f_g is well defined.

Also, $\quad f_g(x) = f_g(y) \quad \Rightarrow \quad gx = gy \quad \Rightarrow \quad x = y$

$\therefore$ f_g is one-one.

Now $x \in G$ implies $g^{-1} x \in G$

and $\quad f_g\ (g^{-1} x) = g(g^{-1} x) = (gg^{-1})\ x = ex = x$

$\therefore$ f_g is onto

$\therefore$ f_g is a permutation of G *i.e.*, $f_g \in A(G)$.

Define $\phi : G \to A(G)$ by $\phi(g) = f_g \quad \forall\ g \in G$.

ϕ is well defined. Let $g_1, g_2 \in G$.

$$g_1 = g_2 \quad \Rightarrow \quad g_1x = g_2x \quad \forall\ x \in G$$

$$\Rightarrow \quad f_{g_1}(x) = f_{g_2}(x) \quad \forall\ x \in G$$

$$\Rightarrow \quad f_{g_1} = f_{g_2} \quad \Rightarrow \quad \phi(g_1) = \phi(g_2).$$

$\therefore \quad \phi$ is well defined.

ϕ is a homomorphism. Let $g_1, g_2 \in$ G.

$$\phi(g_1 g_2) = f_{g_1 g_2} \text{ and } \phi(g_1) o\, \phi(g_2) = f_{g_1} o\, f_{g_2}.$$

For $x \in$ G, we have

$$f_{g_1 g_2}(x) = (g_1 g_2)\, x = g_1(g_2 x) = f_{g_1}(g_2 x) = f_{g_1}(f_{g_2}(x)) = (f_{g_1} o\, f_{g_2})(x).$$

$\therefore \quad f_{g_1 g_2} = f_{g_1} o\, f_{g_2}$

$\therefore \quad \phi(g_1 g_2) = \phi(g_1) o\, \phi(g_2)$

$\therefore$ ϕ is a homomorphism.

ϕ is one-one. Let $g_1, g_2 \in$ G.

$$\phi(g_1) = \phi(g_2) \Rightarrow f_{g_1} = f_{g_2} \Rightarrow f_{g_1}(x) = f_{g_1}(x)\ \forall\, x \in G$$

$$\Rightarrow \quad g_1 x = g_2 x\ \forall\, x \in G \Rightarrow g_1 = g_2.$$

$\therefore$ ϕ is one-one.

$\therefore$ ϕ is a one-one homomorphism of G into A(G).

$\therefore$ G is isomorphic to ϕ(G) which, being a subgroup of A(G), is a permutation group.

$\therefore$ Every group is isomorphic to a permutation group.

5. PERMUTATIONS OF FINITE SETS

Let S be a finite set having n elements. There are n ! ways of arranging these n elements.

$\therefore$ Number of one-one mappings of S onto itself = n !

$\therefore$ Number of permutations of S = n !

$\therefore$ The set A(S) of all permutations of S contains n ! elements.

Let S = $\{a_1, a_2, \ldots\ldots, a_n\}$.

Let $f \in$ A(S) and $f(a_1) = a_{k_1}, f(a_2) = a_{k_2}, \ldots\ldots, f(a_n) = a_{k_n}$, where $k_1, k_2, \ldots\ldots, k_n$ are distinct integers from 1 to n.

The permutation f is denoted by the symbol $\begin{pmatrix} a_1 & a_2 \ldots\ldots a_n \\ a_{k_1} & a_{k_2} \ldots\ldots a_{k_n} \end{pmatrix}$.

The first row contains the elements of S. In the second row, the element below a_i represents the image of a_i under f. The columns in the above notation may be inter changed. This notation is called **2-rowed notation** for f.

For example, let S = $\{a_1, a_2, a_3\}$ and f : S $\to$ S be a permutation such that $f(a_1) = a_2$, $f(a_2) = a_3$ and $f(a_3) = a_1$.

In 2-rowed notation, we write f as $\begin{pmatrix} a_1 & a_2 & a_3 \\ a_2 & a_3 & a_1 \end{pmatrix}$.

f can also be written as $\begin{pmatrix} a_1 & a_3 & a_2 \\ a_2 & a_1 & a_3 \end{pmatrix}$ or as $\begin{pmatrix} a_2 & a_3 & a_1 \\ a_3 & a_1 & a_2 \end{pmatrix}$.

In short, we also write f as $\begin{pmatrix} 1 & 2 & 3 \\ 2 & 3 & 1 \end{pmatrix}$.

Thus if $g = \begin{pmatrix} 1 & 2 & 3 \\ 3 & 1 & 2 \end{pmatrix}$, then it means

$$g(a_1) = a_3,\ g(a_2) = a_1 \text{ and } g(a_3) = a_2.$$

Example 1. *If S = {1, 2, 3} write A(S).*

Sol. We have S = {1, 2, 3}. S contains 3 elements.

∴ A(S) contains 6 (= 3 !) elements. Different arrangements of 1, 2, 3 are 1, 2, 3   1, 3, 2   2, 1, 3   2, 3, 1   3, 1, 2 and 3, 2, 1.

$$\therefore \quad A(S) = \left\{ \begin{pmatrix} 1 & 2 & 3 \\ 1 & 2 & 3 \end{pmatrix}, \begin{pmatrix} 1 & 2 & 3 \\ 1 & 3 & 2 \end{pmatrix}, \begin{pmatrix} 1 & 2 & 3 \\ 2 & 1 & 3 \end{pmatrix}, \begin{pmatrix} 1 & 2 & 3 \\ 2 & 3 & 1 \end{pmatrix}, \begin{pmatrix} 1 & 2 & 3 \\ 3 & 1 & 2 \end{pmatrix}, \begin{pmatrix} 1 & 2 & 3 \\ 3 & 2 & 1 \end{pmatrix} \right\}.$$

6. EQUALITY OF PERMUTATIONS

Two permutations f and g of set S are said to be **equal permutations** if

$$f(s) = g(s) \quad \forall\, s \in S.$$

For example, let S = {1, 2, 3, 4} and $f = \begin{pmatrix} 1 & 2 & 3 & 4 \\ 2 & 3 & 4 & 1 \end{pmatrix}$, $g = \begin{pmatrix} 2 & 1 & 4 & 3 \\ 3 & 2 & 1 & 4 \end{pmatrix}$.

Here $f = g$ because $f(1) = g(1) = 2$, $f(2) = g(2) = 3$,
$f(3) = g(3) = 4$, $f(4) = g(4) = 1$.

7. COMPOSITION OF PERMUTATIONS IN 2-ROWED NOTATION

Let S = $\{a_1, a_2, \ldots\ldots, a_n\}$ be a finite set having n elements.

Let f be a permutation of S defined by $f(a_1) = b_1, f(a_2) = b_2, \ldots\ldots, f(a_n) = b_n$, where $b_1, b_2, \ldots\ldots, b_n$ are distinct element of the set S. *i.e.*, b_i is equal to some a_j.

$$\therefore \qquad f = \begin{pmatrix} a_1 & a_2 \ldots\ldots a_n \\ b_1 & b_2 \ldots\ldots b_n \end{pmatrix}.$$

Let g be another permutation of S defined by $g(b_1) = c_1, g(b_2) = c_2, \ldots\ldots, g(b_n) = c_n$, where $c_1, c_2, \ldots\ldots, c_n$ are also distinct elements of the set S *i.e.*, c_i is equal to some a_k.

$$\therefore \qquad g = \begin{pmatrix} b_1 & b_2 \ldots\ldots b_n \\ c_1 & c_2 \ldots\ldots c_n \end{pmatrix}$$

By definition
$$(gof)(a_1) = g(f(a_1)) = g(b_1) = c_1$$
$$(gof)(a_2) = g(f(a_2)) = g(b_2) = c_2$$
$$\ldots\ldots\ldots\ldots\ldots\ldots\ldots\ldots\ldots\ldots$$
$$\ldots\ldots\ldots\ldots\ldots\ldots\ldots\ldots\ldots\ldots$$
$$(gof)(a_n) = g(f(a_n)) = g(b_n) = c_n$$

∴ In 2-rowed notation, we have

$$gof = \begin{pmatrix} a_1 & a_2 \ldots\ldots a_n \\ c_1 & c_2 \ldots\ldots c_n \end{pmatrix}.$$

In short we write
$$gof = \begin{pmatrix} a_1 & a_2 \ldots\ldots a_n \\ b_1 & b_2 \ldots\ldots b_n \end{pmatrix} \begin{pmatrix} b_1 & b_2 \ldots\ldots b_n \\ c_1 & c_2 \ldots\ldots c_n \end{pmatrix} = \begin{pmatrix} a_1 & a_2 \ldots\ldots a_n \\ c_1 & c_2 \ldots\ldots c_n \end{pmatrix}.$$

Remark. In writing gof, the order is reversed *i.e.*, g is written after writing f.

Example 2. *Let* $f = \begin{pmatrix} 1 & 2 & 3 & 4 \\ 2 & 4 & 3 & 1 \end{pmatrix}$ *and* $g = \begin{pmatrix} 1 & 2 & 3 & 4 \\ 2 & 3 & 4 & 1 \end{pmatrix}$ *be two permutations of the set* $S = \{1, 2, 3, 4\}$. *Find :*

(i) *fof* (ii) *fog*

(iii) *gof* (iv) *gog*.

Sol. (i)
$$fof = \begin{pmatrix} 1 & 2 & 3 & 4 \\ 2 & 4 & 3 & 1 \end{pmatrix} \begin{pmatrix} 1 & 2 & 3 & 4 \\ 2 & 4 & 3 & 1 \end{pmatrix}$$
$$= \begin{pmatrix} 1 & 2 & 3 & 4 \\ 2 & 4 & 3 & 1 \end{pmatrix} \begin{pmatrix} 2 & 4 & 3 & 1 \\ 4 & 1 & 3 & 2 \end{pmatrix} = \begin{pmatrix} \mathbf{1} & \mathbf{2} & \mathbf{3} & \mathbf{4} \\ \mathbf{4} & \mathbf{1} & \mathbf{3} & \mathbf{2} \end{pmatrix}.$$

Remark. Since interchanging columns of a permutation does not make any difference, the permutations $\begin{pmatrix} 1 & 2 & 3 & 4 \\ 2 & 4 & 3 & 1 \end{pmatrix}$ **and** $\begin{pmatrix} 2 & 4 & 3 & 1 \\ 4 & 1 & 3 & 2 \end{pmatrix}$ **are same.**

(ii)
$$fog = \begin{pmatrix} 1 & 2 & 3 & 4 \\ 2 & 3 & 4 & 1 \end{pmatrix} \begin{pmatrix} 1 & 2 & 3 & 4 \\ 2 & 4 & 3 & 1 \end{pmatrix}$$
$$= \begin{pmatrix} 1 & 2 & 3 & 4 \\ 2 & 3 & 4 & 1 \end{pmatrix} \begin{pmatrix} 2 & 3 & 4 & 1 \\ 4 & 3 & 1 & 2 \end{pmatrix} = \begin{pmatrix} \mathbf{1} & \mathbf{2} & \mathbf{3} & \mathbf{4} \\ \mathbf{4} & \mathbf{3} & \mathbf{1} & \mathbf{2} \end{pmatrix}.$$

(iii)
$$gof = \begin{pmatrix} 1 & 2 & 3 & 4 \\ 2 & 4 & 3 & 1 \end{pmatrix} \begin{pmatrix} 1 & 2 & 3 & 4 \\ 2 & 3 & 4 & 1 \end{pmatrix}$$
$$= \begin{pmatrix} 1 & 2 & 3 & 4 \\ 2 & 4 & 3 & 1 \end{pmatrix} \begin{pmatrix} 2 & 4 & 3 & 1 \\ 3 & 1 & 4 & 2 \end{pmatrix} = \begin{pmatrix} \mathbf{1} & \mathbf{2} & \mathbf{3} & \mathbf{4} \\ \mathbf{3} & \mathbf{1} & \mathbf{4} & \mathbf{2} \end{pmatrix}.$$

(iv)
$$gog = \begin{pmatrix} 1 & 2 & 3 & 4 \\ 2 & 3 & 4 & 1 \end{pmatrix} \begin{pmatrix} 1 & 2 & 3 & 4 \\ 2 & 3 & 4 & 1 \end{pmatrix}$$
$$= \begin{pmatrix} 1 & 2 & 3 & 4 \\ 2 & 3 & 4 & 1 \end{pmatrix} \begin{pmatrix} 2 & 3 & 4 & 1 \\ 3 & 4 & 1 & 2 \end{pmatrix} = \begin{pmatrix} \mathbf{1} & \mathbf{2} & \mathbf{3} & \mathbf{4} \\ \mathbf{3} & \mathbf{4} & \mathbf{1} & \mathbf{2} \end{pmatrix}.$$

Example 3. *Let* $f = \begin{pmatrix} 1 & 2 & 3 \\ 2 & 3 & 1 \end{pmatrix}$, $g = \begin{pmatrix} 1 & 2 & 3 \\ 1 & 3 & 2 \end{pmatrix}$ *and* $h = \begin{pmatrix} 1 & 2 & 3 \\ 2 & 1 & 3 \end{pmatrix}$ *be three elements of A(S), where S = {1, 2, 3}. Show that fo(goh) = (fog)oh.*

Sol.
$$goh = \begin{pmatrix} 1 & 2 & 3 \\ 2 & 1 & 3 \end{pmatrix} \begin{pmatrix} 1 & 2 & 3 \\ 1 & 3 & 2 \end{pmatrix} = \begin{pmatrix} 1 & 2 & 3 \\ 2 & 1 & 3 \end{pmatrix} \begin{pmatrix} 2 & 1 & 3 \\ 3 & 1 & 2 \end{pmatrix} = \begin{pmatrix} 1 & 2 & 3 \\ 3 & 1 & 2 \end{pmatrix}$$

$$\therefore \quad fo(goh) = \begin{pmatrix} 1 & 2 & 3 \\ 3 & 1 & 2 \end{pmatrix} \begin{pmatrix} 1 & 2 & 3 \\ 2 & 3 & 1 \end{pmatrix} = \begin{pmatrix} 1 & 2 & 3 \\ 3 & 1 & 2 \end{pmatrix} \begin{pmatrix} 3 & 1 & 2 \\ 1 & 2 & 3 \end{pmatrix} = \begin{pmatrix} 1 & 2 & 3 \\ 1 & 2 & 3 \end{pmatrix}$$

$$fog = \begin{pmatrix} 1 & 2 & 3 \\ 1 & 3 & 2 \end{pmatrix} \begin{pmatrix} 1 & 2 & 3 \\ 2 & 3 & 1 \end{pmatrix} = \begin{pmatrix} 1 & 2 & 3 \\ 1 & 3 & 2 \end{pmatrix} \begin{pmatrix} 1 & 3 & 2 \\ 2 & 1 & 3 \end{pmatrix} = \begin{pmatrix} 1 & 2 & 3 \\ 2 & 1 & 3 \end{pmatrix}$$

$$\therefore \quad (fog)oh = \begin{pmatrix} 1 & 2 & 3 \\ 2 & 1 & 3 \end{pmatrix} \begin{pmatrix} 1 & 2 & 3 \\ 2 & 1 & 3 \end{pmatrix} = \begin{pmatrix} 1 & 2 & 3 \\ 2 & 1 & 3 \end{pmatrix} \begin{pmatrix} 2 & 1 & 3 \\ 1 & 2 & 3 \end{pmatrix} = \begin{pmatrix} 1 & 2 & 3 \\ 1 & 2 & 3 \end{pmatrix}$$

$\therefore$ **fo(goh) = (fog)oh.**

8. CYCLIC PERMUTATIONS

Let S be a finite set. A permutation f of S is called a **cyclic permutation** or a **cycle** if there exists elements $i_1, i_2, \ldots\ldots, i_t$ in S such that $f(i_1) = i_2$, $f(i_2) = i_3$, ..., $f(i_{t-i}) = i_t$, $f(i_t) = i_1$ and for any $j \in S$ different from $i_1, i_2, \ldots\ldots, i_t$ we have $f(j) = j$. The cyclic permutation f is denoted by the symbol $(i_1\ i_2 \ldots\ldots i_t)$. This notation of f is called **one-rowed notation.** t is called the **length** of the cycle f. The cycle f is also called a **t-cycle.**

Illustrations. *(i)* Let S = {1, 2, 3, 4}.

Let $f \in A(S)$ and $f = \begin{pmatrix} 1 & 2 & 3 & 4 \\ 2 & 3 & 1 & 4 \end{pmatrix}$.

Here $f(1) = 2, f(2) = 3, f(3) = 1$ and $f(4) = 4$.

$\therefore \quad f = (1\ 2\ 3)$.

$\therefore$ f is a cyclic permutation of length 3.

Remark. The cycle (1 2 3) can also be written as (2 3 1) or as (3 1 2).

(ii) Let S = {1, 2, 3, 4, 5}.

Let $f \in A(S)$ and $f = \begin{pmatrix} 1 & 2 & 3 & 4 & 5 \\ 1 & 3 & 2 & 4 & 5 \end{pmatrix}$

Here $f(2) = 3, f(3) = 2$ and $f(1) = 1, f(4) = 4, f(5) = 5$.

$\therefore \quad f = (2, 3)$.

$\therefore$ f is a cyclic permutation of length 2.

Example 4. *Let S = {1, 2, 3, 4, 5, 6}. Represents the following cycles of S by 2-rowed notation :*

(i) (2 3 4) *(ii)* (6 1)

(iii) (2 3 1 4 5) *(iv)* (1 6 3).

Sol. We have S = {1, 2, 3, 4, 5, 6}.

(i) Let $f = (2\ 3\ 4)$

$\therefore \quad f(2) = 3, f(3) = 4, f(4) = 2, f(1) = 1, f(5) = 5, f(6) = 6.$

$\therefore \quad f = \begin{pmatrix} \mathbf{1} & \mathbf{2} & \mathbf{3} & \mathbf{4} & \mathbf{5} & \mathbf{6} \\ \mathbf{1} & \mathbf{3} & \mathbf{4} & \mathbf{2} & \mathbf{5} & \mathbf{6} \end{pmatrix}$.

(ii) Let $g = (6\ 1)$.

$\therefore \quad g(6) = 1, g(1) = 6, g(2) = 2, g(3) = 3, g(4) = 4, g(5) = 5.$

$\therefore \quad g = \begin{pmatrix} \mathbf{1} & \mathbf{2} & \mathbf{3} & \mathbf{4} & \mathbf{5} & \mathbf{6} \\ \mathbf{6} & \mathbf{2} & \mathbf{3} & \mathbf{4} & \mathbf{5} & \mathbf{1} \end{pmatrix}$.

(iii) Let $\phi = (2\ 3\ 1\ 4\ 5)$.

$\therefore \quad \phi(2) = 3, \phi(3) = 1, \phi(1) = 4, \phi(4) = 5, \phi(5) = 2, \phi(6) = 6.$

$\therefore \quad \phi = \begin{pmatrix} \mathbf{1} & \mathbf{2} & \mathbf{3} & \mathbf{4} & \mathbf{5} & \mathbf{6} \\ \mathbf{4} & \mathbf{3} & \mathbf{1} & \mathbf{5} & \mathbf{2} & \mathbf{6} \end{pmatrix}$.

(iv) Let $\psi = (1\ 6\ 3)$.

$\therefore \quad \psi(1) = 6, \psi(6) = 3, \psi(3) = 1, \psi(2) = 2, \psi(4) = 4, \psi(5) = 5.$

$\therefore \quad \psi = \begin{pmatrix} \mathbf{1} & \mathbf{2} & \mathbf{3} & \mathbf{4} & \mathbf{5} & \mathbf{6} \\ \mathbf{6} & \mathbf{2} & \mathbf{1} & \mathbf{4} & \mathbf{5} & \mathbf{3} \end{pmatrix}$.

Example 5. *Let S = {1, 2, 3, 4, 5, 6, 7}. Which of the following elements of A(S) are cycles of S ?*

(i) $\begin{pmatrix} 1 & 2 & 3 & 4 & 5 & 6 & 7 \\ 1 & 3 & 4 & 2 & 5 & 6 & 7 \end{pmatrix}$ *(ii)* $\begin{pmatrix} 1 & 2 & 3 & 4 & 5 & 6 & 7 \\ 1 & 2 & 4 & 3 & 5 & 7 & 6 \end{pmatrix}$.

Sol. We have $S = \{1, 2, 3, 4, 5, 6, 7\}$.

(*i*) Let $f = \begin{pmatrix} 1 & 2 & 3 & 4 & 5 & 6 & 7 \\ 1 & 3 & 4 & 2 & 5 & 6 & 7 \end{pmatrix}$

$\therefore$ $2 \to 3, 3 \to 4, 4 \to 2,$
$1 \to 1, 5 \to 5, 6 \to 6, 7 \to 7.$

$\therefore$ f **is a cycle of length 3 and is equal to (2 3 4).**

(*ii*) Let $g = \begin{pmatrix} 1 & 2 & 3 & 4 & 5 & 6 & 7 \\ 1 & 2 & 4 & 3 & 5 & 7 & 6 \end{pmatrix}$

$\therefore$ $3 \to 4, 4 \to 3,$
$6 \to 7, 7 \to 6,$
and $1 \to 1, 2 \to 2, 5 \to 5.$

$\therefore$ **g is not a cycle.**

Example 6. *Let S = {1, 2, 3, 4, 5, 6}. Let (1 2 3) and (5 6 4 1) be two cyclic permutations of S. Find (1 2 3)o(5 6 4 1) and (5 6 4 1) o(1 2 3).*

Sol. We have $S = \{1, 2, 3, 4, 5, 6\}$.

$(1 \ 2 \ 3) \ o(5 \ 6 \ 4 \ 1)$

$$= \begin{pmatrix} 1 & 2 & 3 & 4 & 5 & 6 \\ 2 & 3 & 1 & 4 & 5 & 6 \end{pmatrix} o \begin{pmatrix} 1 & 2 & 3 & 4 & 5 & 6 \\ 5 & 2 & 3 & 1 & 6 & 4 \end{pmatrix}$$

$$= \begin{pmatrix} 1 & 2 & 3 & 4 & 5 & 6 \\ 5 & 2 & 3 & 1 & 6 & 4 \end{pmatrix} \begin{pmatrix} 1 & 2 & 3 & 4 & 5 & 6 \\ 2 & 3 & 1 & 4 & 5 & 6 \end{pmatrix}$$

$$= \begin{pmatrix} 1 & 2 & 3 & 4 & 5 & 6 \\ 5 & 2 & 3 & 1 & 6 & 4 \end{pmatrix} \begin{pmatrix} 5 & 2 & 3 & 1 & 6 & 4 \\ 5 & 3 & 1 & 2 & 6 & 4 \end{pmatrix}$$

$$= \begin{pmatrix} 1 & 2 & 3 & 4 & 5 & 6 \\ 5 & 3 & 1 & 2 & 6 & 4 \end{pmatrix} = \mathbf{(1\ 5\ 6\ 4\ 2\ 3).}$$

$(5 \ 6 \ 4 \ 1) \ o(1 \ 2 \ 3)$

$$= \begin{pmatrix} 1 & 2 & 3 & 4 & 5 & 6 \\ 5 & 2 & 3 & 1 & 6 & 4 \end{pmatrix} o \begin{pmatrix} 1 & 2 & 3 & 4 & 5 & 6 \\ 2 & 3 & 1 & 4 & 5 & 6 \end{pmatrix}$$

$$= \begin{pmatrix} 1 & 2 & 3 & 4 & 5 & 6 \\ 2 & 3 & 1 & 4 & 5 & 6 \end{pmatrix} \begin{pmatrix} 1 & 2 & 3 & 4 & 5 & 6 \\ 5 & 2 & 3 & 1 & 6 & 4 \end{pmatrix}$$

$$= \begin{pmatrix} 1 & 2 & 3 & 4 & 5 & 6 \\ 2 & 3 & 1 & 4 & 5 & 6 \end{pmatrix} \begin{pmatrix} 2 & 3 & 1 & 4 & 5 & 6 \\ 2 & 3 & 5 & 1 & 6 & 4 \end{pmatrix}$$

$$= \begin{pmatrix} 1 & 2 & 3 & 4 & 5 & 6 \\ 2 & 3 & 5 & 1 & 6 & 4 \end{pmatrix} = \mathbf{(1\ 2\ 3\ 5\ 6\ 4).}$$

9. TRANSPOSITION

A cyclic permutation of length 2 of a finite set is called a **transposition.**

10. DISJOINT CYCLES

Two cyclic permutations of a finite set are called **disjoint cycles** if they have no common element.

For example, the cycles (2 1 3) and (5 6) of the finite set {1, 2, 3, 4, 5, 6} are disjoint cycles. It can be proved that the composition of disjoint cycles is commutative.

Example 7. *Let S = {1, 2, 3, 4, 5}. Let f = (2 3), g = (1 4). Show that fog = gof.*

Sol. We have S = {1, 2, 3, 4, 5}.

$$f = (2 \;\; 3) = \begin{pmatrix} 1 & 2 & 3 & 4 & 5 \\ 1 & 3 & 2 & 4 & 5 \end{pmatrix} \text{ and } g = (1 \;\; 4) = \begin{pmatrix} 1 & 2 & 3 & 4 & 5 \\ 4 & 2 & 3 & 1 & 5 \end{pmatrix}.$$

Now
$$fog = \begin{pmatrix} 1 & 2 & 3 & 4 & 5 \\ 4 & 2 & 3 & 1 & 5 \end{pmatrix} \begin{pmatrix} 1 & 2 & 3 & 4 & 5 \\ 1 & 3 & 2 & 4 & 5 \end{pmatrix}$$

$$= \begin{pmatrix} 1 & 2 & 3 & 4 & 5 \\ 4 & 2 & 3 & 1 & 5 \end{pmatrix} \begin{pmatrix} 4 & 2 & 3 & 1 & 5 \\ 4 & 3 & 2 & 1 & 5 \end{pmatrix} = \begin{pmatrix} 1 & 2 & 3 & 4 & 5 \\ 4 & 3 & 2 & 1 & 5 \end{pmatrix}.$$

$$gof = \begin{pmatrix} 1 & 2 & 3 & 4 & 5 \\ 1 & 3 & 2 & 4 & 5 \end{pmatrix} \begin{pmatrix} 1 & 2 & 3 & 4 & 5 \\ 4 & 2 & 3 & 1 & 5 \end{pmatrix}$$

$$= \begin{pmatrix} 1 & 2 & 3 & 4 & 5 \\ 1 & 3 & 2 & 4 & 5 \end{pmatrix} \begin{pmatrix} 1 & 3 & 2 & 4 & 5 \\ 4 & 3 & 2 & 1 & 5 \end{pmatrix} = \begin{pmatrix} 1 & 2 & 3 & 4 & 5 \\ 4 & 3 & 2 & 1 & 5 \end{pmatrix}.$$

∴ **fog = gof.**

11. THEOREM

Every permutation can be expressed as a product of disjoint cycles.

Note. The proof of this theorem is beyond the scope of this book.

We can write (1 2 3 ...... *k*) as (1 *k*)*o* ...... *o*(1 3) *o*(1 2).

∴ Every *k*-cycle can be expressed as a product of transpositions.

∴ By the above theorem, every permutation can be expressed as a product of transpositions.

12. EVEN AND ODD PERMUTATIONS

A permutation is called an **even permutation** if it can be expressed as a product of even number of transpositions.

A permutation is called an **odd permutation** if it can be expressed as a product of odd number of transpositions.

Properties :

1. A permutation is either even or odd.
2. Product of two even permutations is an even permutation.
3. Product of an even permutation and an odd permutation is an odd permutation.
4. Product of an odd permutation and an even permutation is an odd permutation.
5. Product of two odd permutations is an even permutation.
6. If number of elements in S is *n*, then the number of even permutations of S is (n !)/2.

13. ALTERNATING GROUP

Let S be a finite set containing *n* elements. The set of all even permutations of S is denoted by A_n. We shall show that A_n is a group with composition of permutations as the binary operation.

A_n is a non-empty subset of the set A(S) and A(S) is a finite group with composition of permutations as the binary operation.

Let $f, g \in A_n$.

$\therefore$ f and g are even permutations of the set S.

$\therefore$ Their composition (*i.e.*, product) fog is also an even permutation.

$\therefore \quad fog \in A_n$

$\therefore$ A_n is a subgroup of A(S).

$\therefore$ A_n is also a group with composition of permutations as the binary operation.

The group A_n is called the **alternating group.**

IMPORTANT RESULTS

1. *The set A(S) of all permutations of a non-empty set S is a group with composition of mappings as the binary operation.*
2. *Every group is isomorphic to a permutation group (Cayley).*
3. *Every permutation can be expressed as a product of transpositions.*
4. *The set of even permutations on a finite set is a group with composition of mappings as the binary operation.*

EXERCISE

1. If S = {α, β, γ}, write A(S).
2. $f = \begin{pmatrix} 1 & 2 & 3 \\ 2 & 1 & 3 \end{pmatrix}, g = \begin{pmatrix} 2 & 3 & 1 \\ 1 & 3 & 2 \end{pmatrix}, h = \begin{pmatrix} 3 & 2 & 1 \\ 1 & 2 & 3 \end{pmatrix}$ be permutations of the set S = {1, 2, 3}. Show that $f = g, f \neq h, g \neq h$.
3. Let $f = \begin{pmatrix} 1 & 2 & 3 \\ 2 & 3 & 1 \end{pmatrix}$ and $g = \begin{pmatrix} 1 & 2 & 3 \\ 3 & 1 & 2 \end{pmatrix}$ be two permutations of the set S = {1, 2, 3}. Find *fof, fog, gof, gog.*
4. Let $f = \begin{pmatrix} a & b & c & d & e \\ b & c & d & a & e \end{pmatrix}$ and $g = \begin{pmatrix} a & c & d & b & e \\ a & e & b & c & d \end{pmatrix}$ be two permutations of the set S = {*a, b, c, d, e*}. Find the permutations *fof, fog, gof, gog.*
5. Let $f = \begin{pmatrix} 1 & 2 & 3 & 4 \\ 2 & 3 & 4 & 1 \end{pmatrix}, g = \begin{pmatrix} 2 & 3 & 1 & 4 \\ 1 & 2 & 3 & 4 \end{pmatrix}$ and $h = \begin{pmatrix} 1 & 3 & 2 & 4 \\ 2 & 1 & 3 & 4 \end{pmatrix}$ be three permutations of the set S = {1, 2, 3, 4}. Verify that $fo(goh) = (fog)oh$.
6. Let S = {1, 2, 3, 4}. Write the following cycles of S by using 2-rowed notation :

   (*i*) (2 3) (*ii*) (1 3 4) (*iii*) (2 1 4 3).
7. Let S = {1, 2, 3, 4, 5, 6, 7}. Write the following cycles of S by using 2 rowed notation :

   (*i*) (2 7) (*ii*) (1 3 2 5)

   (*iii*) (2 1 7 3 5) (*iv*) (2 1 3 4 6 7).
8. Let S = {1, 2, 3, 4, 5, 6}. Which of the following elements of A(S) are cycles of S ?

   (*i*) $\begin{pmatrix} 1 & 2 & 3 & 4 & 5 & 6 \\ 3 & 2 & 5 & 4 & 1 & 6 \end{pmatrix}$ (*ii*) $\begin{pmatrix} 1 & 2 & 3 & 4 & 5 & 6 \\ 5 & 1 & 4 & 3 & 2 & 6 \end{pmatrix}$

   (*iii*) $\begin{pmatrix} 1 & 2 & 3 & 4 & 5 & 6 \\ 2 & 1 & 4 & 3 & 5 & 6 \end{pmatrix}$ (*iv*) $\begin{pmatrix} 1 & 2 & 3 & 4 & 5 & 6 \\ 2 & 3 & 4 & 5 & 6 & 1 \end{pmatrix}$.

9. Let S = {1, 2, 3, 4, 5, 6, 7}. If $f = (2\ 3)$, $g = (1\ 4\ 5)$, find the elements *fog* and *gof* of A(S).

10. Let S = {1, 2, 3, 4, 5, 6, 7} and $f = (1\ 3)$, $g = (2\ 4\ 6)$. Show that *fog* = *gof*.

ANSWERS

1. $\left\{\begin{pmatrix}\alpha & \beta & \gamma\\ \alpha & \beta & \gamma\end{pmatrix}, \begin{pmatrix}\alpha & \beta & \gamma\\ \alpha & \gamma & \beta\end{pmatrix}, \begin{pmatrix}\alpha & \beta & \gamma\\ \beta & \alpha & \gamma\end{pmatrix}, \begin{pmatrix}\alpha & \beta & \gamma\\ \beta & \gamma & \alpha\end{pmatrix}, \begin{pmatrix}\alpha & \beta & \gamma\\ \gamma & \alpha & \beta\end{pmatrix}, \begin{pmatrix}\alpha & \beta & \gamma\\ \gamma & \beta & \alpha\end{pmatrix}\right\}$

3. $\begin{pmatrix}1 & 2 & 3\\ 3 & 1 & 2\end{pmatrix}, \begin{pmatrix}1 & 2 & 3\\ 1 & 2 & 3\end{pmatrix}, \begin{pmatrix}1 & 2 & 3\\ 1 & 2 & 3\end{pmatrix}, \begin{pmatrix}1 & 2 & 3\\ 2 & 3 & 1\end{pmatrix}$

4. $\begin{pmatrix}a & b & c & d & e\\ c & d & a & b & e\end{pmatrix}, \begin{pmatrix}a & b & c & d & e\\ b & d & e & c & a\end{pmatrix}, \begin{pmatrix}a & b & c & d & e\\ c & e & b & a & d\end{pmatrix}, \begin{pmatrix}a & b & c & d & e\\ a & e & d & c & b\end{pmatrix}$

6. (*i*) $\begin{pmatrix}1 & 2 & 3 & 4\\ 1 & 3 & 2 & 4\end{pmatrix}$ (*ii*) $\begin{pmatrix}1 & 2 & 3 & 4\\ 3 & 2 & 4 & 1\end{pmatrix}$ (*iii*) $\begin{pmatrix}1 & 2 & 3 & 4\\ 4 & 1 & 2 & 3\end{pmatrix}$

7. (*i*) $\begin{pmatrix}1 & 2 & 3 & 4 & 5 & 6 & 7\\ 1 & 7 & 3 & 4 & 5 & 6 & 2\end{pmatrix}$ (*ii*) $\begin{pmatrix}1 & 2 & 3 & 4 & 5 & 6 & 7\\ 3 & 5 & 2 & 4 & 1 & 6 & 7\end{pmatrix}$

(*iii*) $\begin{pmatrix}1 & 2 & 3 & 4 & 5 & 6 & 7\\ 7 & 1 & 5 & 4 & 2 & 6 & 3\end{pmatrix}$ (*iv*) $\begin{pmatrix}1 & 2 & 3 & 4 & 5 & 6 & 7\\ 3 & 1 & 4 & 6 & 5 & 7 & 2\end{pmatrix}$

8. (*i*), (*iv*) **9.** $\begin{pmatrix}1 & 2 & 3 & 4 & 5 & 6 & 7\\ 4 & 3 & 2 & 5 & 1 & 6 & 7\end{pmatrix}, \begin{pmatrix}1 & 2 & 3 & 4 & 5 & 6 & 7\\ 4 & 3 & 2 & 5 & 1 & 6 & 7\end{pmatrix}.$

15
Rings

1. DEFINITION OF A RING

A non-empty set R with two binary operations denoted by '+' and '.' and called addition and multiplication respectively is said to be a **ring** if :

1. $a + (b + c) = (a + b) + c \quad \forall\, a, b, c \in R$

2. There exists an element $0 \in R$ such that

$$a + 0 = a = 0 + a \quad \forall\, a \in R$$

3. For $a \in R$, there exists an element $-a$ in R such that

$$a + (-a) = 0 = (-a) + a.$$

4. $a + b = b + a \quad \forall\, a, b \in R$

5. $a \,.\, (b \,.\, c) = (a \,.\, b) \,.\, c \quad \forall\, a, b, c \in R$

6. $a \,.\, (b + c) = a \,.\, b + a \,.\, c$ and $(b + c) \,.\, a = b \,.\, a + c \,.\, a \quad \forall\, a, b, c \in R.$

The element '0' is called the **additive identity** of the ring (R, +, .). The element '$-a$' is called the **additive inverse** of a.

The additive identity of the ring (R, +, .) is represented by '0' and it has nothing to do with the number zero.

In place of '+' and '.', the binary operations of the ring can very well be denoted by other symbols like $*$, o, $\oplus$, $\odot$ etc.

Axioms **1—4** shows that (R, +) is an abelian group.

Axiom **5** states that the binary operation '.' is associative.

Axiom **6** states that the multiplication is distributive over addition.

Remark. The binary operations of a ring are generally taken as '+' and '.' but these operations have no bearing with the usual addition and multiplication of numbers.

2. COMMUTATIVE RING

A ring (R, +, .) is called a **commutative ring** if $a \,.\, b = b \,.\, a \quad \forall\, a, b \in R$.

3. RING WITH UNIT ELEMENT

A ring (R, +, .) is called a **ring with unit element** if there exists an element '1' in R such that $a \,.\, 1 = a = 1 \,.\, a \quad \forall\, a \in R$.

The unit element of a ring is also called its **multiplication identity.**

4. RING WITHOUT ZERO DIVISORS

A ring (R, +, .) is called a **ring without zero divisors** if $a \,.\, b = 0$ for $a, b \in R$ then either $a = 0$ or $b = 0$.

5. INTEGRAL DOMAIN

A ring (R, +, .) is called an **integral domain** if :

(*i*) the ring (R, +, .) is commutative

(*ii*) the ring (R, +, .) is without zero divisors.

In short, we can say that a commutative ring without zero divisors is an integral domain.

6. MULTIPLICATIVE INVERSE OF AN ELEMENT

Let (R, +, .) be a ring with unit element '1' and let a be any element of R. An element $b \in$ R is called the **multiplicative inverse** of a if

$$a \,.\, b = 1 = b \,.\, a.$$

The multiplicative inverse of an element a is denoted by a^{-1}.

7. FIELD

A ring (R, +, .) is called a **field** if :

(*i*) the ring (R, +, .) is commutative.

(*ii*) the ring (R, +, .) has unit element.

(*iii*) every non-zero element of R has multiplicative inverse.

Equivalently we can say that a non-empty set R with two binary operations '+' and '.' is a field if :

(*i*) (R, +) is an abelian group.

(*ii*) (R – {0}, .) is an abelian group.

(*iii*) Multiplication '.' distributes over addition '+'.

Thus a non-empty set R with two binary operations '+' and '.' is a field if :

1. $a + (b + c) = (a + b) + c \quad \forall\, a, b, c \in \text{R}$

2. There exist an element $0 \in$ R such that

$$a + 0 = a = 0 + a \quad \forall\, a \in \text{R}$$

'0' is called additive identity of R.

3. For $a \in$ R, there exists an element $-a$ in R such that

$$a + (-a) = 0 = (-a) + a$$

$-a$ is called the additive inverse of a.

4. $a + b = b + a \quad \forall\, a, b \in \text{R}.$

5. $a \,.\, (b \,.\, c) = (a \,.\, b) \,.\, c \quad \forall\, a, b, c \in \text{R}.$

6. There exist an element $1 \in$ R such that

$$a \,.\, 1 = a = 1 \,.\, a \quad \forall\, a \in \text{R}$$

'1' is called the multiplicative identity or unit element of R.

7. For $a\ (\neq 0) \in$ R, there exist an element a^{-1} in R such that

$$a \,.\, a^{-1} = 1 = a^{-1} \,.\, a$$

a^{-1} is called the multiplicative inverse of a.

8. $a \,.\, b = b \,.\, a \quad \forall\, a, b \in \text{R}.$

9. $a \,.\, (b + c) = a \,.\, b + a \,.\, c$ and $(b + c) \,.\, a = b \,.\, a + c \,.\, a \quad \forall\, a, b, c \in \text{R}.$

Remark. In axiom **9**, we need writing only one equality because the other follows by using axiom **8**.

8. DIVISION RING

A ring (R, +, .) is called a division ring if :

(*i*) the ring (R, +, .) has unit element

(*ii*) every non-zero element of R has multiplicative inverse.

Remark 1. A division ring is also called a **skew field.**

Remark 2. A commutative division ring is a field.

Example 1. *Let R = **Z**, the set of integers and binary operations '+' and '.' be respectively the usual addition and multiplication of integers. Show that (R, +, .) is a ring.*

Sol. We have :

1. $a + (b + c) = (a + b) + c \quad \forall\, a, b, c \in \mathbf{Z}$

2. $0 \in \mathbf{Z}$ and $a + 0 = a = 0 + a \quad \forall\, a \in \mathbf{Z}$

3. For $a \in \mathbf{Z}$, we have $-a \in \mathbf{Z}$ and $a + (-a) = 0 = (-a) + a$

4. $a + b = b + a \quad \forall\, a, b \in \mathbf{Z}$

5. $a . (b . c) = (a . b) . c \quad \forall\, a, b, c \in \mathbf{Z}$

6. $a . (b + c) = a . b + a . c$ and $(b + c) . a = b . a + c . a \quad \forall\, a, b, c \in \mathbf{Z}$

$\therefore$ (**Z**, +, .) is a ring.

Note 1. (**Z**, +, .) is also an integral domain because :

(*i*) $a . b = b . a \quad \forall\, a, b \in \mathbf{Z}$

(*ii*) $a . b = 0$ for $a, b \in \mathbf{Z}$ implies either $a = 0$ or $b = 0$.

Note 2. (**Z**, +, .) is not a field because $4\ (\neq 0) \in \mathbf{Z}$ and there is no integer k such that

$$4 . k = 1 = k . 4.$$

Example 2. *Let **C** be the set of complex numbers and binary operations '+' and '.' be respectively the usual addition and multiplication of complex numbers. Show that (**C**, +, .) is a ring.*

Sol. We have :

1. $z_1 + (z_2 + z_3) = (z_1 + z_2) + z_3 \quad \forall\, z_1, z_2, z_3 \in \mathbf{C}$

2. $0\ (= 0 + 0\,i) \in \mathbf{C}$ and $z + 0 = z = 0 + z \quad \forall\, z \in \mathbf{C}$

3. For $z\ (= a + ib) \in \mathbf{C}$, we have $-z\ (= -a - ib) \in \mathbf{C}$

and $$z + (-z) = 0 = (-z) + z.$$

4. $z_1 + z_2 = z_2 + z_1 \quad \forall\, z_1, z_2 \in \mathbf{C}$

5. $z_1 . (z_2 . z_3) = (z_1 . z_2) . z_3 \quad \forall\, z_1, z_2, z_3 \in \mathbf{C}$

6. $z_1 . (z_2 + z_3) = z_1 . z_2 + z_1 . z_3$ and $(z_2 + z_3) . z_1 = z_2 . z_1 + z_3 . z_1 \quad \forall\, z_1, z_2, z_3 \in \mathbf{C}$

$\therefore$ (**C**, +, .) is a ring.

Note 1. (**C**, +, .) is also an integral domain because

(*i*) $z_1 . z_2 = z_2 . z_1 \quad \forall\, z_1, z_2 \in \mathbf{C}$

(*ii*) $z_1 . z_2 = 0$ for $z_1, z_2 \in \mathbf{C}$ implies either $z_1 = 0$ or $z_2 = 0$.

Note 2. (**C**, +, .) is also a field because :

(*i*) $z_1 . z_2 = z_2 . z_1 \quad \forall\, z_1, z_2 \in \mathbf{C}$.

(*ii*) $1 (= 1 + 0i) \in \mathbf{C}$ and $z . 1 = z = 1 . z \quad \forall\, z \in \mathbf{C}$

(*iii*) Let $z\ (= a + ib)$ be any non-zero complex numbers

$\therefore \quad a^2 + b^2 \neq 0$

$\therefore \quad \dfrac{a}{a^2+b^2} + \dfrac{(-b)}{a^2+b^2}\, i \in \mathbf{C}$ and

$$(a+ib)\left(\frac{a}{a^2+b^2} + \frac{(-b)}{a^2+b^2}i\right) = 1 = \left(\frac{a}{a^2+b^2} + \frac{(-b)}{a^2+b^2}i\right)(a+ib).$$

$\therefore$ **Every non-zero element of C has a multiplicative inverse.**

Example 3. *Let ⊕ and ⊙ be binary operations on* **Z** *defined by*

$$a \oplus b = a + b - 1 \quad \forall\, a, b \in \mathbf{Z} \text{ and } a \odot b = a + b - ab \quad \forall\, a, b \in \mathbf{Z}.$$

Show that (**Z**, ⊕, ⊙) *is a commutative ring with unit element. Also show that* (**Z**, ⊕, ⊙) *is an integral domain but not a field.*

Sol. 1. Let $a, b, c \in \mathbf{Z}$.

$$a \oplus (b \oplus c) = a \oplus (b + c - 1) = a + (b + c - 1) - 1 = a + b + c - 2$$

Also $\quad (a \oplus b) \oplus c = (a + b - 1) \oplus c = (a + b - 1) + c - 1 = a + b + c - 2$

$\therefore \quad (a \oplus b) \oplus c = a \oplus (b \oplus c) \quad \forall\, a, b, c \in \mathbf{Z}.$

2. $1 \in \mathbf{Z}$ and for $a \in \mathbf{Z}$,

$$a \oplus 1 = a + 1 - 1 = a \text{ and } 1 \oplus a = 1 + a - 1 = a$$

$\therefore \quad a \oplus 1 = a = 1 \oplus a \quad \forall\, a \in \mathbf{Z}.$

3. Let $a \in \mathbf{Z}$. $\therefore$ $2 - a \in \mathbf{Z}$ and

$$a \oplus (2 - a) = a + (2 - a) - 1 = 1$$

and $\quad (2 - a) \oplus a = (2 - a) + a - 1 = 1$

$\therefore \quad a \oplus (2 - a) = 1 = (2 - a) \oplus a$

$\therefore$ $2 - a$ is additive inverse of a.

4. Let $a, b \in \mathbf{Z}$.

$$a \oplus b = a + b - 1 = b + a - 1 = b \oplus a$$

$\therefore \quad a \oplus b = b \oplus a \quad \forall\, a, b \in \mathbf{Z}.$

5. Let $a, b, c \in \mathbf{Z}$.

$$\begin{aligned} a \odot (b \odot c) &= a \odot (b + c - bc) \\ &= a + (b + c - bc) - a(b + c - bc) = a + b + c - bc - ab - ac + abc \end{aligned}$$

Also
$$\begin{aligned} (a \odot b) \odot c &= (a + b - ab) \odot c \\ &= (a + b - ab) + c - (a + b - ab)\, c = a + b + c - ab - ac - bc + abc \end{aligned}$$

$\therefore \quad (a \odot b) \odot c = a \odot (b \odot c) \;\forall\, a, b, c \in \mathbf{Z}.$

6. Let $a, b, c \in \mathbf{Z}$.

$$\begin{aligned} a \odot (b \oplus c) &= a \odot (b + c - 1) \\ &= a + (b + c - 1) - a(b + c - 1) = 2a + b + c - ab - ac - 1 \end{aligned}$$

$$\begin{aligned} (a \odot b) \oplus (a \odot c) &= (a + b - ab) \oplus (a + c - ac) \\ &= (a + b - ab) + (a + c - ac) - 1 = 2a + b + c - ab - ac - 1 \end{aligned}$$

$\therefore \quad a \odot (b \oplus c) = (a \odot b) \oplus (a \odot c)$

Similarly, $\quad (b \oplus c) \odot a = (b \odot a) \oplus (c \odot a).$

$\therefore$ (**Z**, ⊕, ⊙) is a ring.

7. Let $a, b \in \mathbf{Z}$.

$$a \odot b = a + b - ab = b + a - ba = b \odot a$$

$\therefore \quad a \odot b = b \odot a \quad \forall\, a, b \in \mathbf{Z}$

$\therefore$ $(\mathbf{Z}, \oplus, .)$ is a commutative ring .

The additive identity is '1'. Let $a \odot b = 1$.

$\Rightarrow \quad a + b - ab = 1 \quad \Rightarrow \quad 1 - a - b + ab = 0$

$\Rightarrow \quad (1 - a)(1 - b) = 0 \quad \Rightarrow \quad a = 1 \text{ or } b = 1$

$\therefore \quad a \odot b = 1 \quad \Rightarrow \quad a = 1 \text{ or } b = 1.$

$\therefore$ The ring $(\mathbf{Z}, \oplus, \odot)$ is without zero divisors.

$\therefore$ $(\mathbf{Z}, \oplus, \odot)$ is an integral domain.

We have $0 \in \mathbf{Z}$. Let $a \in \mathbf{Z}$

$\therefore \quad a \odot 0 = a + 0 + a(0) = a$

and $\quad 0 \odot a = 0 + a + 0(a) = a$

$\therefore \quad a \odot 0 = a = 0 \odot a \quad \forall\, a \in \mathbf{Z}.$

$\therefore$ 0 is the unit element.

Let $a (\neq 1) \in \mathbf{Z}$.

$$a \odot b = 0 \;\Rightarrow\; a + b - ab = 0 \;\Rightarrow\; b = \frac{a}{a-1}$$

For $a\ (\neq 1) \in \mathbf{Z}$, the number b *i.e.*, $\dfrac{a}{a-1}$ may not be in $\mathbf{Z}$.

$\therefore$ Non-zero elements of $\mathbf{Z}$ may not have multiplicative inverses in $\mathbf{Z}$.

$\therefore$ $(\mathbf{Z}, \oplus, \odot)$ is not a field.

Note 1. $a \oplus e = a \;\Rightarrow\; a + e - 1 = a \;\Rightarrow\; e = 1$

This gives us the idea of taking '1' as the additive identity.

Note 2. $a \odot e = a \;\Rightarrow\; a + e - ae = a \;\Rightarrow\; e(1 - a) = 0$

This prompted us to take the multiplicative identity as '0'.

Example 4. *Let M_n be the set of all $n \times n$ matrices with elements from the set of integers (or rational numbers or real numbers or complex numbers). Show that $(M_n, +, .)$ is a non-commutative ring with unit element.*

Sol. Sum and product of two $n \times n$ matrices are $n \times n$ matrices.

$\therefore$ '+' and '.' are binary operations on M_n.

1. Let $A, B, C \in M_n$.

We have $\quad A + (B + C) = (A + B) + C.$

2. The null matrix O is in M_n and for $A \in M_n$, we have $A + O = A = O + A$.

$\therefore$ O is the additive identity.

3. Let $A \in M_n$. $\therefore$ $-A \in M_n$ and $A + (-A) = O = (-A) + A$.

$\therefore$ $-A$ is the additive inverse of A.

4. Let $A, B \in M_n$.

We have $\quad A + B = B + A.$

5. Let A, B, C $\in$ M_n

We have A . (B . C) = (A . B) . C.

6. Let A, B, C $\in$ M_n.

We have A . (B + C) = A . B + A . C and (B + C) . A = B . A + C . A.

$\therefore$ (M_n, +, .) is a ring.

For A, B $\in$ M_n, we may not have A . B = B . A.

$\therefore$ (M_n, +, .) is not a commutative ring.

We have $I_n \in M_n$ and for A $\in$ M_n,

$$A \,.\, I_n = A = I_n \,.\, A.$$

$\therefore$ I_n is the unit element of the ring (M_n, +, .).

Example 5. *Let binary operations* $\oplus$ *and* $\odot$ *be defined on* **Z** *by*

$$a \oplus b = ab \quad \forall\, a, b \in \mathbf{Z}$$

$$a \odot b = a + b \quad \forall\, a, b \in \mathbf{Z}.$$

Show that (**Z**, $\oplus$, $\odot$) *is not a ring.*

Sol. 1. Let $a, b, c \in \mathbf{Z}$.

$$a \oplus (b \oplus c) = a \oplus bc = a(bc)$$

and $$(a \oplus b) \oplus c = ab \oplus c = (ab)\, c = a(bc)$$

$\therefore$ $$a \oplus (b \oplus c) = (a \oplus b) \oplus c.$$

2. $1 \in \mathbf{Z}$ and for $a \in \mathbf{Z}$, we have

$$a \oplus 1 = a(1) = a \text{ and } 1 \oplus a = 1(a) = a$$

$\therefore$ $$a \oplus 1 = a = 1 \oplus a \quad \forall\, a \in \mathbf{Z}$$

$\therefore$ 1 is the additive identity.

3. Let $a \in \mathbf{Z}$.

For $b \in \mathbf{Z}$, $a \oplus b = 1 = b \oplus a$ implies $ab = 1 = ba$ *i.e.*, $b = 1/a$.

But for $a \in \mathbf{Z}$, the number $1/a$ may not be in **Z**

$\therefore$ Additive inverse does not exist in **Z**.

$\therefore$ (**Z**, $\oplus$, $\odot$) is not a ring.

Example 6. *Let* $R = \left\{ \begin{bmatrix} a & b \\ 0 & 0 \end{bmatrix} : a, b \in \mathbf{R} \right\}$. *Show that R is a non-commutative ring and without multiplicative identity under usual addition and multiplication.*

Sol. Let $$\begin{bmatrix} a & b \\ 0 & 0 \end{bmatrix}, \begin{bmatrix} c & d \\ 0 & 0 \end{bmatrix} \in \text{R}.$$

$\therefore$ $$\begin{bmatrix} a & b \\ 0 & 0 \end{bmatrix} + \begin{bmatrix} c & d \\ 0 & 0 \end{bmatrix} = \begin{bmatrix} a+c & b+d \\ 0 & 0 \end{bmatrix} \in \text{R}$$

and $$\begin{bmatrix} a & b \\ 0 & 0 \end{bmatrix} \begin{bmatrix} c & d \\ 0 & 0 \end{bmatrix} = \begin{bmatrix} ac+0 & ad+0 \\ 0+0 & 0+0 \end{bmatrix} = \begin{bmatrix} ac & ad \\ 0 & 0 \end{bmatrix} \in \text{R}$$

$\therefore$ '+' and '.' are binary operations on R.

1. Let $$\text{A} = \begin{bmatrix} a & b \\ 0 & 0 \end{bmatrix}, \text{B} = \begin{bmatrix} c & d \\ 0 & 0 \end{bmatrix}, \text{C} = \begin{bmatrix} e & f \\ 0 & 0 \end{bmatrix} \in \text{R}$$

$$A + (B + C) = \begin{bmatrix} a & b \\ 0 & 0 \end{bmatrix} + \left(\begin{bmatrix} c & d \\ 0 & 0 \end{bmatrix} + \begin{bmatrix} e & f \\ 0 & 0 \end{bmatrix} \right)$$

$$= \begin{bmatrix} a & b \\ 0 & 0 \end{bmatrix} + \begin{bmatrix} c+e & d+f \\ 0 & 0 \end{bmatrix} = \begin{bmatrix} a+(c+e) & b+(d+f) \\ 0 & 0 \end{bmatrix}$$

$$= \begin{bmatrix} (a+c)+e & (b+d)+f \\ 0 & 0 \end{bmatrix} = \begin{bmatrix} a+c & b+d \\ 0 & 0 \end{bmatrix} + \begin{bmatrix} e & f \\ 0 & 0 \end{bmatrix}$$

$$= \left(\begin{bmatrix} a & b \\ 0 & 0 \end{bmatrix} + \begin{bmatrix} c & d \\ 0 & 0 \end{bmatrix} \right) + \begin{bmatrix} e & f \\ 0 & 0 \end{bmatrix} = (A + B) + C$$

$\therefore \quad A + (B + C) = (A + B) + C \quad \forall A, B, C \in R.$

2. $O = \begin{bmatrix} 0 & 0 \\ 0 & 0 \end{bmatrix} \in R$ **and for** $A = \begin{bmatrix} a & b \\ 0 & 0 \end{bmatrix}$**, we have**

$$A + O = \begin{bmatrix} a & b \\ 0 & 0 \end{bmatrix} + \begin{bmatrix} 0 & 0 \\ 0 & 0 \end{bmatrix} = \begin{bmatrix} a+0 & b+0 \\ 0+0 & 0+0 \end{bmatrix} = \begin{bmatrix} a & b \\ 0 & 0 \end{bmatrix} = A$$

and
$$O + A = \begin{bmatrix} 0 & 0 \\ 0 & 0 \end{bmatrix} + \begin{bmatrix} a & b \\ 0 & 0 \end{bmatrix} = \begin{bmatrix} 0+a & 0+b \\ 0+0 & 0+0 \end{bmatrix} = \begin{bmatrix} a & b \\ 0 & 0 \end{bmatrix} = A$$

$\therefore \quad A + O = A = O + A \quad \forall A \in R.$

3. Let $A = \begin{bmatrix} a & b \\ 0 & 0 \end{bmatrix} \in R$. $\therefore \quad -A = \begin{bmatrix} -a & -b \\ 0 & 0 \end{bmatrix} \in R$

$$A + (-A) = \begin{bmatrix} a & b \\ 0 & 0 \end{bmatrix} + \begin{bmatrix} -a & -b \\ 0 & 0 \end{bmatrix} = \begin{bmatrix} 0 & 0 \\ 0 & 0 \end{bmatrix} = O$$

and
$$(-A) + A = \begin{bmatrix} -a & -b \\ 0 & 0 \end{bmatrix} + \begin{bmatrix} a & b \\ 0 & 0 \end{bmatrix} = \begin{bmatrix} 0 & 0 \\ 0 & 0 \end{bmatrix} = O$$

$\therefore \quad A + (-A) = O = (-A) + A.$

4. Let $\quad A = \begin{bmatrix} a & b \\ 0 & 0 \end{bmatrix}, B = \begin{bmatrix} c & d \\ 0 & 0 \end{bmatrix} \in R$

$$A + B = \begin{bmatrix} a & b \\ 0 & 0 \end{bmatrix} + \begin{bmatrix} c & d \\ 0 & 0 \end{bmatrix} = \begin{bmatrix} a+c & b+d \\ 0 & 0 \end{bmatrix}$$

$$= \begin{bmatrix} c+a & d+b \\ 0 & 0 \end{bmatrix} = \begin{bmatrix} c & d \\ 0 & 0 \end{bmatrix} + \begin{bmatrix} a & d \\ 0 & 0 \end{bmatrix} = B + A$$

$\therefore \quad A + B = B + A \quad \forall A, B \in R.$

5. Let $\quad A = \begin{bmatrix} a & b \\ 0 & 0 \end{bmatrix}, B = \begin{bmatrix} c & d \\ 0 & 0 \end{bmatrix}, C = \begin{bmatrix} e & f \\ 0 & 0 \end{bmatrix} \in R$

$$A \,.\, (B \,.\, C) = \begin{bmatrix} a & b \\ 0 & 0 \end{bmatrix} \left(\begin{bmatrix} c & d \\ \xrightarrow{} & \\ 0 & 0 \end{bmatrix} . \begin{bmatrix} e & \downarrow & f \\ 0 & & 0 \end{bmatrix} \right) = \begin{bmatrix} a & b \\ 0 & 0 \end{bmatrix} \begin{bmatrix} ce+0 & cf+0 \\ 0+0 & 0+0 \end{bmatrix}$$

$$= \begin{bmatrix} a & b \\ \xrightarrow{} & \\ 0 & 0 \end{bmatrix} \begin{bmatrix} ce & \downarrow & cf \\ 0 & & 0 \end{bmatrix} = \begin{bmatrix} a(ce)+0 & a(cf)+0 \\ 0+0 & 0+0 \end{bmatrix} = \begin{bmatrix} a(ce) & a(cf) \\ 0 & 0 \end{bmatrix}$$

$$(A \,.\, B) \,.\, C = \left(\begin{bmatrix} a & b \\ 0 & 0 \end{bmatrix} . \begin{bmatrix} c & d \\ 0 & 0 \end{bmatrix}\right) . \begin{bmatrix} e & f \\ 0 & 0 \end{bmatrix}$$

$$= \begin{bmatrix} ac+0 & ad+0 \\ 0+0 & 0+0 \end{bmatrix} . \begin{bmatrix} e & f \\ 0 & 0 \end{bmatrix}$$

$$= \begin{bmatrix} ac & ad \\ 0 & 0 \end{bmatrix} . \begin{bmatrix} e & f \\ 0 & 0 \end{bmatrix} = \begin{bmatrix} (ac)e+0 & (ac)f+0 \\ 0+0 & 0+0 \end{bmatrix} = \begin{bmatrix} a(ce) & a(cf) \\ 0 & 0 \end{bmatrix}$$

$\therefore$ A . (B . C) = (A . B) . C $\forall$ A, B, C $\in$ R.

6. Let

$$A = \begin{bmatrix} a & b \\ 0 & 0 \end{bmatrix}, B = \begin{bmatrix} c & d \\ 0 & 0 \end{bmatrix}, C = \begin{bmatrix} e & f \\ 0 & 0 \end{bmatrix} \in R$$

$$A \,.\, (B + C) = \begin{bmatrix} a & b \\ 0 & 0 \end{bmatrix}\left(\begin{bmatrix} c & d \\ 0 & 0 \end{bmatrix} + \begin{bmatrix} e & f \\ 0 & 0 \end{bmatrix}\right) = \begin{bmatrix} a & b \\ 0 & 0 \end{bmatrix} . \begin{bmatrix} c+e & d+f \\ 0 & 0 \end{bmatrix}$$

$$= \begin{bmatrix} a(c+e)+0 & a(d+f)+0 \\ 0+0 & 0+0 \end{bmatrix} = \begin{bmatrix} a(c+e) & a(d+f) \\ 0 & 0 \end{bmatrix}$$

$$A \,.\, B + A \,.\, C = \begin{bmatrix} a & b \\ 0 & 0 \end{bmatrix}\begin{bmatrix} c & d \\ 0 & 0 \end{bmatrix} + \begin{bmatrix} a & b \\ 0 & 0 \end{bmatrix} . \begin{bmatrix} e & f \\ 0 & 0 \end{bmatrix}$$

$$= \begin{bmatrix} ac+0 & ad+0 \\ 0+0 & 0+0 \end{bmatrix} + \begin{bmatrix} ae+0 & af+0 \\ 0+0 & 0+0 \end{bmatrix}$$

$$= \begin{bmatrix} ac+ae & ad+af \\ 0 & 0 \end{bmatrix} = \begin{bmatrix} a(c+e) & a(d+f) \\ 0 & 0 \end{bmatrix}$$

$\therefore$ A . (B + C) = A . B + A . C $\forall$ A, B, C $\in$ R.

Similarly, (B + C) . A = B . A + C . A $\forall$ A, B, C $\in$ R.

$\therefore$ (R, +, .) is a ring.

Let

$$A = \begin{bmatrix} a & b \\ 0 & 0 \end{bmatrix}, B = \begin{bmatrix} c & d \\ 0 & 0 \end{bmatrix} \in R.$$

$$A \,.\, B = \begin{bmatrix} a & b \\ 0 & 0 \end{bmatrix}\begin{bmatrix} c & d \\ 0 & 0 \end{bmatrix} = \begin{bmatrix} ac+0 & ad+0 \\ 0+0 & 0+0 \end{bmatrix} = \begin{bmatrix} ac & ad \\ 0 & 0 \end{bmatrix}$$

and

$$B \,.\, A = \begin{bmatrix} c & d \\ 0 & 0 \end{bmatrix}\begin{bmatrix} a & b \\ 0 & 0 \end{bmatrix} = \begin{bmatrix} ca+0 & cb+0 \\ 0+0 & 0+0 \end{bmatrix} = \begin{bmatrix} ca & cb \\ 0 & 0 \end{bmatrix}$$

$\therefore$ A . B $\neq$ B . A

$\therefore$ The ring (R, +, .) is not commutative.

Let $\begin{bmatrix} x & y \\ 0 & 0 \end{bmatrix} \in$ R be such that for $\begin{bmatrix} a & b \\ 0 & 0 \end{bmatrix} \in$ R, we have

$$\begin{bmatrix} a & b \\ 0 & 0 \end{bmatrix}\begin{bmatrix} x & y \\ 0 & 0 \end{bmatrix} = \begin{bmatrix} a & b \\ 0 & 0 \end{bmatrix} = \begin{bmatrix} x & y \\ 0 & 0 \end{bmatrix}\begin{bmatrix} a & b \\ 0 & 0 \end{bmatrix}$$

$$\Rightarrow \quad \begin{bmatrix} ax & ay \\ 0 & 0 \end{bmatrix} = \begin{bmatrix} a & b \\ 0 & 0 \end{bmatrix} = \begin{bmatrix} xa & xb \\ 0 & 0 \end{bmatrix}$$

$$\Rightarrow \quad ax = a, ay = b = xb$$

$$\Rightarrow \quad x = 1, ay = b \;\Rightarrow\; x = 1, y = b/a$$

$\therefore \begin{bmatrix} x & y \\ 0 & 0 \end{bmatrix} = \begin{bmatrix} 1 & b/a \\ 0 & 0 \end{bmatrix}$, which is not independent of the matrix $\begin{bmatrix} a & b \\ 0 & 0 \end{bmatrix}$

$\therefore$ The ring (R, +, .) does not have multiplicative identity.

Example 7. *Let R = {0, 1, 2, 3, 4}. Let the symbols ⊕ and ⊙ denote the operations 'addition modulo 5' and 'multiplication modulo 5' respectively. Show that (R, ⊕, ⊙) is a commutative ring with unit element, an integral domain, a field.*

Sol. **Composition table of ⊕**

⊕	*0*	*1*	*2*	*3*	*4*
0	0	1	2	3	4
1	1	2	3	4	0
2	2	3	4	0	1
3	3	4	0	1	2
4	4	0	1	2	3

Composition table of ⊙

⊙	*0*	*1*	*2*	*3*	*4*
0	0	0	0	0	0
1	0	1	2	3	4
2	0	2	4	1	3
3	0	3	1	4	2
4	0	4	3	2	1

The composition tables of ⊕ and ⊙ show that these are binary operations on R.

Let $a, b, c \in$ R, we have

1. $a \oplus (b \oplus c) = (a \oplus b) \oplus c$.

2. $0 \in$ R and $a \oplus 0 = a = 0 \oplus a$.

3. For $0 \in$ R, $\quad 0 \oplus 0 = 0 = 0 \oplus 0$

For $1 \in$ R, $\quad 1 \oplus 4 = 0 = 4 \oplus 1$

For $2 \in$ R, $\quad 2 \oplus 3 = 0 = 3 \oplus 2$

For $3 \in$ R, $\quad 3 \oplus 2 = 0 = 2 \oplus 3$

For $4 \in$ R, $\quad 4 \oplus 1 = 0 = 1 \oplus 4$

$\therefore$ Every element of R has its additive inverse in R.

4. $a \oplus b = b \oplus a$.

5. $a \odot (b \odot c) = (a \odot b) \odot c$.

6. $a \odot (b \oplus c) = (a \odot b) \oplus (a \odot c)$ and $(b \oplus c) \odot a = (b \odot a) \oplus (c \odot a)$.

$\therefore$ (R, ⊕, ⊙) is a ring.

7. $1 \in$ R and we have

$$0 \odot 1 = 0 = 1 \odot 0, \quad 1 \odot 1 = 1 = 1 \odot 1, \quad 2 \odot 1 = 2 = 1 \odot 2,$$
$$3 \odot 1 = 3 = 1 \odot 3 \quad 4 \odot 1 = 4 = 1 \odot 4.$$

$\therefore$ 1 is the unit element of the ring (R, ⊕, ⊙).

8. From the composition table of ⊙, we have

$$a \odot b = b \odot a \quad \forall a, b \in R$$

$\therefore$ (R, ⊕, .) is a commutative ring.

From the composition table, we see that $a \odot b = 0$ implies that at least one of a and b is zero.

$\therefore$ (R, ⊕, ⊙) has no zero divisors.

$\therefore$ (R, ⊕, ⊙) is an integral domain.

For $1 \in$ R, $\quad 1 \odot 1 = 1 = 1 \odot 1$

For $2 \in$ R, $\quad 2 \odot 3 = 1 = 3 \odot 2$

For $3 \in$ R, $\quad 3 \odot 2 = 1 = 2 \odot 3$

For $4 \in$ R, $\quad 4 \odot 4 = 1 = 4 \odot 4$

$\therefore$ Every non-zero element of (R, ⊕, ⊙) has multiplicative inverse.

$\therefore$ (R, ⊕, ⊙) is a commutative ring with unit element and every non-zero element having multiplicative inverse.

$\therefore$ (R, ⊕, ⊙) is a field.

Remark. '(R, ⊕, ⊙) has no zero divisors' can also be proved like this :

$$a \odot b = 0 \Rightarrow 5/ab \Rightarrow 5/a \text{ or } 5/b \qquad (\because \text{ 5 is a prime number})$$

$\Rightarrow a = 0$ or $b = 0$.

Example 8. *Give an example of a ring which is not an integral domain.*

Sol. Let R = {0, 1, 2, 3, 4, 5}. Let symbols $+_6$ and $\times_6$ denote the operations addition modulo 6 and multiplication modulo 6′. We can show that (R, $+_6$, $\times_6$) is a ring.

Here 2, 3 ∈ R and $2 \times_6 3 = 0$.

$\therefore$ R is not free from zero divisors. $\quad\therefore$ R is not an integral domain.

Example 9. *Prove that ({0, 1, 2, 3, 4, 5, 6} $+_7$, $\times_7$) is a field*

Sol. Let R = {0, 1, 2, 3, 4, 5}.

Composition table for $+_7$

$+_7$	*0*	*1*	*2*	*3*	*4*	*5*	*6*
0	0	1	2	3	4	5	6
1	1	2	3	4	5	6	0
2	2	3	4	5	6	0	1
3	3	4	5	6	0	1	2
4	4	5	6	0	1	1	3
5	5	6	0	1	2	3	4
6	6	0	1	2	3	4	5

Composition table for $\times_7$

$\times_7$	0	1	2	3	4	5	6
0	0	0	0	0	0	0	0
1	0	1	2	3	4	5	6
2	0	2	4	6	1	3	5
3	0	3	6	2	5	1	4
4	0	4	1	5	2	6	3
5	0	5	3	1	6	4	2
6	0	6	5	4	3	2	1

The composition tables of $+_7$ and $\times_7$ show that these are binary operations on R.

Let $a, b, c \in$ R. We have

1. $a +_7 (b +_7 c) = (a +_7 b) +_7 c.$

2. $0 \in$ R and $a +_7 0 = a = 0 +_7 a.$

3. For $0 \in$ R, $\quad 0 +_7 0 = 0 = 0 +_7 0$

For $1 \in$ R, $\quad 1 +_7 6 = 0 = 6 +_7 1$

For $2 \in$ R, $\quad 2 +_7 5 = 0 = 5 +_7 2$

For $3 \in$ R, $\quad 3 +_7 4 = 0 = 4 +_7 3$

For $4 \in$ R, $\quad 4 +_7 3 = 0 = 3 +_7 4$

For $5 \in$ R, $\quad 5 +_7 2 = 0 = 2 +_7 5$

For $6 \in$ R, $\quad 6 +_7 1 = 0 = 1 +_7 6.$

$\therefore$ Every element of R has its additive inverse in R.

4. $a +_7 b = b +_7 a.$

5. $a \times_7 (b \times_7 c) = (a \times_7 b) \times_7 c.$

6. From the composition table of $\times_7$, we have

$a \times_7 1 = a = 1 \times_7 a.$

7. For $1 \in$ R, $\quad 1 \times_7 1 = 1 = 1 \times_7 1$

For $2 \in$ R, $\quad 2 \times_7 4 = 1 = 4 \times_7 2$

For $3 \in$ R, $\quad 3 \times_7 5 = 1 = 5 \times_7 3$

For $4 \in$ R, $\quad 4 \times_7 2 = 1 = 2 \times_7 4$

For $5 \in$ R, $\quad 5 \times_7 3 = 1 = 3 \times_7 5$

For $6 \in$ R, $\quad 6 \times_7 6 = 1 = 6 \times_7 6$

$\therefore$ Every non-zero element of R has its multiplicative inverse in R.

8. $a \times_7 b = b \times_7 a.$

9. $a \times_7 (b +_7 c) = (a \times_7 b) +_7 (a \times_7 c)$ and $(b +_7 c) \times_7 a = (b \times_7 a) +_7 (c \times_7 a).$

$\therefore$ (R, $+_7$, $\times_7$) is a field.

EXERCISE 1

1. Show that the set of rational number with usual addition and multiplication is (*i*) a ring (*ii*) an integral domain (*iii*) a field.
2. Show that (**R**, +, .) is (*i*) a ring (*ii*) an integral domain (*iii*) a field, where the binary operations '+' and '.' are respectively usual addition and multiplication of real numbers.

3. Let R = $\{4n : n \in \mathbf{Z}\}$. Let the binary operations 'usual addition' and 'usual multiplication' be denoted by '+' and '.' respectively. Show that :

(*i*) (R, +, .) is a commutative ring. (*ii*) (R, +, .) has no unit element.

(*iii*) (R, +, .) is an integral domain. (*iv*) (R, +, .) is not a field.

4. Prove that the set I of all even integers is a commutative ring without unity, the addition and multiplication of integers being the two ring compositions.

5. Prove that if :

$$S = \{a + b\sqrt{3} : a, b \in \mathbf{Q}\}$$

then (S, +, .) is a field.

6. Let R = $\{a + \sqrt{2}\, b : a, b \in \mathbf{Q}\}$. Under usual addition and multiplication, show that :

(*i*) (R, +, .) is a ring (*ii*) (R, +, .) is an integral domain

(*iii*) (R, +, .) is a field.

7. Let R = $\{a + b\sqrt[3]{2} : a, b \in \mathbf{Q}\}$. Under usual addition and multiplication, show that (R, +, .) is not a ring.

8. Let R = $\{a + b\sqrt{5} : a, b \in \mathbf{Z}\}$. Under usual addition and multiplication, show that :

(*i*) (R, +, .) is commutative ring with unit element.

(*ii*) (R, +, .) is an integral domain.

(*iii*) (R, +, .) is not a field.

9. Let R = {0, 1, 2, 3, 4, 5}. Let $\oplus$ and $\odot$ denote the operations 'addition modulo 6' and 'multiplication modulo 6' respectively. Show that :

(*i*) (R, $\oplus$, $\odot$) is a commutative ring with unit element.

(*ii*) (R, $\oplus$, $\odot$) is not an integral domain.

(*iii*) (R, $\oplus$, $\odot$) is not a field.

10. Let R = $\mathbf{Z} \times \mathbf{Z}$. Define operations $\oplus$ and $\odot$ on R as :

$$(a, b) \oplus (c, d) = (a + c, b + d)$$

and $$(a, b) \odot (c, d) = (ac, bd) \quad \forall\ (a, b), (c, d) \in R.$$

Show that (R, $\oplus$, $\odot$) is a commutative ring with unit element. Is (R, $\oplus$, $\odot$) an integral domain ?

11. Let F be the set of all continuous mappings from **R** into **R.** Show that F is a commutative ring with unit element under usual addition and multiplication of functions.

12. Show that (**N**, +, ×) is not a ring.

ANSWERS

10. No.

HINTS

7. Usual multiplication is not a binary operation on R.

8. (*ii*) Let $(a + b\sqrt{5})(c + d\sqrt{5}) = 0$, where $a + b\sqrt{5}$ and $c + d\sqrt{5}$ are non-zero elements of R.

$$\Rightarrow \qquad (ac + 5bd) + (ad + bc)\sqrt{5} = 0.$$

$$\Rightarrow \qquad ac + 5bd = 0 \text{ and } ad + bc = 0 \Rightarrow \frac{a}{b} = -\frac{5d}{c} \text{ and } \frac{a}{b} = -\frac{c}{d}$$

$$\Rightarrow \qquad -\frac{5d}{c} = -\frac{c}{d} \Rightarrow c = \pm\sqrt{5}\,d$$

This is impossible, because $\pm\sqrt{5}\,d \notin \mathbf{Z}$.

(*iii*) Let $a + b\sqrt{5}\ (\neq 0) \in$ R.

$$\therefore \quad \left(\frac{a}{a^2 - 5b^2} + \left(\frac{-b}{a^2 - 5b^2}\right)\sqrt{5}\right)(a + b\sqrt{5}) = 1 \text{ but } \frac{a}{a^2 - 5b^2} \text{ and } -\frac{b}{a^2 - 5b^2} \text{ may not be integers.}$$

10. (R, ⊕, ⊙) is not an integral domain because (4, 0) ⊙ (0, 2) = (0, 0).

---

9. A PROPERTY OF FIELDS

Theorem 1. *Every field is an integral domain.*

Proof. Let (F, +, .) be a field.

∴ (F, +, .) is a commutative ring.

∴ It is sufficient to show that the ring (F, +, .) is without zero divisors.

Let $a, b \in$ F and $a \,.\, b = 0$.

If $a = 0$, then we have nothing to prove.

Let $a \neq 0$. Since (F, +, .) is a field, $\exists\, a^{-1} \in$ F : $a \,.\, a^{-1} = 1 = a^{-1} \,.\, a$

$$\therefore \quad a \,.\, b = 0 \Rightarrow a^{-1} \,.\, (a \,.\, b) = a^{-1} \,.\, 0 \Rightarrow (a^{-1} \,.\, a) \,.\, b = 0$$

$$\Rightarrow \quad 1 \,.\, b = 0 \quad \Rightarrow \quad b = 0.$$

∴ The ring (F, +, .) is without zero divisors.

∴ (F, +, .) is an integral domain.

Remark. The converse of above theorem is not true. For example, (**Z**, +, .) is an integral domain and it is not a field.

10. SCALAR MULTIPLE AND POWERS OF ELEMENTS OF A RING

Let (R, +, .) be a ring.

For $a \in$ R, we define

$$2a = a + a,\ 3a = a + 2a,\ \ldots\ldots,\ na = a + (n - 1)a \text{ etc.}$$

and
$$a^2 = a \,.\, a,\quad a^3 = a \,.\, a^2,\ \ldots\ldots,\ a^n = a \,.\, a^{n-1} \text{ etc.}$$

Remark 1. If (R, +, .) is a ring then for the sake of simplicity, the product $a \,.\, b$ is written as ab for $a, b \in$ R.

Remark 2. If (R, +, .) is a ring then for $a, b \in$ R, the element $a + (-b)$ is written as $a - b$.

Example 1. *Let (R, +, .) be a ring. Show that the ring is commutative if and only if*

$$(a + b)^2 = a^2 + 2ab + b^2 \quad \forall\, a, b \in R.$$

Sol. Let the ring (R, +, .) be commutative.

$$\therefore \quad xy = yx \quad \forall\, x, y \in \text{R}$$

Now
$$(a + b)^2 = (a + b)(a + b)$$
$$= (a + b)a + (a + b)b = (aa + ba) + (ab + bb)$$
$$= a^2 + ab + ab + b^2 = a^2 + 2ab + b^2$$

$$\therefore \quad (a + b)^2 = a^2 + 2ab + b^2 \quad \forall\, a, b \in \text{R}.$$

Conversely, let

$$(a + b)^2 = a^2 + 2ab + b^2 \quad \forall\, a, b \in R.$$

Let $x, y \in R$.

$\therefore \quad (x + y)^2 = x^2 + 2xy + y^2 \quad \Rightarrow \quad (x + y)(x + y) = x^2 + xy + xy + y^2$

$\Rightarrow \quad (x + y)x + (x + y)y = x^2 + xy + xy + y^2 \quad \Rightarrow \quad xx + yx + xy + yy = xx + xy + xy + yy$

$\Rightarrow \quad yx = xy.$

$\therefore$ (R, +, .) is a commutative ring.

Example 2. *Let R be a non-empty set with binary operations '+' and '.' satisfying all the axioms for a ring except $a + b = b + a \quad \forall\, a, b \in R$. If there exists an element $c \in R$ such that $a \, . \, c = b \, . \, c \;\Rightarrow\; a = b \quad \forall\, a, b \in R$, show that (R, +, .) is a ring.*

Sol. $(a + b) \, . \, (c + c) = (a + b) \, . \, c + (a + b) \, . \, c$

$= (a \, . \, c + b \, . \, c) + (a \, . \, c + b \, . \, c) = a \, . \, c + (b \, . \, c + a \, . \, c) + b \, . \, c$

Also $\quad (a + b) \, . \, (c + c) = a \, . \, (c + c) + b \, . \, (c + c)$

$= (a \, . \, c + a \, . \, c) + (b \, . \, c + b \, . \, c) = a \, . \, c + (a \, . \, c + b \, . \, c) + b \, . \, c$

$\therefore \quad a \, . \, c + (b \, . \, c + a \, . \, c) + b \, . \, c = a \, . \, c + (a \, . \, c + b \, . \, c) + b \, . \, c$

$\Rightarrow \quad b \, . \, c + a \, . \, c = a \, . \, c + b \, . \, c$

$\Rightarrow \quad (b + a) \, . \, c = (a + b) \, . \, c$

$\Rightarrow \quad b + a = a + b. \qquad (\because \; x \, . \, c = y \, . \, c \Rightarrow x = y \quad \forall\, x, y \in R)$

$\therefore$ (R, +, .) is a ring.

Example 3. *If (R, +, .) is a ring with unity and $(xy)^2 = x^2y^2 \;\forall\, x, y \in R$ then show that the ring R is commutative.*

Sol. Let $x, y \in R$. $\therefore\; x, y + 1 \in R$

$\Rightarrow \quad [x(y + 1)]^2 = x^2(y + 1)^2 \quad \Rightarrow \quad (xy + x)^2 = x^2(y^2 + 2y + 1)$

$\Rightarrow \quad (xy)^2 + (xy)x + x(xy) + x^2 = x^2y^2 + 2x^2y + x^2$

$\Rightarrow \quad x^2y^2 + (xy)x + x^2y = x^2y^2 + x^2y + x^2y$

$\Rightarrow \quad (xy)x = x^2y \qquad \text{...(1)}$

Replacing x by $x + 1$, we get $\quad [(x + 1)y](x + 1) = (x + 1)^2\, y$

$\Rightarrow \quad (xy + y)(x + 1) = (x^2 + 2x + 1)y$

$\Rightarrow \quad xyx + xy + yx + y = x^2y + 2xy + y \quad \Rightarrow \quad x^2y + yx = x^2y + xy \qquad$ [Using (1)]

$\Rightarrow \quad yx = xy.$

$\therefore$ Ring (R, +, .) is commutative.

EXERCISE 2

1. If (R, +, .) is a ring and $a, b, c, d \in R$, show that :
   $(a + b) \, . \, (c + d) = a \, . \, c + a \, . \, d + b \, . \, c + b \, . \, d.$
2. If a, b, c, d are any four elements of a ring R, show that :
   $(a - b)(c - d) = (ac + bd) - (ad + bc).$
3. Let (R, +, .) be a ring such that $x^2 = x \quad \forall\, x \in R$. Show that :
   (*i*) $x + x = 0 \quad \forall\, x \in R$ $\qquad$ (*ii*) $x + y = 0 \;\Rightarrow\; x = y$
   (*iii*) R is a commutative ring.

HINTS

3. (i) $(x + x)^2 = x + x \Rightarrow x^2 + x^2 + x^2 + x^2 = x + x \Rightarrow x + x + x + x = x + x$
(ii) $x + y = 0 \Rightarrow x + y = x + x$
(iii) $(x + y)^2 = x + y \Rightarrow x + xy + yx + y = x + y \Rightarrow xy + yx = yx + yx.$

---

11. SUBRINGS

Let (R, +, .) be a ring. A non-empty subset S of R is called a **subring** of the ring (R, +, .) if S itself is a ring under the binary operations '+' and '.' of R.

A subring of a commutative ring is always commutative.

Example 1. If (R, +, .) is a ring, then {0} and R are always subrings of R. These are called **improper subrings** of R. Other subrings, if any, are called **proper subrings** of R.

Example 2. Z is a subring of the ring (**Q**, +, .).

Example 3. Q is a subring of the ring (**R**, +, .).

12. TEST FOR A SUBRING

Theorem. *Let (R, +, .) be a ring. A non-empty subset S of R is a subring of R if and only if* $a - b \in S$ *and* $ab \in S \quad \forall\, a, b \in S$.

Proof. Let S be a subring of the ring (R, +, .).

$\therefore$ (S, +, .) is a ring. Let $a, b \in S$.

$\therefore$ $-b$, being the additive inverse of b, is in S.

$\therefore$ $a + (-b)$ *i.e.*, $a - b$ is in S.

Also $ab \in S$ ($\because$ '.' is binary operation of S)

$\therefore$ $a - b \in S, ab \in S \quad \forall\, a, b, \in S.$

Conversely, let $a - b \in S$, $ab \in S \quad \forall\, a, b \in S$.

'+' is associative because elements of S are also elements of R.

Let $a \in S$. $\therefore$ $a - a \in S \Rightarrow a + (-a) \in S \Rightarrow 0 \in S$

Let $x \in S$. $\therefore$ $x \in R$.

$\therefore$ $x + 0 = x = 0 + x$ ($\because$ R is a ring)

Let $a \in S$. $\therefore$ $0 - a = -a \in S$

$\therefore$ For $a \in S, \exists -a \in S$:

$$a + (-a) = 0 = (-a) + a$$

$$a, b \in S \Rightarrow a, -b \in S \Rightarrow a - (-b) \in S \Rightarrow a + b \in S.$$

$\therefore$ '+' is a binary operation on S.

'.' is a binary operation on S because $ab \in S \quad \forall\, a, b \in S$.

'.' is associative because elements of S are also elements of R.

Distributive laws holds in S because elements of S are also elements of R.

$\therefore$ **(S, +, .) is a ring.**

$\therefore$ **S is a subring of the ring (R, +, .).**

Example 1. *Show that the set of all even integers is a subring of* (**Z**, +, .).

Sol. Let S = set of all even integers.

$S \neq \phi$ because $0\ (= 2(0)) \in S$.

Let $2a, 2b \in S$.

$\therefore \quad 2a - 2b = 2(a - b) \in S$

and $\quad (2a)(2b) = 4ab = 2(2ab) \in S$.

$\therefore \quad 2a - 2b, (2a)(2b) \in S \quad \forall\, 2a, 2b \in S$.

$\therefore$ S is a subring of the ring $(\mathbf{Z}, +, .)$.

Example 2. *Show that the set of matrices* $\begin{bmatrix} a & b \\ 0 & c \end{bmatrix}$ *is a subring of the ring* $(M_2, +, .)$, *where* $M_2 = \left\{\begin{bmatrix} a & b \\ c & d \end{bmatrix} : a, b, c, d \in \mathbf{Z}\right\}$.

Sol. The set M_2 is a ring with usual addition and multiplication as the binary operations.

Let $\quad S = \left\{\begin{bmatrix} a & b \\ 0 & c \end{bmatrix} : a, b, c \in \mathbf{Z}\right\}$.

$S \neq \phi$ because $\begin{bmatrix} 0 & 0 \\ 0 & 0 \end{bmatrix} \in S$.

Let $A = \begin{bmatrix} a & b \\ 0 & c \end{bmatrix}$ and $B = \begin{bmatrix} d & e \\ 0 & f \end{bmatrix}$ be any two elements of S.

Now $\quad A - B = \begin{bmatrix} a & b \\ 0 & c \end{bmatrix} - \begin{bmatrix} d & e \\ 0 & f \end{bmatrix} = \begin{bmatrix} a-d & b-e \\ 0 & c-f \end{bmatrix} \in S$

Also $\quad AB = \begin{bmatrix} a & b \\ 0 & c \end{bmatrix}\begin{bmatrix} d & e \\ 0 & f \end{bmatrix} = \begin{bmatrix} ad+0 & ae+bf \\ 0+0 & 0+cf \end{bmatrix} = \begin{bmatrix} ad & ae+bf \\ 0 & cf \end{bmatrix} \in S$.

$\therefore \quad A - B, AB \in S \quad \forall\, A, B \in S$

$\therefore$ S is a subring of the ring $(M_2, +, .)$.

Example 3. *Let (R, +, .) be a ring. Show that*

$S = \{x \in R : xy = yx \quad \forall\, y \in R\}$ *is a subring of R.*

Sol. $\quad S = \{x \in R : xy = yx \quad \forall\, y \in R\}$

$0 \in S$ because $\quad 0y = y0 = 0 \quad \forall\, y \in R$

$\therefore \quad S \neq \phi$.

Let $x_1, x_2 \in S$

$\therefore \quad x_1y = yx_1$ and $x_2y = yx_2 \quad \forall\, y \in R$.

Now $\quad (x_1 - x_2)y = x_1y - x_2y = yx_1 - yx_2 = y(x_1 - x_2)$.

$\therefore \quad x_1 - x_2 \in S$

Also $\quad (x_1x_2)y = x_1(x_2y) = x_1(yx_2) = (x_1y)x_2 = (yx_1)x_2 = y(x_1x_2)$.

$\therefore \quad x_1x_2 \in S$

$\therefore \quad x_1 - x_2, x_1x_2 \in S \quad \forall\, x_1, x_2 \in S$.

$\therefore$ S is subring of (R, +, .).

13. INTERSECTION OF SUBRINGS

Theorem. *The intersection of two subrings of a ring is also a subring of the ring.*

Proof. Let S_1 and S_2 be two subrings of a ring (R, +, .).

Let '0' be additive identity of the ring R.

$\therefore \quad 0 \in S_1, 0 \in S_2$

$\therefore \quad 0 \in S_1 \cap S_2. \qquad \therefore \quad S_1 \cap S_2 \neq \phi$

Let $a, b \in S_1 \cap S_2$

$\therefore \quad a, b \in S_1, a, b \in S_2$

$\Rightarrow \quad a - b, ab \in S_1$ and $a - b, ab \in S_2$

$\Rightarrow \quad a - b, ab \in S_1 \cap S_2$

$\therefore \quad S_1 \cap S_2$ is a subring of the ring (R, +, .).

IMPORTANT RESULTS

1. *Let (R, +, .) be a ring. A non-empty subset S of R is a subring of R if and only if $a - b \in S, ab \in S \quad \forall\, a, b \in S$.*
2. *The intersection of two subrings of a ring is also a subring of the ring.*

EXERCISE 3

1. Show that the set **R** of real numbers is a subring of the ring (**C**, +, .).
2. Show that the set of matrices $\begin{bmatrix} a & b \\ 0 & 0 \end{bmatrix}$, where $a, b \in \mathbf{Z}$ is a subring of the ring of all 2 × 2 matrices over integers under usual addition and multiplication of matrices.
3. Show that the set of matrices $\begin{bmatrix} a & 4 \\ 0 & 0 \end{bmatrix}$, where $a \in \mathbf{Z}$ is not a subring of the ring of all 2 × 2 matrices over integers under usual addition and multiplication of matrices.
4. Show that $S = \{ma : a \in \mathbf{Z}\}$ is a subring of the ring (**Z**, +, .).
5. If (R, +, .) is a ring and $a \in R$, show that $N(a) = \{r \in R : ar = ra\}$ is a subring of R.
6. Show that the union of two subrings of a ring may not be a subring of the ring.
7. Prove that an arbitrary intersection of subrings is a subring.

HINT

6. $S_1 = \{3n : n \in \mathbf{Z}\}$ and $S_2 = \{4n : n \in \mathbf{Z}\}$ are subrings of the ring (**Z**, +, .).
   $3 \in S_1, 4 \in S_2$ implies $3, 4 \in S_1 \cup S_2$.
   Now $3 + 4 = 7$ and $7 \notin S_1 \cup S_2$
   $\therefore \quad S_1 \cup S_2$ is not closed w.r.t. addition.

14. CHARACTERISTIC OF A RING

Let (R, +, .) be a ring. Let '0' denote the additive identity of this ring. If there exists a positive integer n such that $na = 0 \quad \forall\, a \in R$, then the smallest such positive integer is called the **characteristic** of the ring R.

If no such positive integer 'n' exists, then R is said to have characteristic zero.

The characteristics of ring R is written as '*char R*'.

Example 1. There is no positive integer 'n' such that $na = 0 \quad \forall\, a \in \mathbf{Z}$.

$\therefore$ The characteristic of the ring $(\mathbf{Z}, +, .)$ is zero.

Example 2. Let $R = \{0, 1, 2, 3, 4\}$. $(R, \oplus, \odot)$ is a ring, where $\oplus$ and $\odot$ denote 'addition modulo 5' and 'multiplication modulo 5' respectively.

For each $x \in R$, we have $5x = 0$

i.e., $$x \oplus x \oplus x \oplus x \oplus x = 0.$$

$\therefore$ Char $R \le 5$. $\therefore$ Char R is either 5 or 4 or 3 or 2 or 1.

Now
$$4(2) = (2 \oplus 2) \oplus 2 \oplus 2 = (4 \oplus 2) \oplus 2 = 1 \oplus 2 = 3 \neq 0$$
$$3(2) = (2 \oplus 2) \oplus 2 = 4 \oplus 2 = 1 \neq 0$$
$$2(2) = 2 \oplus 2 = 4 \neq 0$$
$$1(2) = 2 \neq 0$$

$\therefore$ Char R is neither 4 nor 3 nor 2 nor 1.

$\therefore$ Char R = **5.**

Theorem 1. *Let (R, +, .) be a ring. If there is an element in R of infinite order when regarded as an element of the additive group (R, +) then the ring (R, +, .) is of characteristic zero.*

Proof. Let $a \in R$ such that the order of a as an element of the group (R, +) is infinite.

$\therefore$ There does not exist any positive integer 'n' such that $na = 0$.

By definition, char R cannot be any positive integer.

$\therefore$ Char R = 0.

Theorem 2. *Let (R, +, .) be a ring with unit element. The characteristic of this ring is a positive integer n (resp. zero) if the order of the unit element regarded as an element of the additive group (R, +) is n (resp. infinite).*

Proof. Let '1' be the unit element of the ring (R, +, .).

Case I. Let order of '1' as an element of the group (R, +) be infinite.

$\therefore$ There does not exist any positive integer 'm' such that $m(1) = 0$.

$\therefore$ By definition, char R cannot be any positive integer.

$\therefore$ Char R = 0.

Case II. Let order of '1' as an element of the group (R, +) be the positive integer n.

$\therefore$ n is the smallest positive integer satisfying $n(1) = 0$.

Let a be any element of R.

$$\begin{aligned} \therefore \qquad na &= a + a + a + \ldots\ldots n \text{ terms} \\ &= 1a + 1a + 1a + \ldots\ldots n \text{ terms} \\ &= (1 + 1 + 1 + \ldots\ldots n \text{ terms})\, a \qquad \text{(By right distributive law)} \\ &= (n(1))a = 0a = 0 \end{aligned}$$

$$\therefore \qquad na = 0 \quad \forall\, a \in R$$

$\therefore$ Char $R \le n$

If possible, let char $R = k$ and $k < n$.

$$\therefore \qquad ka = 0 \quad \forall\, a \in R$$

$\therefore$ In particular, $k(1) = 0$.

This is impossible, because $o(1) = n$. $\therefore$ Our supposition is wrong.

$\therefore$ Char R = n.

Theorem 3. *Let (D, +, .) be an integral domain. The characteristic of the integral domain is a positive integer n (resp. zero) if the order of any non-zero element of the additive group (D, +) is n (resp. infinite).*

Proof. Let a be any non-zero element of the integral domain (D, +, .).

Case I. Let order of 'a' as an element of the group (D, +) be infinite.

$\therefore$ There does not exist any positive integer 'm' such that $ma = 0$.

$\therefore$ By definition, char D cannot be any positive integer.

$\therefore$ Char D = 0.

Case II. Let order of 'a' as an element of the group (D, +) be the positive integer n.

$\therefore$ n is the smallest positive integer satisfying $na = 0$.

Let b be any element of D.

$\Rightarrow$ $(na)b = 0b = 0$

$\Rightarrow$ $(a + a + a + \ldots\ldots n \text{ times})\, b = 0$

$\Rightarrow$ $ab + ab + ab + \ldots\ldots n \text{ times} = 0$

$\Rightarrow$ $a(b + b + b + \ldots\ldots n \text{ times}) = 0$

$\Rightarrow$ $a(nb) = 0$

$\therefore$ $nb = 0 \quad \forall\, b \in \text{R}$ $\quad$ ($\because$ $a \neq 0$ and D is an integral domain)

$\therefore$ Char D $\leq n$.

If possible, let char D = k and $k < n$.

$$kc = 0 \quad \forall\, c \in \text{D}$$

In particular, $ka = 0$

This is impossible, because $o(a) = n$. $\therefore$ Our supposition is wrong.

$\therefore$ Char D = n.

Theorem 4. *Let (D, +, .) be an integral domain. The order of each non-zero element of the group (D, +) is same.*

Proof. Let a be any non-zero element of D. Let order of a as an element of the group (D, +) be n. Let b be any other non-zero element of D.

Case I. n is infinite.

$\therefore$ There does not exist any positive integer 'm' such that $ma = 0$.

If possible let $o(b) = k$ and $k > 0$.

$\therefore$ $kb = 0$ $\quad\Rightarrow\quad$ $a(kb) = a0 = 0$

$\Rightarrow$ $a(b + b + b + \ldots\ldots k \text{ times}) = 0$ $\quad\Rightarrow\quad$ $ab + ab + ab + \ldots.. k \text{ times} = 0$

$\Rightarrow$ $(a + a + a + \ldots\ldots k \text{ times})\, b = 0$ $\quad\Rightarrow\quad$ $(ka)b = 0$

$\Rightarrow$ $ka = 0$ $\quad$ ($\because$ $b \neq 0$ and D is an integral domain)

$\therefore$ Order of a is not infinite

This is impossible.

$\therefore$ $o(b)$ is infinite.

Case II. n > 0.

$\therefore$ n is the smallest positive integer such that $na = 0$.

$\therefore$ $(na)b = 0b = 0$

$\Rightarrow$ $(a + a + a + \ldots\ldots n \text{ times})b = 0$

$\Rightarrow$ $ab + ab + ab + \ldots\ldots n \text{ times} = 0$

$\Rightarrow$ $a(b + b + b + \ldots\ldots n \text{ times}) = 0$

$\Rightarrow$ $a(nb) = 0$

$\Rightarrow$ $nb = 0$ $\quad (\because\ a \neq 0$ and D is an integral domain)

$\therefore$ $o(b) \leq n$

If possible, let $o(b) = k$ and $k < n$.

$\therefore$ $kb = 0$

$\therefore$ $a(kb) = a0 = 0$

$\Rightarrow$ $a(b + b + b + \ldots\ldots k \text{ times}) = 0$

$\Rightarrow$ $ab + ab + ab + \ldots\ldots k \text{ times} = 0$

$\Rightarrow$ $(a + a + a + \ldots\ldots k \text{ times})b = 0$

$\Rightarrow$ $(ka)b = 0$

$\Rightarrow$ $ka = 0$ $\quad (\because\ b \neq 0$ and D is an integral domain)

$\therefore$ $o(a) \leq k$ *i.e.*, $n \leq k$

This is impossible.

$\therefore$ $o(b) = n$.

Theorem 5. *The characteristic of an integral domain is either zero or a prime number.*

Proof. Let (D, +, .) be an integral domain.

Let a be any non-zero element of D.

$\therefore$ Char D is a positive integer n (resp. zero) if the order of a as an element of the group (D, +) is n (resp. infinite). If a is of infinite order as an element of the group (D, +), then char D = 0 and we have nothing to prove.

Let n be a positive integer.

We shall show that n is a prime number. If possible, let n be not a prime number and $n = q_1 q_2$, where $1 < q_1, q_2 < n$.

If possible, let $a^2 = 0$.

$\Rightarrow$ $aa = 0$ $\Rightarrow$ $a = 0$ $\quad (\because$ D is an integral domain)

This is impossible because $a \neq 0$.

$\therefore$ $a^2 \neq 0$

$\therefore$ $o(a^2) = n$ $\quad (\because$ order of non-zero elements of integral domain are same)

$\Rightarrow$ $na^2 = 0$ $\Rightarrow$ $q_1 q_2 a^2 = 0$ $\Rightarrow$ $(q_1 a)(q_2 a) = 0$

$\Rightarrow$ Either $q_1 a = 0$ or $q_2 a = 0$.

$\Rightarrow$ Either $n \leq q_1$ or $n \leq q_2$.

This is impossible because $q_1, q_2 < n$. $\therefore$ Our supposition is wrong.

$\therefore$ n must be a prime number.

IMPORTANT RESULTS

1. *Let (R, +, .) be a ring. If there is an element in R of infinite order when regarded as an element of the additive group (R, +) then the ring (R, +, .) is of characteristic zero.*
2. *Let (R, +, .) be a ring with unit element. The characteristic of the ring is a positive integer n (resp. zero) if the order of the unit element regarded as an element of the additive group (R, +) is n (resp. infinite).*
3. *Let (D, +, .) be an integral domain. The characteristic of the integral domain is a positive integer n (resp. zero) if the order of any non-zero element of the additive group (D, +) is n (resp. infinite).*
4. *Let (D, +, .) be an integral domain. The order of each non-zero element of the group (D, +) is same.*
5. *The characteristic of an integral domain is either zero or a prime number.*

EXERCISE 4

1. Find the characteristic of the following rings :
   (*i*) (**Q**, +, ×) (*ii*) (**R**, +, ×).
2. Let R = {0, 1, 2, 3}. Find the characteristic of the ring (R, ⊕, ⊙), where ⊕ and ⊙ respectively denote 'addition modulo 4' and 'multiplication modulo 4'.
3. Let R = {0, 1, 2, 3, 4, 5, 6}. Find the characteristic of the ring (R, ⊕, ⊙), where ⊕ and ⊙ respectively denote 'addition modulo 7' and 'mutiplication modulo 7'.

ANSWERS

1. (*i*) 0 (*ii*) 0 **2.** 4 **3.** 7.